T0338044

Computational Pharmaceutics

Advances in Pharmaceutical Technology
A Wiley Book Series

Series Editors:
Dennis Douroumis, University of Greenwich, UK
Alfred Fahr, Friedrich–Schiller University of Jena, Germany
Jűrgen Siepmann, University of Lille, France
Martin Snowden, University of Greenwich, UK
Vladimir Torchilin, Northeastern University, USA

Titles in the Series:

Hot-Melt Extrusion: Pharmaceutical Applications
Edited by Dionysios Douroumis

Drug Delivery Strategies for Poorly Water-Soluble Drugs
Edited by Dionysios Douroumis and Alfred Fahr

Forthcoming titles:

Novel Delivery Systems for Transdermal and Intradermal Drug Delivery
Edited by Ryan F. Donnelly and Thakur Raghu Raj Singh

Pulmonary Drug Delivery: Advances and Challenges by Ali Nokhodhi and Gary P. Martin

Computational Pharmaceutics

Application of Molecular Modeling in Drug Delivery

Edited by

DEFANG OUYANG AND SEAN C. SMITH

WILEY

This edition first published 2015

Registered Office
John Wiley & Sons, Ltd, The Atrium, Southern Gate, Chichester, West Sussex, PO19 8SQ, United Kingdom

For details of our global editorial offices, for customer services and for information about how to apply for permission to reuse the copyright material in this book please see our website at www.wiley.com.

Library of Congress Cataloging-in-Publication Data

Computational pharmaceutics : application of molecular modeling in drug delivery / edited by Defang Ouyang and Sean C. Smith.
p. ; cm. – (Advances in pharmaceutical technology)
Includes bibliographical references and index.
ISBN 978-1-118-57399-0 (cloth)
I. Ouyang, Defang, editor. II. Smith, Sean C., editor. III. Series: Advances in pharmaceutical technology.
[DNLM: 1. Drug Delivery Systems–methods. 2. Computational Biology–methods. 3. Drug Design. 4. Models, Molecular. QV 785]
RM301.25
615.1′9–dc23

2014050194

A catalogue record for this book is available from the British Library.

Set in 10/12pt Times by SPi Global, Pondicherry, India
Printed and bound in Singapore by Markono Print Media Pte Ltd

1 2015

Contents

List of Contributors

Amanda S. Barnard, *CSIRO Virtual Nanoscience Laboratory, Australia*

Alex Bunker, *Centre for Drug Research, Faculty of Pharmacy, University of Helsinki, Finland*

Robin Curtis, *School of Chemical Engineering and Analytical Sciences, Manchester Institute of Biotechnology, University of Manchester, UK*

Dennis E. Discher, *Department of Chemical and Biomolecular Engineering, University of Pennsylvania, USA*

Zhiwei Feng, *Institute of Functional Nano and Soft Materials (FUNSOM), Soochow University, China*

Yong Gan, *Shanghai Institute of Materia Medica, Chinese Academy of Sciences, China*

Jingkai Gu, *Research Center for Drug Metabolism, Jilin University, China*

Zhen Guo, *Shanghai Institute of Materia Medica, Chinese Academy of Sciences, China; and Institute of Pharmaceutical Innovation, University of Bradford, UK*

You He, *Shanghai Synchrotron Radiation Facility, Shanghai Institute of Applied Physics, Chinese Academy of Sciences, China*

Tingjun Hou, *Institute of Functional Nano and Soft Materials (FUNSOM), Soochow University, China*

Myungshim Kang, *Department of Chemistry, College of Staten Island, City University of New York, USA*

Peng Ke, *Merck & Co, Inc., UK*

Lin Lai, *CSIRO Virtual Nanoscience Laboratory, Australia*

Dennis Lam, *Department of Chemistry, College of Staten Island, City University of New York, USA*

Youyong Li, *Institute of Functional Nano and Soft Materials (FUNSOM), Soochow University, China*

Sharon M. Loverde, *Department of Chemistry, College of Staten Island, City University of New York, USA*

Vinuthaa Murthy, *School of Psychological and Clinical Sciences, Charles Darwin University, Australia*

Sang Young Noh, *Department of Chemistry and Centre for Scientific Computing, University of Warwick, UK; and Molecular Organization and Assembly in Cells Doctoral Training Centre, University of Warwick, UK*

Rebecca Notman, *Department of Chemistry and Centre for Scientific Computing, University of Warwick, UK*

David William O'Neill, *Department of Chemistry and Centre for Scientific Computing, University of Warwick, UK*

Defang Ouyang, *Institute of Chinese Medical Sciences, University of Macau, Macau*

Harendra S. Parekh, *School of Pharmacy, The University of Queensland, Australia*

Sarah L. Price, *Department of Chemistry, University College London, UK*

Sheng Qi, *School of Pharmacy, University of East Anglia, UK*

Xiaohong Ren, *Shanghai Institute of Materia Medica, Chinese Academy of Sciences, China; and Research Center for Drug Metabolism, Jilin University, China*

Dorota Roberts, *School of Chemical Engineering and Analytical Sciences, Manchester Institute of Biotechnology, University of Manchester, UK*

Gabriele Sadowski, *Department of Biochemical and Chemical Engineering, TU Dortmund University, Germany*

Carl H. Schwable, *School of Life and Health Sciences, Aston University, UK*

Qun Shao, *Institute of Pharmaceutical Innovation, University of Bradford, UK*

Raj K. Singh Badhan, *Medicines Research Unit, School of Life and Health Sciences, Aston University, UK*

Sean C. Smith, *School of Chemical Engineering, University of New South Wales, Australia*

Xiaotian Sun, *Institute of Functional Nano and Soft Materials (FUNSOM), Soochow University, China*

Sachin S. Thakur, *School of Pharmacy, The University of Queensland, Australia*

Jim Warwicker, *Faculty of Life Sciences, Manchester Institute of Biotechnology, University of Manchester, UK*

Li Wu, *Shanghai Institute of Materia Medica, Chinese Academy of Sciences, China; and Research Center for Drug Metabolism, Jilin University, China*

Tiqiao Xiao, *Shanghai Synchrotron Radiation Facility, Shanghai Institute of Applied Physics, Chinese Academy of Sciences, China*

Zhi Ping Xu, *Australian Institute for Bioengineering and Nanotechnology, University of Queensland, Australia*

Xianzhen Yin, *Shanghai Institute of Materia Medica, Chinese Academy of Sciences, China; and Institute of Pharmaceutical Innovation, University of Bradford, UK*

Peter York, *Institute of Pharmaceutical Innovation, University of Bradford, UK*

Jiwen Zhang, *Shanghai Institute of Materia Medica, Chinese Academy of Sciences, China; and Research Center for Drug Metabolism, Jilin University, China*

Series Preface

The series *Advances in Pharmaceutical Technology* covers the principles, methods, and technologies that the pharmaceutical industry use to turn a candidate molecule or new chemical entity into a final drug form and hence a new medicine. The series will explore means of optimizing the therapeutic performance of a drug molecule by designing and manufacturing the best and most innovative of new formulations. The processes associated with the testing of new drugs, the key steps involved in the clinical trials process, and the most recent approaches utilized in the manufacture of new medicinal products will all be reported. The focus of the series will very much be on new and emerging technologies and the latest methods used in the drug development process.

The topics covered by the series include:

- *Formulation*: the manufacture of tablets in all forms (caplets, dispersible, fast-melting) will be described, as will capsules, suppositories, solutions, suspensions and emulsions, aerosols and sprays, injections, powders, ointments and creams, sustained release, and the latest transdermal products. The developments in engineering associated with fluid, powder and solids handling, solubility enhancement, and colloidal systems including the stability of emulsions and suspensions will also be reported within the series. The influence of formulation design on the bioavailability of a drug will be discussed and the importance of formulation with respect to the development of an optimal final new medicinal product will be clearly illustrated.
- *Drug delivery*: The use of various excipients and their role in drug delivery will be reviewed. Amongst the topics to be reported and discussed will be a critical appraisal of the current range of modified-release dosage forms currently in use and also those under development. The design and mechanism(s) of controlled release systems, including macromolecular drug delivery, microparticulate controlled drug delivery, the delivery of biopharmaceuticals, delivery vehicles created for gastro-intestinal tract targeted delivery, transdermal delivery, and systems designed specifically for drug delivery to the lung will all be reviewed and critically appraised. Further site-specific systems used for the delivery of drugs across the blood–brain barrier will be reported, including dendrimers, hydrogels, and new innovative biomaterials.
- *Manufacturing*: The key elements of the manufacturing steps involved in the production of new medicines will be explored in this series. The importance of crystallization; batch and continuous processing, seeding; mixing including a description of the key engineering principles relevant to the manufacture of new medicines will all be reviewed and reported. The fundamental processes of quality control including good laboratory practice (GLP), good manufacturing practice (GMP), quality by design (QbD), the Deming cycle, regulatory requirements, and the design of appropriate robust statistical sampling procedures for the control of raw materials will all be an integral part of this book series.

An evaluation of the current analytical methods used to determine drug stability, the quantitative identification of impurities, contaminants, and adulterants in pharmaceutical materials will be described as will the production of therapeutic biomacromolecules, bacteria, viruses, yeasts, moulds, prions, and toxins through chemical synthesis and emerging synthetic/molecular biology techniques. The importance of packaging including the compatibility of materials in contact with drug products and their barrier properties will also be explored.

Advances in Pharmaceutical Technology is intended as a comprehensive one stop shop for those interested in the development and manufacture of new medicines. The series will appeal to those working in the pharmaceutical and related industries, both large and small, and will also be valuable to those who are studying and learning about the drug development process and the translation of those drugs into new life-saving and life-enriching medicines.

Dennis Douroumis
Alfred Fahr
Juergen Siepmann
Martin Snowden
Vladimir Torchilin

Preface

While computer-aided drug design (or rational drug design) has been practised for half a century, the application of computational modeling to drug delivery and pharmaceutical formulations "computational pharmaceutics" has emerged only in recent years. In combination with existing branches of pharmaceutics, it offers rapidly growing potential for developing rational, deductive, and knowledge-based strategies in pharmaceutics. Thus, this discipline has emerged and grown enormously in importance. Exploiting the exponential growth in power of high performance computing systems, computational pharmaceutics has the ability to provide multiscale lenses to pharmaceutical scientists, revealing mechanistic details ranging across chemical reactions, small drug molecules, proteins, nucleic acids, nanoparticles, and powders to the human body.

Given the advances in the field, it is becoming increasingly important that pharmaceutical scientists have cognizance of computational modeling, which may be anticipated to become a more common part of postgraduate curricula. While the fundamental theories and technical implementations of many of the methods in this book are complex, the computational software is becoming much more readily accessible and usable. Hence, it is no longer a prerequisite to understand in detail how the methods work at the theory and computer coding levels in order to gain substantial insights and benefits from carrying out simulations. Written for an audience with little experience in molecular modeling, with a focus on applications, this book will prove an excellent resource for researchers and students working in pharmaceutical sciences. We anticipate it will be useful not only for pharmaceutical scientists but more broadly for computational chemists looking to move into the pharmaceutical domain, for those working in medicinal chemistry, materials science, and nanotechnology.

Defang Ouyang
Sean C. Smith
March 2015

Editors' Biographies

Defang Ouyang

Dr Defang Ouyang is an Assistant Professor in the University of Macau, China. He has a multidisciplinary background in pharmaceutics and computer modeling, with experience in academia and industry. He completed his PhD in pharmacy at The University of Queensland, Australia, in 2010 and progressed directly to a faculty position at Aston University, UK in January 2011. At the end of 2014, he moved to the University of Macau. Since 2011, he has pioneered the integration of molecular modeling techniques and experimental approaches in the field of drug delivery – "computational pharmaceutics". He has published over 20 refereed journal papers in this area. He chaired the Computational Pharmaceutics Workshop at the 2014 Controlled Release Society Annual Meeting (Chicago, USA). Currently he supervises one PhD student and two MSc students. He has also successfully supervised one PhD student and 18 Master research projects.

Sean C. Smith

Professor Sean C. Smith received his PhD in theoretical chemistry at the University of Canterbury, New Zealand, in 1989. Following an Alexander von Humboldt Fellowship at the University of Göttingen, Germany and postdoctoral research at the University of California, Berkeley, USA he accepted a faculty position at The University of Queensland (UQ), Australia in 1993. He became Professor and Founding Director of the UQ Centre for Computational Molecular Science in 2002 and in 2006 his laboratory moved to the Australian Institute for Bioengineering and Nanotechnology at UQ. He was a founding chief investigator, computational program leader, and Deputy Director in the Australian Research Council Center of Excellence for Functional Nanomaterials 2002–2010. In 2011, he undertook the role of Director of the Department of Energy funded Center for Nanophase Materials Sciences (CNMS) at Oak Ridge National Laboratory, USA. In 2014 he moved to the University of New South Wales (UNSW)

Australia as Professor of Computational Nanomaterials Science and Engineering and Foundation Director of the Integrated Materials Design Centre. His specific research involves theoretical and computational studies of structure, reactivity and catalysis, photochemistry, electrochemistry and transport phenomena within nanomaterials, proteins, and hybrid nano-bio systems as well as his long-time passions of chemical kinetics and reaction dynamics. He has published over 230 refereed journal papers. In 2006 he was recipient of a Bessel Research Award from the Alexander von Humboldt Foundation in Germany and in 2012 he was elected Fellow of the American Association for the Advancement of Science (AAAS).

1

Introduction to Computational Pharmaceutics

Defang Ouyang[1] and Sean C. Smith[2]

[1] Institute of Chinese Medical Sciences, University of Macau, Macau
[2] School of Chemical Engineering, University of New South Wales, Australia

1.1 What Is Computational Pharmaceutics?

It is a given that active pharmaceutical ingredients (APIs) should be made into safe and effective dosage forms or formulations before administration to patients. Pharmaceutics is the discipline to make an API into the proper dosage form or medicine, which may then be safely and effectively used by patients [1]. Pharmaceutics also relates to the absorption, distribution, metabolism, and excretion of medicines in the body. Branches of pharmaceutics include formulation development, pharmaceutical manufacture and associated technologies, dispersing pharmacy, physical pharmacy, pharmacokinetics, and biopharmaceutics [1–4]. Today there are various dosage forms, such as tablets, capsules, solutions, suspensions, creams, inhalations, patches, and recently nanomedicines (e.g., liposomes, nanoparticles, nanopatches). Although numerous new techniques have been developed for the form of dosage, current development of drug formulations still strongly relies on personal experience of pharmaceutical scientists by trial and error in the laboratory [1]. The process of formulation development is laborious, time-consuming, and costly. Therefore, the simplification of formulation development becomes more and more important in pharmaceutical research. *Computational pharmaceutics* involves the application of computational modeling to drug delivery and pharmaceutical

Computational Pharmaceutics: Application of Molecular Modeling in Drug Delivery, First Edition.
Edited by Defang Ouyang and Sean C. Smith.

nanotechnology. In combination with existing branches of pharmaceutics, it offers rapidly growing potential for developing rational, deductive and knowledge-based strategies in pharmaceutics.

With stunningly rapid advances in hardware, theory and software algorithms, computer simulation is now able to model complex systems, which may be difficult, costly, or even impractical to measure or monitor directly by experiment [5, 6]. The first example of computer modeling was the simulation of the nuclear bomb process in the Manhattan Project, World War II. With the development of high performance computing, multiscale modeling techniques have been widely pursued, from quantum mechanics (QM) and molecular dynamics (MDs) to stochastic Monte Carlo methods, coarse grained dynamics, discrete element methods (DEMs), finite element methods as well as advanced analytical modeling. In principle, all properties of all systems are able to be described by QM. However, first principle calculations are limited to small systems, <1000 atoms, which is impractical for solving applications of large molecules or systems [5, 6]. MD simulations mimic the physical motion of atoms and molecules under Newton's laws of physics, which is applicable to larger systems containing millions of atoms [5, 6]. MD simulation is based on molecular mechanics, which models the interactions between atoms with force fields. Monte Carlo (MC) simulation uses the same empirical force field as MD simulation. However, MC simulation features playing and recording the results in casino-like conditions by repeated random sampling [5, 6]. Thus, unlike MD simulation, MC simulation cannot offer dynamical information with time evolution of the system in a form suitable for viewing. MC methods are especially useful for modeling systems with significant uncertainty and high degrees of freedom, such as polymer chains and protein membranes. For much larger systems, coarse-grain models do further classical approximations by treating functional groups as rigid bodies of constrained particles [5, 6]. The DEM is one of the numerical methods for computing the motion and effect of a large number of small particles, which is widely used in the pharmaceutical process and manufacturing [5, 6].

In the past three decades, the application of computational modeling approaches in the field of drug design (e.g., QSAR, ligand docking) has been intensively developed to the point of being a mature field [7]. Pharmaceutical research is, however, a far broader field than drug design alone. Proceeding beyond drug design, the application of computational modeling to drug delivery and pharmaceutical nanotechnology, *computational pharmaceutics*, is a very new field with great potential for growth [8]. As shown in Figure 1.1, computational pharmaceutics has the ability to provide multiscale lenses to pharmaceutical scientists, revealing mechanistic details ranging across the chemical reactions of small drug molecules, proteins, nucleic acids, nanoparticles, and powders with the human body. The aim of this book is to provide a contemporary overview of the application of computational modeling techniques to problems relating to pharmaceutics (drug delivery and formulation development) that will be of great relevance for pharmaceutical scientists and computational chemists in both industry and academia. Contributions from leading researchers cover both computational modeling methodologies and various examples where these methods have been applied successfully in this field.

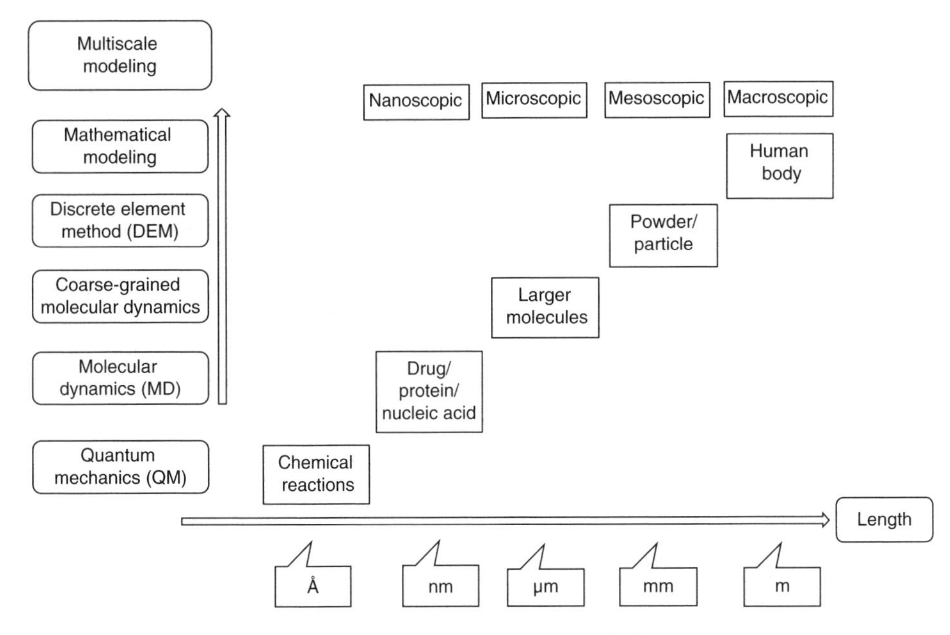

Figure 1.1 *Application of computational pharmaceutics.*

1.2 Application of Computational Pharmaceutics

Polymorphism of small drug molecules (e.g., crystal, hydrate, solvate, salt, cocrystal) plays a very important role in pharmaceutical research because it greatly influences the dissolution behavior and the bioavailability of pharmaceutical products. Thus, crystal structure prediction (CSP) methods have gained wide attention from pharmaceutical scientists. Chapter 2 discusses the general principles of CSP and recent progress in solid form screening by the crystal energy landscape method.

Cyclodextrins are a family of cyclic oligosaccharides, which are widely used in drug delivery and formulations for the solubilization of poorly soluble drugs. In Chapter 3, the physicochemical principles of cyclodextrin/drug complexation, experimental characterization and recent theoretical progress of drug/cyclodextrin modeling are discussed.

Polymeric-based micellar vehicles have been widely used in pharmaceutics for the delivery of both hydrophilic and lipophilic drugs. Multiscale modeling for polymeric-based vehicles for drug delivery is discussed in Chapter 4, including different computational approaches, micellar self-assembly and stability, the interaction of Taxol with model cellular membranes, and Taxol–tubulin association.

Solid dispersion refers to the dispersion of drug molecules in carriers in a solid state, prepared by the hot melting method. Poor physical stability has strongly hindered the commercialization of this technique. Chapter 5 discusses the possible molecular structure of amorphous solid dispersions and the mechanism of physical stability of this technology.

Biological lipid membranes are the key to drug absorption and bioavailability. Liposomes, artificially prepared vesicles with a phospholipid bilayer, are widely employed in drug delivery for cancer and other diseases. Chapter 6 reviews the theoretical progress of lipid membrane models, small-molecular uptake and permeation across lipid membranes, nanoparticle–membrane interaction, and the mechanism of chemical penetration enhancers.

Protein/peptide drugs are becoming increasingly important in the pharmaceutical market. However, the development of stable and effective formulations for biopharmaceuticals is still quite challenging for pharmaceutical scientists. Chapter 7 summarizes the diverse modeling results on the solution behavior of protein formulation, including protein aggregation pathways in liquid formulations, protein-cosolvent interactions, and protein–protein interactions.

Inorganic nanoparticles had been increasingly utilized for drug/gene delivery in recent decades. One of the main advantages of nanoparticle drug delivery systems is the targeting effect to specific organs and tissues. Chapter 8 discusses recent progress in computational modeling of inorganic nanoparticle drug delivery systems: carbon nanotubes, graphene/graphene oxide, silica, and gold nanoparticles.

Although the concept of nanodiamonds (diamond nanoparticles) for drug delivery is still in its infancy, recently nanodiamonds have been widely studied for bio-imaging and drug targeting for improved chemotherapeutic effect. Chapter 9 reviews the structure of the individual nanodiamond, its surface chemistry and interactions, and nanodiamond drug delivery systems.

Layered double hydroxides (LDHs) are composed of nanoscale cationic brucite-like layers and exchangeable interlayer anions. LDH nanoparticles are an efficient drug delivery system for anionic chemicals, such as small drug molecules (e.g., methotrexate, heparin) and nucleic acids (RNA and DNA). In Chapter 10, different computational approaches are discussed to investigate the properties and interactions of LDH/anion systems.

The structure of particles or powders plays an important role in many dosage forms, such as tablets, granules, and capsules. However, the microstructure of these dosage forms is less well investigated. Chapter 11 reviews the principles of synchrotron radiation-based microtomography (SR-μCT) and its application to determine the particulate architecture of granules, osmotic pump tablets. and HPMC matrix tablets.

Physiology-based pharmacokinetics plays an important role in pre-clinical drug development and formulation development. Chapter 12 discusses the principles of pharmacokinetic modeling and simulation and commercially available models for pharmaceutical scientists.

1.3 Future Prospects

"*Today the computer is just as important a tool for chemists as the test tube*" (Karplus, Levitt, and Warshel, Nobel Prize in Chemistry 2013). Analogous to the paradigm shift of drug development in the past three decades by computer-aided drug design, computational pharmaceutics also has great potential to shift the paradigm of drug delivery research in the near future [8].

References

[1] Aulton, M.E. and K. Taylor, *Alton's Pharmaceutics – The Design and Manufacture of Medicines* 4th edn. 2013, Elsevier.

[2] Sinko, P.J., *Martin's Physical Pharmacy and Pharmaceutical Sciences* 6th edn. 2010, Philadelphia, PA: Lippincott Williams and Wilkins.

[3] Florence, A.T. and D. Attwood, *Physicochemical Principles of Pharmacy* 5th edn 2011, London: Pharmaceutical Press.
[4] Allen, L.V., N.G. Popovich, and H.C. Ansel, *Ansel's Pharmaceutical Dosage Forms and Drug Delivery Systems* 9th edn. 2010, Philadelphia, PA: Lippincott Williams and Wilkins.
[5] Leach, A.R., *Molecular Modelling: Principles and Applications*. 2nd edn 2001, New York: Prentice Hall.
[6] Young, D.C., *Computational Chemistry: A Practical Guide for Applying Techniques to Real World Problems*. 2001, New York: John Wiley & Sons, Inc.
[7] Seddon, G., V. Lounnas, R. McGuire, *et al.*, Drug design for ever, from hype to hope. *Journal of Computer-Aided Molecular Design*, 2012. **26**(1): 137–150.
[8] Ouyang, D., A. Bunker, A. Rostami, and J. Zhang, Computational pharmaceutics: the application of computer modelling to drug delivery. *CRS Newsletter*, 2014. **31**(2): 16–18.

2

Crystal Energy Landscapes for Aiding Crystal Form Selection

Sarah L. Price

Department of Chemistry, University College London, UK

2.1 Introduction

The fundamental interrelationships between crystal structure, properties, performance and processing of a drug form the basis of "turning the development of pharmaceutical products from an art into a science" [1]. Thus, the development of an active pharmaceutical ingredient (API) into a pharmaceutical product requires detailed knowledge of its solid state forms and their physical properties. Even if the drug is not going to be delivered in a solid tablet, for example from a transdermal patch, there is a need to avoid crystallisation in the product. The manufacturing process will usually involve crystallisation as a purification step, possibly for separating out enantiomers as well as less similar synthetic impurities. Hence the design of the manufacturing process requires information about the solid forms that could be involved, including any hydrates or solvates [2].

An API rarely has only one crystalline phase. There may be polymorphs, hydrates, solvates, salts or cocrystals, though only cases where the solvent, counterion or coformer is pharmaceutically acceptable can be considered for development into a drug product. Salts and cocrystal forms can be used to improve solubility and dissolution rate [3]. Hence, a careful consideration of all alternatives is needed to select the physical form that will be used in the product, and the stability of multicomponent crystalline forms relative to their components is a key property [4].

Computational Pharmaceutics: Application of Molecular Modeling in Drug Delivery, First Edition.
Edited by Defang Ouyang and Sean C. Smith.

Polymorphism [5, 6], the ability of a molecule (or defined stoichiometry hydrate, salt or cocrystal) to adopt more than one crystal structure, plays an important role in pharmaceutical development, because the physical properties of a crystal, most critically the solubility and dissolution rates, can differ between polymorphs. Thus, the pharmaceutical product must contain just the approved polymorphic form. A drug is most often formulated using the most stable polymorph to avoid potential problems of phase transformations during storage. However, since metastable polymorphs or even amorphous forms may have better properties, particularly being more soluble, a metastable form may be developed provided that transformation to the more stable form during production and storage can be excluded. Unfortunately, molecules do not necessarily readily crystallise in their most stable form: indeed Ostwald's Rule of Stages [7] encapsulates the observation that metastable forms tend to be the first observed. The problem is that, once a more stable form has been nucleated, it can prove very difficult if not practically impossible to produce the metastable crystal structure, giving rise to the phenomenon of disappearing polymorphs [8]. There was a huge expansion of industrial and academic research into organic polymorphism after the Norvir® crisis, when Abbott Laboratories had to reformulate ritonavir at considerable expense after a novel, thermodynamically more stable, considerably less soluble polymorph emerged two years after the launch [9]. This led to the development of many solid form screening and characterisation methods [10]. However, polymorphism still causes problems within the industry, such as the crystallisation of rotigotine within the transdermal patches being used to deliver it to sufferers of Parkinson's disease in 2008 [11]. How can you ensure experimentally that the most stable crystalline form is known? One rare example of knowing that the most stable form has not yet been crystallised has been reported for a melatonin agonist, because the inactive enantiomer has two polymorphs and despite extensive efforts only the corresponding metastable form can be found for the active enantiomer [12]. The rationalisation is that the inactive enantiomer has come into contact with a chiral substance or surface capable of nucleating its most stable polymorph. It is the uncertainty in knowing that the most stable form has been crystallised that led to considerable practical interest in developing a computational method of predicting the most stable crystal structure of a given API [4].

Crystal structure prediction (CSP) methods gained their name decades ago, when people were trying to predict the crystal structure of a molecule, assuming that polymorphism was unlikely. The challenge was to predict the crystal structure of a molecule from just the chemical diagram. Success would indicate that all the factors which determined which crystal structure was adopted had been incorporated into the model. Most methods assumed that this involved searching for the most thermodynamically stable crystal structure, assuming that thermodynamics was the major determinant. The hope behind the first CSP programs was that there would be a significant energy difference between the observed structures and the hypothetical structures generated, a hope that was quickly dashed [13, 14]. The growing realisation that the prediction of crystal structures was industrially relevant, as well as being a fundamental scientific challenge, led to the Cambridge Crystallographic Data Centre organising blind tests of CSP [15–19]. In these tests, groups developing CSP methods are sent the chemical diagrams of the molecules whose solved crystal structures are kept confidential until the groups have submitted three predictions by a deadline. The accounts of the discussion of the results form an excellent review of both the variety of methodologies and the progress in the field, as well as showing the low success rate (Figure 2.1). In the fourth blind test (CSP2007), one group had the remarkable success of correctly predicting

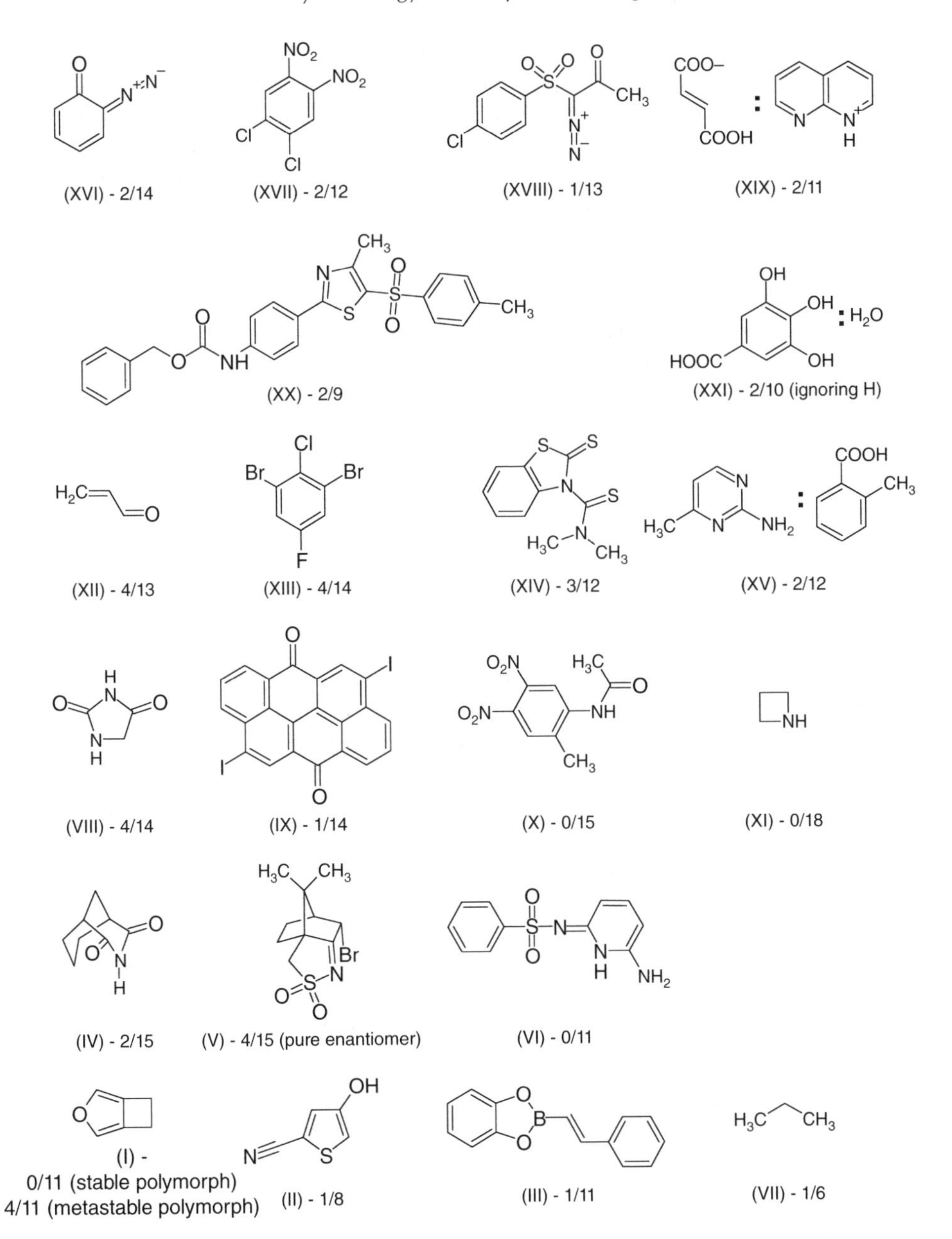

Figure 2.1 *Molecules used in the CCDC blind tests of crystal structure prediction. The Roman numberals denoting the targets are approximately in year order: I–III and VII CSP1999 [15], IV–VI CSP2001 [16], VIII–XI CSP2004 [17], XII–XV CSP2007 [18], and XVI–XXI CSP2010 [19]. The success rate is given as x/y, where x of the y groups which submitted three predictions included the experimental structure. For XXI the challenge was to predict further polymorphs.*

all four targets [20]. In the following test, the variety of target systems was increased to include a molecular salt, polymorphs of a hydrate and a molecule of a size that is more relevant to pharmaceutical development ($C_{25}H_{22}N_2O_4S_2$, XX in Figure 2.1). The crystal structure of this "model pharmaceutical" required a significant adaption of the methodology and was successfully predicted by two groups [21] but not by the approach which had been successful for the smaller molecules. Hence, the methodology of CSP studies is still actively evolving and is a challenge to computational chemistry methods. Future blind tests will help to evaluate progress, with computational costs (currently measured in CPU years for larger molecules) being only part of the challenge.

Polymorphism has always been acknowledged as a complicating factor in the interpretation of blind test results. Indeed further experimentation after blind tests found novel polymorphs for XXI [22], IV [23] and VI [24]. Whilst the thermodynamically most stable structure has to be found, which of the structures that are competitive in energy should correspond to polymorphs? CSP methods usually generate more plausible crystal structures than observed polymorphs, raising many questions about organic crystallisation [25]. This has led to the concept of a crystal energy landscape, a set of structures that are sufficiently low in energy to be potential polymorphs [26], and the question as to how these energy landscapes relate to the experimental crystal structures that the molecule could adopt. More practically, how can the calculated crystal energy landscapes help establish the true diversity of solid forms and act as a complement to other polymorph screening techniques?

Thus this chapter is divided into two sections. First, the general principles behind the computational approaches [27] that have been successful in blind tests will be outlined, with particular emphasis on the challenges for pharmaceutical molecules with their greater structural complexity. Pharmaceutical applications of CSP differ from those of computational materials design, where the aim is to suggest molecules whose crystals will have particular properties and avoid the synthesis of unpromising molecules. For example, an isolated molecule can have a large nonlinear optical coefficient, but if it crystallises in a centrosymmetric space group, the crystalline material will not be optically active. Molecular electronics, energetics, pigments and porous frameworks are other classes of functional materials where CSP methods have be applied. The blind tests are more closely aligned with the materials design problem, whereas for pharmaceutical applications, there may already be relevant experimental information that can be used to focus computational efforts, or known problems that need to be understood at the molecular level. The second section will therefore review some recent work where the crystal energy landscapes have been used as a complement to detailed experimental work on model pharmaceuticals, which indicates the potential use of CSP studies to complement experimental solid form screening.

2.2 CSP Methods for Generating Crystal Energy Landscapes

CSP methods [28] based on searching for the most thermodynamically stable crystal structures have to compromise between the number and range of crystal structures considered and the approximations used in evaluating the crystal energy. Current methods are based on the lattice energy as an approximation to the crystal energy, that is the energy of a completely ordered crystal at 0 K ignoring zero-point vibrational effects. This is only seeking the group of energetically favourable crystal structures and does not necessarily give the correct relative ranking of enantiotropically related polymorphs, whose relative stability order changes with temperature. The lattice energy is defined as the energy of an infinite, perfect, static crystal relative to the molecules in their lowest energy conformation at infinite separation.

Early work on CSP concentrated on simple rigid molecules, and in blind tests the molecular diagram is given (Figure 2.1). In applications, it is necessary to check whether you are modelling the appropriate molecules, and not their isomers or tautomers. The crystal energy landscapes are very different for different isomers [29], structural analogues or chemically closely related molecules [30]. A proton position can make a huge difference, for example the expected hydrogen bonding motif between a pyridine and a carboxylic acid is the same as between the corresponding pyridinium ion and carboxylate group, such that it would be reasonable to expect the crystal structures to be essentially the same apart from the acidic proton. However, comparing the crystal energy landscapes for three pyridinium carboxylate salts with those of the corresponding pyridine:carboxylic acid cocrystals showed considerable differences in the lower energy structures, except in the case where the proton is found to be disordered [31].

2.2.1 Assessment of Flexibility Required in Molecular Model

The definition of the lattice energy leads naturally to its separation into an intermolecular part, U_{inter}, and the intramolecular energy penalty paid for changing the conformation of the molecule to improve the molecular packing, ΔE_{intra}, giving $E_{latt} = U_{inter} + \Delta E_{intra}$. There is an important distinction between minor conformational changes and those that change the shape of the molecule and correspond to different conformers [32]. Minor changes of a few degrees in torsion angles, or amide pyramidalisation, methyl rotation or similar changes in proton positions for given covalent bonding can have a significant effect on the lattice energy but are usually refined in later stages of generating the crystal energy landscapes. Major changes, such as rotations about torsions which result in a significant change in the molecular shape or the relative positions of hydrogen bonding donors and acceptors, need to be considered explicitly within the search to ensure that the possibility of any of these conformations generating low energy crystal structures is considered. In general, molecules adopt low energy conformations in their crystal structures [32, 33]. However, there are exceptions, for example when the low energy conformation of the isolated molecule has an intramolecular hydrogen bond, which does not appear in the crystal structure because an intermolecular hydrogen bond is more favourable. If the molecule has different conformations that are in deep energy wells, then separate searches with each conformation held rigid are adequate. However, when there are low energy barrier torsions which change the shape of the molecule, such as between aromatic rings, then the flexibility to change shape has to be included in the first stages of the search. One of the important lessons from the crystal structure of XX in the 2010 blind test (Figure 2.1) is that the conformations adopted by larger molecules, whilst low in energy, can correspond to rather different molecular shapes than the conformational energy minima [21]. Figure 2.2 illustrates which torsions were considered as flexible in the search, and which only refined later, for a range of the molecules whose CSP studies are discussed in this chapter. The range of conformations to be considered in the search is one of the key limitations in the type of molecule that can be tackled or the computer time needed. Hence the development of strategies for CSP studies of flexible molecules depending on both the molecular size and flexibility and purpose of study is an active area, with phenobarbital [34] and the tyrosine kinase inhibitor axitinib (Inlyta®) [4, 35] giving complementary examples to those outlined in this chapter. The critical conformational analysis may well have been carried out in the drug design process for pharmaceuticals. Hence a combination of conformer generators and analysis of the conformations of fragments in known crystal structures of related molecules in the Cambridge Structural Database (CSD) [36] can be used to suggest conformational regions to be included in the search. Care is needed to ensure that the flexibility and relative energies are sufficiently realistic to ensure that the search covers the required region and does not disfavour the likely structures. For example, a conformational energy scan for the low energy torsion of tolfenamic acid (Figure 2.2) performed with a routine *ab initio* molecular orbital method [self–consistent–field calculations with a 6–31+(d) basis set] had a minimum where more accurate methods that represented the correlation in electron motions, and hence modelled the intramolecular dispersion, had a low energy maximum. Although this maximum was only about 3 kJ mol^{-1} for tolfenamic acid or 6 kJ mol^{-1} for unsubstituted fenamic acid, the distribution of this torsion angle in crystal structures of related molecules correlated with the more realistic *ab initio* method [37]. Thus, the CrystalPredictor methodology uses an interpolated grid representing the conformational energy calculated at a sufficient level of theory to be qualitatively correct [38, 39], and the GRACE procedure uses a tailor-made force field [40] for the specific molecule derived by extensive dispersion-corrected density functional calculations [41]. The computational

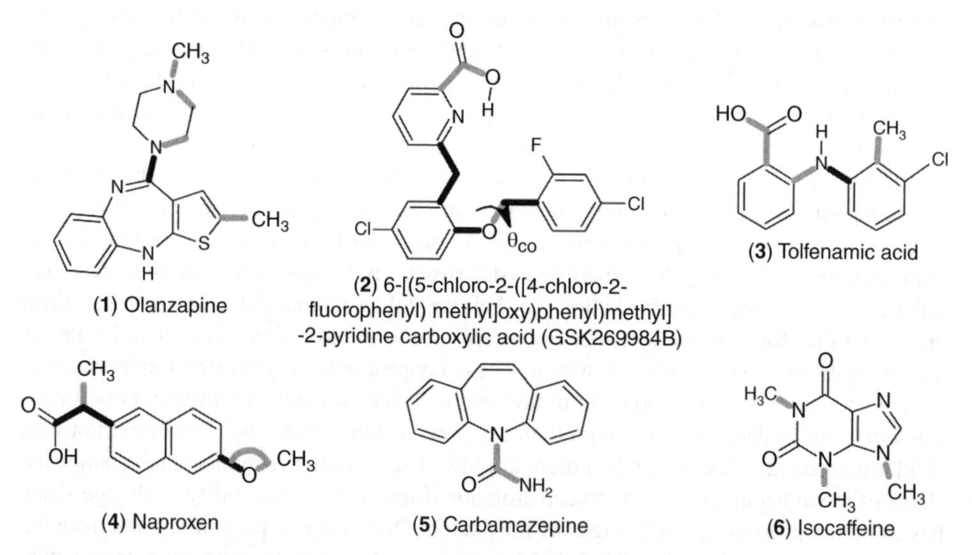

Figure 2.2 *Examples of the degree of flexibility considered during the generation of crystal energy landscapes. Changes to torsion angles marked in bold black are fundamental to the shape of the molecule and need to be considered at the outset of the search. The lighter grey bold angles only need to be refined at later stages of the structure optimisation. For* (**2**) *these approaches are combined in that the search was separated into nine subsearches in which one of three conformations of the CO_2H group and one of the three values of θ_{CO} was initially fixed and refined later.*

cost of treating flexibility at this level of accuracy is quite prohibitive, but has to be balanced by the possibility of the search procedure missing key structures if traditional force fields are used.

2.2.2 Search Method for Generating Putative Structures

A second choice which determines the computational cost and value of the CSP search is the coverage of crystallographic space. The number of independent molecules in the asymmetric unit cell (Z') is a critical choice, as the relative positions of these molecules vastly increase the number of search variables. This greatly adds to the cost of generating crystal energy landscapes for multicomponent systems, such as salts, cocrystals and hydrates. However, the usual procedure of restricting the search to one molecule in the asymmetric unit cell for single component systems ($Z' = 1$) will miss the crystal structures with $Z' > 1$. Many crystal structures with $Z' > 1$ are sufficiently closely related to $Z' = 1$ structures that this restriction is unimportant. Such structures can be generated in the search when the $Z' = 1$ structures are found not to be true minima from examination of the second derivative properties such as elastic tensor or phonon modes. However, there are crystal structures in which different molecules have different hydrogen bonds or different conformations and so intrinsically could only be found in a $Z' > 1$ search. Usually only a $Z' = 1$ search is worthwhile, unless there is evidence, for example from solid state NMR, that a $Z' = 2$ form needs to be structurally characterised. All experimentally known crystal structures can be used to check the validity of the crystal energy landscape by calculating the corresponding lattice energy minima. The lattice energies of all known structures should be among the most stable. This comparison can reveal that the molecule adopts a $Z' > 1$ structure because it has a lower energy than $Z' = 1$ possibilities [29, 37].

The next choice is the range of space groups that are considered in the search. If the molecule is chiral and an enantiomerically pure sample is being crystallised, as is often the case for pharmaceuticals, then it can only adopt a limited number of space groups, the most likely being $P2_12_12_1$, $P2_1$, $P1$, $C2$ and $P2_12_12$. If the molecule is not chiral, or both enantiomers are present, it can adopt a far wider range of space groups that contain an inversion centre, mirror or glide plane. The most popular space group for structures in the CSD is $P2_1/c$, and for a $Z' = 1$ structure in this space group, there are the three cell lengths, one cell angle and the position of the molecule relative to the unit cell axes as variables in the search. The ability to distinguish between an enantiopure and racemic crystal structure means that CSP is being developed to increase our understanding of the resolution of enantiomers by crystallisation, either through spontaneous resolution [42] or by diastereomeric salt formation [43]. For example, CSP was used [44] to solve the crystal structure of the racemic form of the nonsteroidal anti-inflammatory naproxen (Figure 2.2), which required consideration of closely related $Z' = 1$ and 2 structures, as well as confirmation that the known chiral structure was the most stable for the single enantiomer. Restricting the search to the chiral space groups will save a great deal of time for chiral APIs, but may exclude gaining valuable insights into the crystallisation process, particularly in cases when the chiral purity is suspect, for example if the barrier to racemisation is low because the chirality is conformational rather than from chiral sp^3 carbon atoms. In the case of tazofelone, there is a racemic solid solution that is isostructural with the enantiopure crystal that nucleates it, demonstrating that chiral impurities may be readily incorporated [45].

Evaluating the success of different search algorithms in covering defined areas of crystallisation space has been a major benefit of the blind tests [46]. In practical applications, the appropriate method will depend on the purpose of the study and timeline factors. For example, there is a rapidly growing use of CSP to aid the determination of the structure in cases where suitable single crystals cannot be grown, where the powder X-ray diffraction [47], solid state NMR [48] or electron diffraction data [49] may provide information that can be used to reduce the search space drastically. The algorithm choice also depends on the purpose of the search, and the blind test papers give details of the ever-emerging range and the assumptions behind them. For example, MOLPAK is based on density searching and designed for energetic materials research [50], PROM has been designed for easy use by crystallographers for small rigid molecules [51] and PolymorphPredictor was designed for industrial use based on simulated annealing. The programs which are currently being used for published pharmaceutical applications are GRACE [52] and CrystalPredictor [35, 38, 39], with other programs being developed for use on high-performance computer clusters [53]. An indication of the data generation and handling requirements is that a general search for a small rigid molecule should produce at least a few thousand plausible structures to be credible, whereas searches for flexible pharmaceuticals, monohydrates of smaller molecules or cocrystals, can readily generate a million reasonable crystal structures and still not satisfy the criteria for a complete search.

2.2.3 Methods for Computing Relative Crystal Energies

The number of plausible crystal structures that are generated makes a hierarchical system of improving the relative lattice energies, using increasingly expensive and accurate methods on a reducing number of the most promising crystal structures, inevitable. For example, the GRACE procedure uses a tailor-made force field [40] to generate the structures and then uses periodic density functional lattice energy minimisation (with an empirical dispersion correction specifically developed for CSP studies [41]) to refine the structures, working up in energy until it is clear that the reranking produced by calculations on further structures will not generate any more structures within the energy range of likely polymorphism.

This immediately begs the question of what is the energy range of potential polymorphism, as well as what is the accuracy required in evaluating the relative energies of different crystal structures of the same molecule? The longevity and practical importance of a metastable polymorph is mainly determined by the barrier to transformation to the most stable form, as this determines whether or not a crystal form is so long lived that it appears to be stable. Thermal calorimetric measurements on polymorphic transformations, heats of fusion and melting temperatures can be used to construct energy temperature diagrams for known polymorphs, to give the order of stability at 0 K for comparison with lattice energy rankings. Measured enthalpies suggest that lattice energy differences between observed polymorphs are small, usually of order of a few $kJ\,mol^{-1}$. Various reviews of computational methods applied to solid form pharmaceutical modelling reveal that predicting the relative stability of known polymorphs is challenging [54, 55]. Moreover, even for smaller molecules, there is considerable research into developing periodic electronic structure methods capable of the predicting lattice energies within the $kcal\,mol^{-1}$ target that is usually considered “chemical accuracy” [56].

The main alternative approach to electronic structure calculations on the crystal structures requires electronic structure calculations on the molecule in each conformation within the crystal structures. As well as providing an estimate of the intramolecular energy penalty paid for changing the molecular conformation within the crystal structure, ΔE_{intra}, the molecular charge density can be analysed to provide an atomistic model for the intermolecular interactions and evaluate the intermolecular contribution to the lattice energy, U_{inter}, using DMACRYS [57]. The lattice energy is $E_{latt} = U_{inter} + \Delta E_{intra}$. Both approaches are under continual development [58], and their relative accuracy, as well as computational cost, is very dependent on the molecule, its flexibility and the nature of the functional groups and types of interactions in the observed and computationally generated competitive crystal structures. Periodic *ab initio* calculations have the advantage of treating the molecular flexibility well, and automatically include the polarisation of the molecule within the crystal structure, but are limited by the dispersion correction and restricted to poorer quality charge densities. Calculating the molecular structure and wavefunction once can be done with a very high quality method for a rigid molecule, and the realism of this molecular structure and model for the intermolecular forces depends on the extent to which the molecular structure and charge distribution is changed by the crystal packing. It has been shown that using an atomic multipole rather than an atomic charge representation of the molecular charge density to model the electrostatic contribution to the lattice energy considerably improves the proportion of the rigid organic crystal structures which are found at, or near, the global minimum in a search [59]. The use of empirically fitted repulsion–dispersion potentials to model all the nonelectrostatic contributions to the intermolecular lattice energy has proved adequate for many studies, and indeed, the empirical fitting can partially absorb the effects of many approximations such as the neglect of intermolecular polarisation or neglect of thermal expansion. However, for flexible pharmaceuticals, such as those which can form a wide range of hydrogen bonding motifs (particularly if inter- and intramolecular hydrogen bonds are competitive), not only does the wavefunction need recalculating for changing conformations [38], but also the differential polarisation of the molecule within the lattice can be important. For example, this can be modelled in an average way by using the atomic multipoles and conformational energy penalty (ΔE_{intra}) calculated in a polarisable continuum [60]. Indeed, since crystallisation in the pharmaceutical industry is performed from different solvents, COSMO-RS theory can be applied to select the conformations that have the highest populations in a small diverse set of solvents as the input conformations in CSP [4]. However, sometimes it is necessary to model the polarisation of the molecule within specific crystal structures, rather than in a dielectric continuum. Testing a wide range of state of the art models for lattice energies to predict the relative energies of the four polymorphs of methylparaben [61] illustrates the challenge: although most methods have the polymorphs differing in energy by about $5\,\text{kJ mol}^{-1}$, the stability order changes. The polymorphic energy differences are dominated by the differences in the induction (polarisation) energy from the different hydrogen-bonding geometries, but the conformational energy, repulsion, dispersion and electrostatic terms all contribute. Calculating the relative energies of polymorphs is a challenge to theoretical chemistry methods, as the experimental crystal structures represent the balance between all the inter- and intramolecular forces acting on the atoms in the molecule, and polymorphism arises because different arrangements balance these forces in different ways.

Ideally, the final stage of a CSP should be the accurate evaluation of the relative free energies of the different crystal structures at ambient temperature. Until this is possible, we need to be aware of the assumptions that are being made in calculating the relative lattice energies and modify the conclusions as to the relative stability of observed and computer-generated structures according to the sensitivity of the relative energies to different models for the crystal energies.

2.2.4 Comparing Crystal Structures, and Idealised Types of Crystal Energy Landscapes

The simplest crystal energy landscape is one where there is only one structure within the energy range of plausible polymorphism. This energy range is likely to be larger for APIs where polymorphs may be produced by desolvating solvates than for small molecules which can easily transform to the most thermodynamically stable form. A working value is usually taken as being of order 5–10 kJ mol^{-1}. However, clearly monomorphic energy landscapes (Figure 2.3a) appear to be rare – in studies of a few hundred crystal energy landscapes, the largest energy gap we found is for isocaffeine of 6 kJ mol^{-1} [62], though a CSP search for crizotinib, using four conformers predicted to be present in a range of five media, had a lattice energy gap of 7.6 kJ mol^{-1} [4]. A large energy gap requires that the molecule has a unique way of packing with itself that is particularly favourable in all three dimensions. Many molecules have a strongly favoured hydrogen bonded sheet, but there are different ways of stacking this sheet which are similar in energy. The shape of a chiral molecule can define a strong preference for a chiral column, but it is less likely that the interactions between the columns also give a strong preference for a chiral crystal over structures where the columns pack with an inversion centre [42]. Whilst it could be possible that the flexibility of pharmaceutical molecules could help produce a larger energetic advantage for the most stable crystal, experience with the crystal energy landscapes calculated to date [21, 27, 63], as well as the high rate of polymorphism (~50%) found in solid form screening [64], do not suggest that this will be very common.

Hence the crystal energy landscape, the set of crystal structures which are thermodynamically plausible, usually requires interpretation as to whether the structures seem likely to be metastable polymorphs, or are sufficiently similar that it seems unlikely that they could nucleate and grow as distinct polymorphs [25]. We do not yet understand the process of nucleation and growth, responsible for the late appearance of the most stable and the disappearance of metastable polymorphs, well enough to implement this in a computer. Better modelling of thermal effects, such as molecular dynamics simulations, can however eliminate some lattice energy minima as being equivalent free energy minima, but the proportion of structures so eliminated differs considerably between benzene and 5-fluorouracil [65]. Qualitative relationships can give a good guide, and hence the structures on the crystal energy landscape are usually classified by type of packing, being more informative than the space group. This classification can be by types of hydrogen bonding, as given by the graph-set notation, or by an analysis of other supramolecular constructs, using various crystallography software tools [66, 67]. Structures with different hydrogen bonding or large conformational differences are more likely to be able to be trapped as distinct polymorphs, because of the larger barrier to rearrangement in the condensed phase. However, the differences between polymorphs can be very subtle and subject to debate [68], with

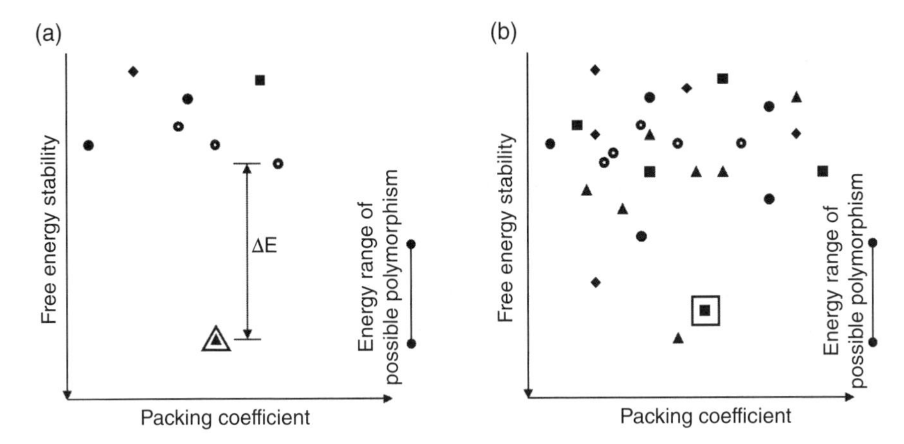

Figure 2.3 *Two idealised plots summarising the lower energy structures generated in a CSP study. Each symbol represents a distinct crystal structure, with different symbols representing unrelated structures (e.g. different hydrogen bonding motifs). The experimental crystal structure is denoted by an open symbol. The comparison with the plausible energy range for polymorphism determines that (a) is a monomorphic system, whereas in (b) there is a clear prediction that a more thermodynamically stable form should exist.*

reports of there being different polymorphs as domains within the same single crystal [69]. Hence, comparisons of CSP studies with experimental polymorph screens are showing us how to interpret the crystal energy landscape [25].

2.2.5 Multicomponent Systems

There are an increasing number of cases where a large number of crystal forms are known and have been structurally characterised, encompassing many solvates and cocrystals, such as over 50 solid forms of carbamazepine [70] or 60 solid forms of olanzapine [71]. In this case the analysis of the relationships between the known multicomponent structures and the hypothetical crystal structures generated by CSP can help interpret the crystallisation behaviour. There are cases where the hydrate [72] or inclusion compound framework structure [73] is found on the single-component crystal energy landscape. The solvent or guest molecule fits in the voids of the CSP generated structure and stabilises the lattice. The relationship can also be with layers; many of the solvates of olanzapine have the same layers as high energy CSP structures, implying that the solvent provides a stabilising "glue" between the layers [71]. Hence a CSP study may help in understanding promiscuous solvate formation or the desolvation behaviour of the system.

CSP studies on multicomponent systems have been applied to pharmaceutical cocrystals [74] and hydrates [75]. The cost of calculations with more than one molecule in the asymmetric unit cell means that these studies are far more useful for understanding the polymorphism of these systems than as a means of predicting whether the multicomponent system would form. The experiments are far quicker [76] and the energy differences between the multicomponent crystal and its components are often small and comparable with the uncertainties in the calculation. A salutary counter-example is the caffeine:benzoic acid cocrystal, which was predicted to be thermodynamically stable, allowing for all the uncertainties in the calculations, and yet persistently failed to form using a broad range of established techniques in at least four geographically diverse laboratories [77]. Once designed heteronuclear seeds facilitated the first formation of this predicted but elusive caffeine:benzoic acid cocrystal, it was very readily formed in subsequent experiments. This example showed how seeds of crystal forms, at such low levels to evade detection, can act as long-lasting laboratory contaminants affecting crystallisation behaviour.

2.3 Examples of the Use of Crystal Energy Landscapes as a Complement to Solid Form Screening

2.3.1 Is the Thermodynamically Stable Form Known?

CSP studies do not often result in such a clear prediction as for the caffeine:benzoic acid cocrystal [77] that a significantly more thermodynamically stable form should exist to justify considerable experimental work to find it. Other curious examples of an extended, but eventually successful, experimental search for the most stable structure on a crystal energy landscape are the racemic crystal structures of progesterone [78] and naproxen [44]. Indeed, if a landscape such as Figure 2.3b is generated, the question arises as to whether the method of evaluating the crystal energies is accurate enough. However, as confidence in the calculations and their interpretation in conjunction with polymorph screening activities increases and is applied more to "hard to crystallise" molecules, then more examples will emerge.

The more nuanced interpretation of the crystal energy landscapes can be illustrated by comparing those of fenamic acid and tolfenamic acid (Figure 2.4) in which all the low energy structures are based on the carboxylic acid dimer, with the phenyl rings able to adopt a wide range of torsion angles [37]. Many fenamates are polymorphic, with the four ordered polymorphs of tolfenamic acid either being found in the search, if they are $Z' = 1$, or having similar lattice energies if they are $Z' > 1$ (Figure 2.4). The crystal energy landscape contains other competitive structures, which could well be found as additional polymorphs, as they are similar in packing to other known fenamate structures, some of which (including tolfenamic acid forms III–V) have been generated by polymer templating. In contrast, fenamic acid is monomorphic with a $Z' = 2$ structure. This structure is closely related to the global minimum in the search (Figure 2.4), and both are related to the next most stable structure, based on identical layers, which is $2\,kJ\,mol^{-1}$ higher in energy. It appears that the stacking of the layers in the second ranked structure is unfavourable, and the energy can be lowered either by shifting the layers (as in the global minimum structure) or by different rotation of the phenyl groups in half the molecules to give the observed $Z' = 2$ structure. It therefore seems unlikely that the global minimum structure could be nucleated and grown as distinct from the known form, and the energy gap suggests that polymorphism is unlikely for fenamic acid. This detailed study of the differences in packing of the known and computer generated thermodynamically competitive structures shows the benefits of CSP studies: there can be general principles that seem to promote polymorphism, such as the conformational flexibility of the fenamate skeleton, but it is the packing of the specific molecule, with the differences caused by the individual substituents, that determines the relative energies of the structures.

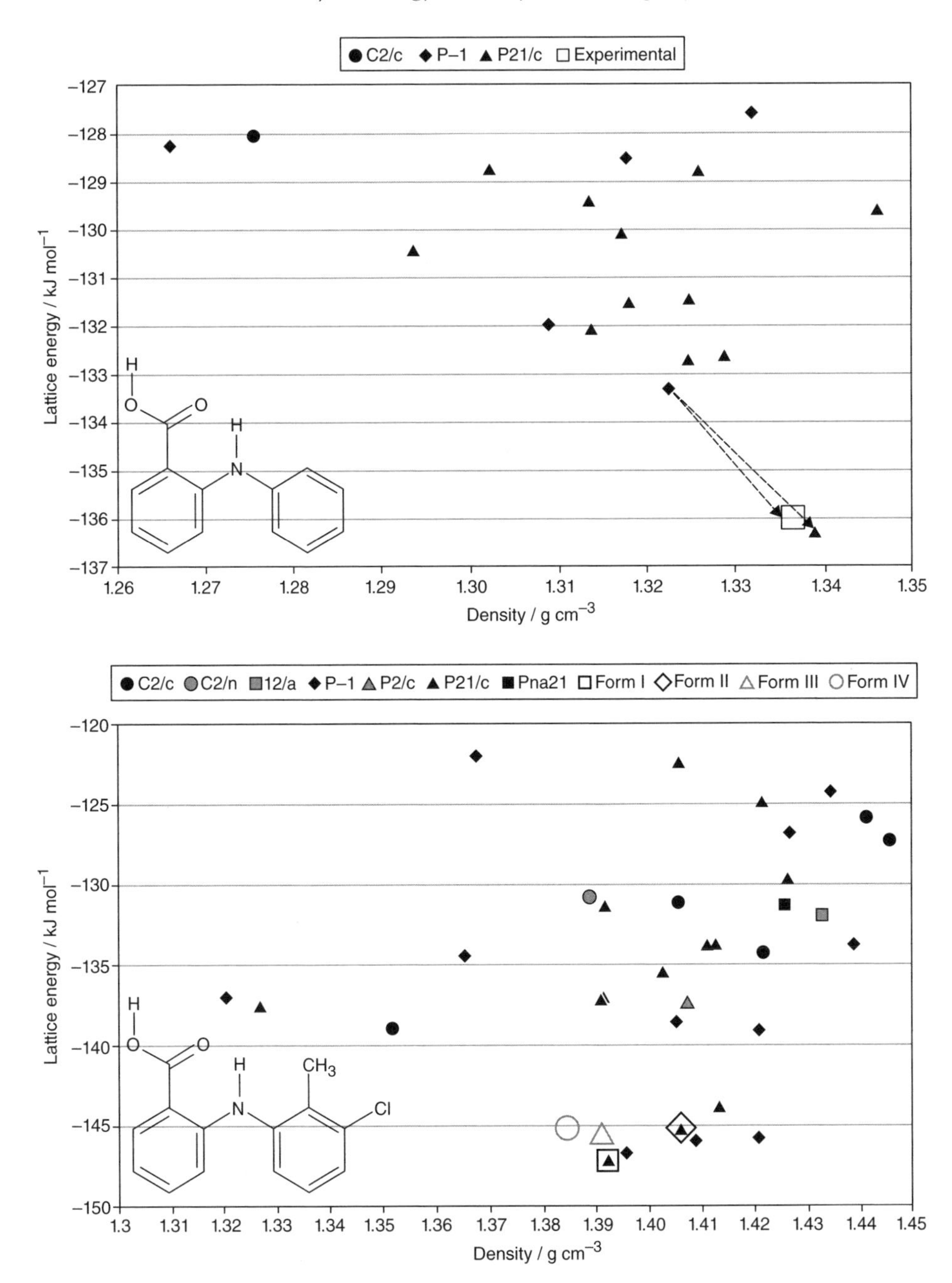

Figure 2.4 *The contrasting CSP searches of (top) fenamic acid and (bottom) tolfenamic acid. The open symbols denote experimental polymorphs: black for $Z' = 1$ structures that could be generated in the search; grey for $Z' > 1$ structures which could not. For fenamic acid, the arrows link closely related structures.*

2.3.2 Supporting and Developing the Interpretation of Experiments

CSP studies have always been seen as having the ability to complement and inform experimental studies. They have long been used to aid the derivation of structures from powder diffraction data, particularly for pigments [79]. The ability to solve structures from powder data has advanced considerably, but it cannot locate protons. These are often placed in chemically reasonable positions, with the positions being optimised by a fixed cell periodic *ab initio* optimisation. This use of computer modelling to correct for the systematic errors in the positions of protons determined from X-ray diffraction experiments is increasing. However, a better check on the validity of the structure would be to do a full optimisation, allowing the cell to also vary to confirm that the structure is close to a lattice energy minimum. Ephedrine d-tartrate provides an example where the initial structure produced from solving the powder diffraction data with protons placed by chemical intuition was far from a lattice energy minimum [80]. This minimum was found to be highly metastable in a CSP search within the experimental space group, but the global minimum structure was essentially the same, apart from the position of two hydroxyl protons. Hence, combining the experimental and computational information gave a much better structural determination [80].

This ability of CSP studies to generate alternative structures, that may only differ in details such as proton positions or stacking of the same layers, can often help in the interpretation of conflicting experimental information. If the energies are very similar, this can lead to disorder, and the CSP generated structures suggest an atomic scale model of the disorder. In the case of phloroglucinol dihydrate, such an analysis of the crystal energy landscape was able to show that variations in crystal surface features and diffuse scattering with growth conditions were not a matter of concern [81].

The use of crystal energy landscapes to provide better insights into pharmaceutical solid form characterisation has been exemplified by a joint screening and computational study on olanzapine [71]. A crystal energy landscape (Figure 2.5) found the structurally characterised forms I and II, and it also showed that there were other structures closely related to form II, differing only in the stacking of the same layers of olanzapine dimers. This interrelationship not only made sense of the inability to grow samples of form III without some form II and/or form I content, but also one of the computer generated structures showed a degree of similarity with form III as indicated by two-phase Pawley-type refinement of the unit cell parameters of the computer generated structure against the best available mixed-phase form III powder diffraction data. The variable proportions of forms II and III with different growth conditions account for some of the confused claims for the polymorphism of olanzapine, although some claimed polymorphs proved to be solvates. The CSP study also explained why olanzapine is such a promiscuous solvate former, with 56 solvates, including isostructural families and mixed solvates. Structures which contain the layers seen in the solvates, corresponding to computationally desolvated solvates, are found at high energy (around $-115\,kJ\,mol^{-1}$) amongst the plethora of structures too unstable to be shown in Figure 2.5. The large number of solid forms found for olanzapine is not matched by the closely related molecules clozapine and amoxapine in analogous experimental screening, consistent with their crystal energy landscapes for the observed conformations being much sparser than Figure 2.5. It is notable that, although the low energy crystal structures of olanzapine include only one implausibly low density structure based on a different conformation, there are many thermodynamically competitive structures that do not contain

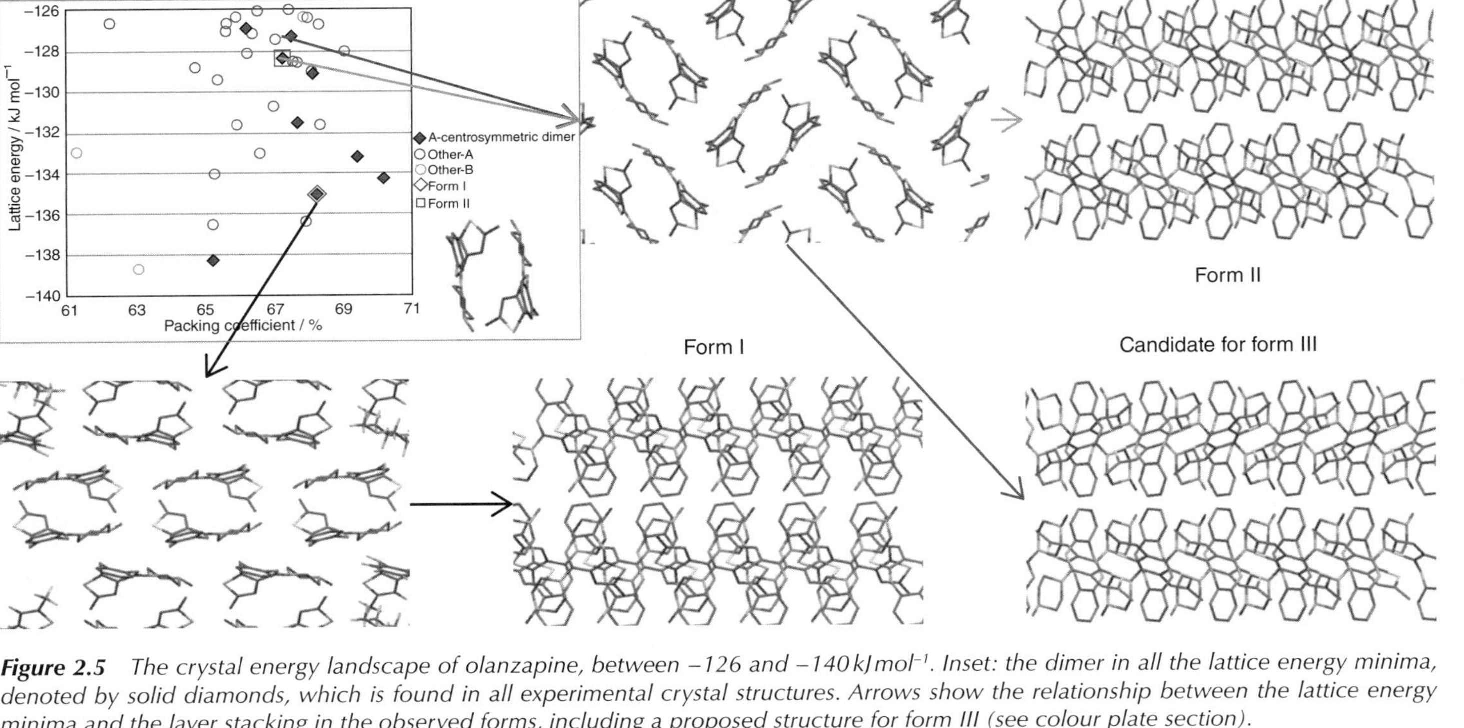

Figure 2.5 *The crystal energy landscape of olanzapine, between −126 and −140 kJ mol⁻¹. Inset: the dimer in all the lattice energy minima, denoted by solid diamonds, which is found in all experimental crystal structures. Arrows show the relationship between the lattice energy minima and the layer stacking in the observed forms, including a proposed structure for form III (see colour plate section).*

the olanzapine dimer motif seen in all the experimentally characterised forms (Figure 2.5). Is this because the kinetics of crystallisation means that the dimer is the growth unit under all crystallisation conditions used? In the case of olanzapine, the screening process included a variety of solution crystallisations using 71 diverse solvents, neat and liquid assisted grinding, crystallisation from the vapour phase, quenching the melt, recrystallisation from the amorphous phase, spray and freeze drying and desolvation of solvates [71], and so the range of crystallisation conditions is particularly large.

2.3.2.1 Design of Experiments to Find New Polymorphs

The tendency of computed crystal energy landscapes to generate more thermodynamically feasible structures than known polymorphs poses a challenge to our understanding of the factors that determine which crystal structures can form. Many lattice energy minima are artefacts of the lack of molecular motion in the calculations, and some polymorphs may not form as they cannot nucleate and grow without transformation to a more stable related structure [25]. However, other structures may inspire the experiment that finds the structure. The most explicit example is the targeting of a polymorph of the anti-epileptic carbamazepine which does not contain the doubly hydrogen-bonded dimer seen in most of its 50 or more solid forms but has catemeric hydrogen bonding. The prediction that this distinctive hydrogen-bonded polymorph was energetically competitive with the known forms led first to intensive solvent screening [82], which generated new solvates both directly and through statistical analysis [83] but no novel polymorphs. The closely related molecule, dihydrocarbamazepine, has a catemeric structure, but adding dihydrocarbamazepine to solutions of carbamazepine leads to a solid solution of the two, proving that carbamazepine could form catemers [84]. Finally, carbamazepine form V was found by subliming carbamazepine onto a crystal of isostructural dihydrocarbamazepine form II [85]. This demonstrates a strategy for finding predicted polymorphs, but much more needs to be known about how heterogeneous surfaces can template the first nucleation of a predicted polymorph to generalise this approach. However, it is well established that heterogeneous surfaces or impurities often produce the first nucleation of polymorphs; specialised screens being developed for difficult molecules are mainly based on providing different conditions for heterogeneous nucleation [10]. Hence CSP studies will increasingly help focus specialised screens by determining the conformations and stronger intermolecular interaction of structures that are thermodynamically plausible as target polymorphs.

2.3.2.2 Understanding the Potential Risks of Solid Form Change

Solid form screening is performed to select the most suitable solid form for development into a product, and to design the optimum crystallisation route and formulation to minimise the risk of any change in the solid form [2]. A study assessing the incorporation of CSP calculations into the industrial screening process was carried out in collaboration with GSK. The molecule GSK269984B was chosen as one where preliminary screening had been done, and a crystal structure was known. A key feature of this molecule (Figure 2.2) was the possibility of inter- or intramolecular hydrogen bonding, with the ortho substituents resulting in a complex conformational energy surface covering a wide range of molecular shapes (Figure 2.6). The crystal energy landscape confirmed that the known form was probably the most thermodynamically stable, but showed that there were thermodynamically

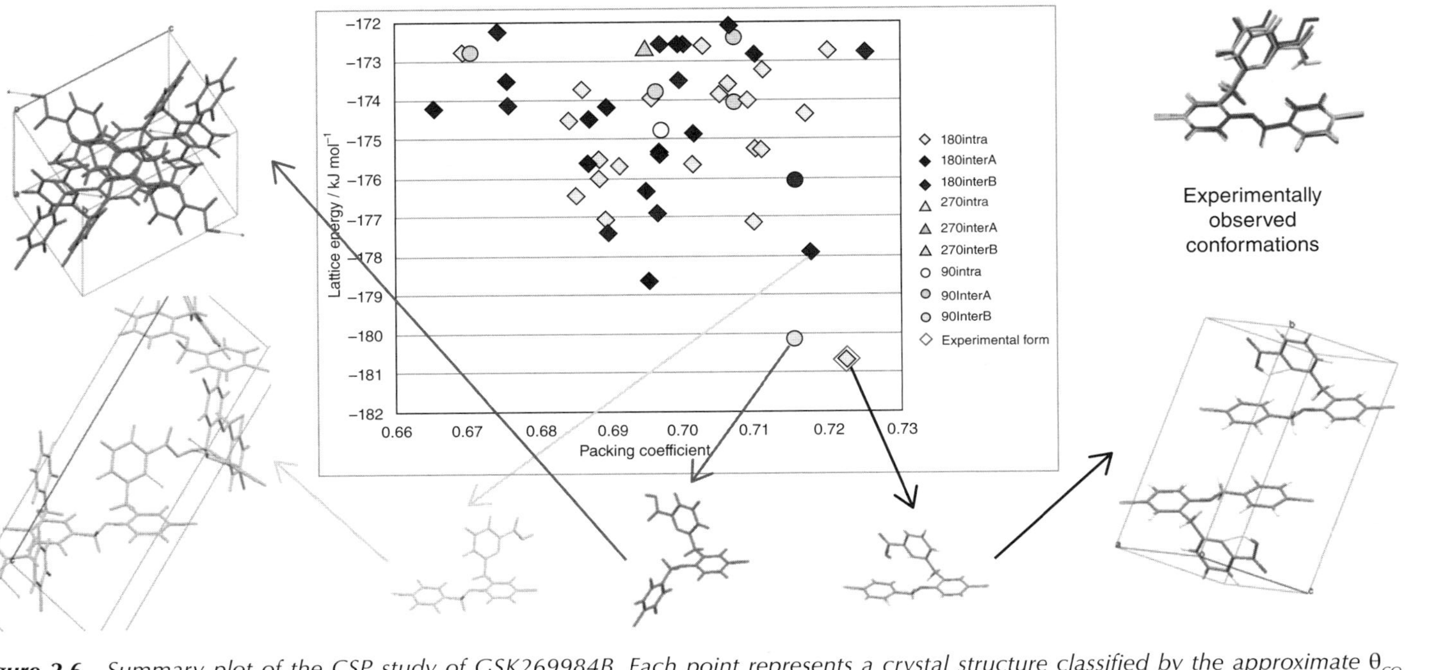

Figure 2.6 *Summary plot of the CSP study of GSK269984B. Each point represents a crystal structure classified by the approximate* θ_{CO} *(defined in Figure 2.2) angle and acid conformation (Intra/InterA/InterB). The three lowest energy structures and their conformations are linked by arrows, and should be contrasted with the overlay of the conformations in the experimental forms: form I and* ab initio *minimum (Intra carboxylic acid conformation, in element colours), the DMSO solvate (InterA, in orange) and the NMP solvate (InterB, in blue; see colour plate section).*

competitive structures with inter- instead of intramolecular hydrogen bonding and very different gross conformations from the known structure (Figure 2.6). More extensive experimental screening, aimed at changing the crystallisation conformation did result in a few metastable solvates. The crystal structures of two solvates could be solved, revealing the surprising result that the solvates were intermolecularly hydrogen bonded but, apart from the hydrogen bonding proton, had the same conformation as the anhydrate form and the molecule in isolation as estimated by *ab initio* calculations (Figure 2.6). The combination of the experimental and computational screening meant that form I could have been developed with considerable confidence. However, there was the nagging result that there could be other polymorphs, if it were possible to crystallise the molecule with a very different conformation. How would it be possible to be sure that there was no risk of a solvent or impurity generating the metastable polymorphs?

2.4 Outlook

There is a considerable challenge to computational chemistry to be able to evaluate the relative energies of polymorphs to a useful accuracy. This is in tension with the challenge for computer scientists and program developers to reduce the computer resources required so that a meaningful CSP study could be performed quickly enough to complement the industrial solid form screening process. There is also the challenge to theory of being able to predict the properties of different polymorphs, such as morphology, mechanical properties, solubility and so on sufficiently well to be able to predict whether any unobserved structure on the crystal energy landscape has a more desirable compromise of physical properties than the known forms. Trying to target finding these polymorphs will be aided by increasing the range of spectral signatures that can be calculated from the CSP generated crystal structures, going beyond easily simulated X-ray powder patterns. This would allow novel polymorphs to be detected more readily, even in unpromising microcrystalline mixed-phase samples.

Fundamental studies on how the conformational flexibility in solvents and heterogeneous surfaces influence the first nucleation of novel forms should lead to strategies to either find CSP generated thermodynamically plausible polymorphs, or conclude that they could never be found. This insight into the kinetic factors which control polymorphism is required before we can hope to reliably predict metastable polymorphs and the conditions to find them. Metastable polymorphs, hydrates and other solid forms are important in developing the appropriate manufacturing process and drug delivery route, and hence it is important to be able to interpret the crystal energy landscape for metastable polymorphs, phase relationships and possible disorder. It is a wasted opportunity to do a CSP study just to confirm that the most stable form is known.

A software code able to predict all polymorphs of any API, and provide a recipe for crystallising them, could be seen as a blue horizons goal that would demonstrate a fundamental understanding of molecular self-assembly, at a level that is also required for computational drug design. Although we are still far from this ambitious goal, we are already at the stage where CSP studies can be usefully combined with solid form screening to help give a molecular level insight into the possible solid form diversity and bring together the experimental results on a system. By showing the range of possible alternative packings to the most easily obtained crystals, CSP studies can help focus experimental solid form work more effectively to deal with complex systems and help reduce the potential problems of late appearing polymorphs.

Acknowledgements

I thank the many people from different disciplines who worked on the CPOSS (Control and Prediction of the Organic Solid State) project, www.cposs.org.uk, and the EPSRC for funding some of them. Dr Louise Price is particularly thanked for her help with producing this chapter.

References

[1] Sun, C.C. (2009) Material science tetrahedron – a useful tool for pharmaceutical research and development. *Journal of Pharmaceutical Sciences*, **98**, 1744–1749.
[2] Tung, H.H. (2013) Industrial perspectives of pharmaceutical crystallization. *Organic Process Research and Development*, **17** (3), 445–454.
[3] Elder, D.P., Holm, R. and de Diego, H.L. (2013) Use of pharmaceutical salts and cocrystals to address the issue of poor solubility. *International Journal of Pharmaceutics*, **453** (1), 88–100.
[4] Abramov, Y.A. (2013) Current computational approaches to support pharmaceutical solid form selection. *Organic Process Research and Development*, **17** (3), 472–485.
[5] Bernstein, J. (2011) Polymorphism – a perspective. *Crystal Growth and Design*, **11** (3), 632–650.
[6] Bernstein, J. (2002) *Polymorphism in Molecular Crystals*, Clarendon Press, Oxford.
[7] Threlfall, T. (2003) Structural and thermodynamic explanations of Ostwald's rule. *Organic Process Research and Development*, **7** (6), 1017–1027.
[8] Dunitz, J.D. and Bernstein, J. (1995) Disappearing polymorphs. *Accounts of Chemical Research*, **28** (4), 193–200.
[9] Chemburkar, S.R., Bauer, J., Deming, K., *et al.* (2000) Dealing with the impact of ritonavir polymorphs on the late stages of bulk drug process development. *Organic Process Research and Development*, **4** (5), 413–417.
[10] Newman, A. (2013) Specialized solid form screening techniques. *Organic Process Research and Development*, **17** (3), 457–471.
[11] Rascol, O. and Perez-Lloret, S. (2009) Rotigotine transdermal delivery for the treatment of Parkinson's disease. *Expert Opinion on Pharmacotherapy*, **10** (4), 677–691.
[12] Stephenson, G.A., Kendrick, J., Wolfangel, C. and Leusen, F.J. (2012) Symmetry breaking: polymorphic form selection by enantiomers of the melatonin agonist and its missing polymorph. *Crystal Growth and Design*, **12** (8), 3964–3976.
[13] van Eijck, B.P., Mooij, W.T.M. and Kroon, J. (1995) Attempted prediction of the crystal-structures of 6 monosaccharides. *Acta Crystallographica Section B: Structural Science*, **51**, 99–103.
[14] Gavezzotti, A. (1994) Are crystal-structures predictable? *Accounts of Chemical Research*, **27** (10), 309–314.
[15] Lommerse, J.P.M., Motherwell, W.D.S., Ammon, H.L., *et al.* (2000) A test of crystal structure prediction of small organic molecules. *Acta Crystallographica Section B: Structural Science*, **56**, 697–714.
[16] Motherwell, W.D.S., Ammon, H.L., Dunitz, J.D., *et al.* (2002) Crystal structure prediction of small organic molecules: a second blind test. *Acta Crystallographica Section B: Structural Science*, **58**, 647–661.
[17] Day, G.M., Motherwell, W.D.S., Ammon, H.L., *et al.* (2005) A third blind test of crystal structure prediction. *Acta Crystallographica Section B: Structural Science*, **61** (5), 511–527.

[18] Day, G.M., Cooper, T.G., Cruz-Cabeza, A.J., *et al.* (2009) Significant progress in predicting the crystal structures of small organic molecules – a report on the fourth blind test. *Acta Crystallographica Section B: Structural Science*, **65** (2), 107–125.

[19] Bardwell, D.A., Adjiman, C.S., Arnautova, Y.A., *et al.* (2011) Towards crystal structure prediction of complex organic compounds – a report on the fifth blind test. *Acta Crystallographica Section B: Structural Science*, **67** (6), 535–551.

[20] Neumann, M.A., Leusen, F.J.J. and Kendrick, J. (2008) A major advance in crystal structure prediction. *Angewandte Chemie International Edition*, **47** (13), 2427–2430.

[21] Kazantsev, A.V., Karamertzanis, P.G., Adjiman, C.S., *et al.* (2011) Successful prediction of a model pharmaceutical in the fifth blind test of crystal structure prediction. *International Journal of Pharmaceutics*, **418** (2), 168–178.

[22] Braun, D.E., Bhardwaj, R.M., Florence, A.J., *et al.* (2013) Complex polymorphic system of gallic acid-five monohydrates, three anhydrates, and over 20 solvates. *Crystal Growth and Design*, **13** (1), 19–23.

[23] Hulme, A.T., Johnston, A., Florence, A.J., *et al.* (2007) Search for a predicted hydrogen bonding motif – a multidisciplinary investigation into the polymorphism of 3-Azabicyclo[3.3.1]nonane-2,4-dione. *Journal of the American Chemical Society*, **129** (12), 3649–3657.

[24] Jetti, R.K.R., Boese, R., Sarma, J., *et al.* (2003) Searching for a polymorph: second crystal form of 6-amino-2-phenylsulfonylimino-1,2-dihydropyridine. *Angewandte Chemie International Edition*, **42**, 1963–1967.

[25] Price, S.L. (2013) Why don't we find more polymorphs? *Acta Crystallographica Section B: Structural Crystallography and Crystal Chemistry*, **69**, 313–328.

[26] Price, S.L. (2009) Computed crystal energy landscapes for understanding and predicting organic crystal structures and polymorphism. *Accounts of Chemical Research*, **42** (1), 117–126.

[27] Price, S.L. (2013) Predicting crystal structures of organic compounds. *Chemical Society Reviews*, **9**, 693–704.

[28] Day, G.M. (2011) Current approaches to predicting molecular organic crystal structures. *Crystallography Reviews*, **17** (1), 3–52.

[29] Barnett, S.A., Johnson, A., Florence, A.J., *et al.* (2008) A systematic experimental and theoretical study of the crystalline state of six chloronitrobenzenes. *Crystal Growth and Design*, **8** (1), 24–36.

[30] Barnett, S.A., Hulme, A.T., Issa, N., *et al.* (2008) The observed and energetically feasible crystal structures of 5-substituted uracils. *New Journal of Chemistry*, **32** (10), 1761–1775.

[31] Mohamed, S., Tocher, D.A. and Price, S.L. (2011) Computational prediction of salt and cocrystal structures – does a proton position matter? *International Journal of Pharmaceutics*, **418** (2), 187–198.

[32] Cruz-Cabeza, A.J. and Bernstein, J. (2013) Conformational polymorphism. *Chemical Reviews*, **6**, 195–205.

[33] Cruz-Cabeza, A.J., Liebeschuetz, J.W. and Allen, F.H. (2012) Systematic conformational bias in small-molecule crystal structures is rare and explicable. *CrystEngComm*, **14** (20), 6797–6811.

[34] Day, G.M., Motherwell, W.D.S. and Jones, W. (2007) A strategy for predicting the crystal structures of flexible molecules: the polymorphism of phenobarbital. *Physical Chemistry Chemical Physics*, **9** (14), 1693–1704.

[35] Vasileiadis, M., Pantelides, C., Adjiman, C. (2015) Prediction of the crystal structures of axitinib, a polymorphic pharmaceutical molecule. *Chemical Engineering Science* **121**, 60–76.

[36] Allen, F. H. (2002) The Cambridge Structural Database: A quarter of a million crystal structures and rising. *Acta Crystallographica Section B: Structural Science*, **58**, 380–388.

[37] Uzoh, O.G., Cruz-Cabeza, A.J. and Price, S.L. (2012) Is the fenamate group a polymorphophore? Contrasting the crystal energy landscapes of fenamic and tolfenamic acids. *Crystal Growth and Design*, **12** (8), 4230–4239.

[38] Vasileiadis, M., Kazantsev, A.V., Karamertzanis, P.G., *et al.* (2012) The polymorphs of ROY: application of a systematic crystal structure prediction technique. *Acta Crystallographica Section B: Structural Science*, **68** (6), 677–685.

[39] Karamertzanis, P.G. and Pantelides, C.C. (2007) Ab initio crystal structure prediction. II. Flexible molecules. *Molecular Physics*, **105** (2/3), 273–291.
[40] Neumann, M.A. (2008) Tailor-made force fields for crystal-structure prediction. *Journal of Physical Chemistry B*, **112** (32), 9810–9829.
[41] Neumann, M.A. and Perrin, M.A. (2005) Energy ranking of molecular crystals using density functional theory calculations and an empirical Van Der Waals correction. *Journal of Physical Chemistry B*, **109** (32), 15531–15541.
[42] D'Oria, E., Karamertzanis, P.G. and Price, S.L. (2010) Spontaneous resolution of enantiomers by crystallization: insights from computed crystal energy landscapes. *Crystal Growth and Design*, **10** (4), 1749–1756.
[43] Karamertzanis, P.G., Anandamanoharan, P.R., Fernandes, P., *et al.* (2007) Toward the computational design of diastereomeric resolving agents: an experimental and computational study of 1-phenylethylammonium-2-phenylacetate derivatives. *Journal of Physical Chemistry B*, **111** (19), 5326–5336.
[44] Braun, D.E., Ardid-Candel, M., D'Oria, E., *et al.* (2011) Racemic naproxen: a multidisciplinary structural and thermodynamic comparison with the enantiopure form. *Crystal Growth and Design*, **11** (12), 5659–5669.
[45] Huang, J., Chen, S., Guzei, I.A. and Yu, L. (2006) Discovery of a solid solution of enantiomers in a racemate-forming system by seeding. *Journal of the American Chemical Society*, **128** (36), 11985–11992.
[46] van Eijck, B.P. (2005) Comparing hypothetical structures generated in the third Cambridge blind test of crystal structure prediction. *Acta Crystallographica Section B: Structural Science*, **61**, 528–535.
[47] Schmidt, M.U., Dinnebier, R.E. and Kalkhof, H. (2007) Crystal engineering on industrial diaryl pigments using lattice energy minimizations and x-ray powder diffraction. *Journal of Physical Chemistry B*, **111** (33), 9722–9732.
[48] Baias, M., Widdifield, C.M., Dumez, J.N., *et al.* (2013) Powder crystallography of pharmaceutical materials by combined crystal structure prediction and solid-state H^{-1} NMR spectroscopy. *Physical Chemistry Chemical Physics*, **15** (21), 8069–8080.
[49] Eddleston, M.D., Hejczyk, K.E., Bithell, E.G., *et al.* (2013) Polymorph identification and crystal structure determination by a combined crystal structure prediction and transmission electron microscopy approach. *Chemistry: A European Journal*, **19** (24), 7874–7882.
[50] Holden, J.R., Du, Z.Y. and Ammon, H.L. (1993) Prediction of possible crystal-structures for C-, H-, N-, O- and F-containing organic compounds. *Journal of Computational Chemistry*, **14** (4), 422–437.
[51] Gavezzotti, A. (2007) *Molecular Aggregation: Structure Analysis and Molecular Simulation of Crystals and Liquids*, Oxford University Press, Oxford.
[52] M. A. Neumann (2007) *GRACE (the Generation, Ranking and Characterisation Engine) [1.0]*. Avant-garde Materials Simulation, Berlin.
[53] Lund, A.M., Orendt, A.M., Pagola, G.I., *et al.* (2013) Optimization of crystal structures of archetypical pharmaceutical compounds: a plane-wave DFT-D study using quantum espresso. *Crystal Growth and Design*, **13** (5), 2181–2189.
[54] Mitchell-Koch, K.R. and Matzger, A.J. (2008) Evaluating computational predictions of the relative stabilities of polymorphic pharmaceuticals. *Journal of Pharmaceutical Sciences*, **97** (6), 2121–2129.
[55] Abramov, Y.A. (2011) QTAIM application in drug development: prediction of relative stability of drug polymorphs from experimental crystal structures. *Journal of Physical Chemistry A*, **115** (45), 12809–12817.
[56] Otero-de-la-Roza, A. and Johnson, E.R. (2012) A benchmark for non-covalent interactions in solids. *Journal of Chemical Physics*, **137** (5), 054103.
[57] Price, S.L., Leslie, M., Welch, G.W.A., *et al.* (2010) Modelling organic crystal structures using distributed multipole and polarizability-based model intermolecular potentials. *Physical Chemistry Chemical Physics*, **12** (30), 8478–8490.
[58] Price, S.L. (2008) Computational prediction of organic crystal structures and polymorphism. *International Reviews in Physical Chemistry*, **27** (3), 541–568.

[59] Day, G.M., Motherwell, W.D.S. and Jones, W. (2005) Beyond the isotropic atom model in crystal structure prediction of rigid molecules: atomic multipoles versus point charges. *Crystal Growth and Design*, **5** (3), 1023–1033.
[60] Cooper, T.G., Hejczyk, K.E., Jones, W. and Day, G.M. (2008) Molecular polarization effects on the relative energies of the real and putative crystal structures of valine. *Journal of Chemical Theory and Computation*, **4** (10), 1795–1805.
[61] Gelbrich, T., Braun, D.E., Ellern, A. and Griesser, U.J. (2013) Four polymorphs of methyl paraben: structural relationships and relative energy differences. *Crystal Growth and Design*, **13** (3), 1206–1217.
[62] Habgood, M. (2011) Form II caffeine: a case study for confirming and predicting disorder in organic crystals. *Crystal Growth and Design*, **11** (8), 3600–3608.
[63] Ismail, S.Z., Anderton, C.L., Copley, R.C., *et al.* (2013) Evaluating a crystal energy landscape in the context of industrial polymorph screening. *Crystal Growth and Design*, **13** (6), 2396–2406.
[64] Stahly, G.P. (2007) Diversity in single- and multiple-component crystals. the search for and prevalence of polymorphs and cocrystals. *Crystal Growth and Design*, **7** (6), 1007–1026.
[65] Karamertzanis, P.G., Raiteri, P., Parrinello, M., *et al.* (2008) The thermal stability of lattice energy minima of 5-fluorouracil: metadynamics as an aid to polymorph prediction. *Journal of Physical Chemistry B*, **112** (14), 4298–4308.
[66] Gelbrich, T. and Hursthouse, M.B. (2005) A versatile procedure for the identification, description and quantification of structural similarity in molecular crystals. *CrystEngComm*, **7**, 324–336.
[67] Macrae, C.F., Bruno, I.J., Chisholm, J.A., *et al.* (2008) Mercury CSD 2.0 – new features for the visualization and investigation of crystal structures. *Journal of Applied Crystallography*, **41**, 466–470.
[68] Gavezzotti, A. (2007) A solid-state chemist's view of the crystal polymorphism of organic compounds. *Journal of Pharmaceutical Sciences*, **96** (9), 2232–2241.
[69] Bond, A.D., Boese, R. and Desiraju, G.R. (2007) On the polymorphism of aspirin: crystalline aspirin as intergrowths of two "polymorphic" domains. *Angewandte Chemie International Edition*, **46** (4), 618–622.
[70] Childs, S.L., Wood, P.A., Rodriguez-Hornedo, N., *et al.* (2009) Analysis of 50 crystal structures containing carbamazepine using the materials module of mercury CSD. *Crystal Growth and Design*, **9** (4), 1869–1888.
[71] Bhardwaj, R.M., Price, L.S., Price, S.L., *et al.* (2013) Exploring the experimental and computed crystal energy landscape of olanzapine. *Crystal Growth and Design*, **13** (4), 1602–1617.
[72] Braun, D.E., Bhardwaj, R.M., Arlin, J.B., *et al.* (2013) Absorbing a little water: the structural, thermodynamic and kinetic relationship between pyrogallol and its tetarto-hydrate. *Crystal Growth and Design*, **13** (9), 4071–4083.
[73] Cruz-Cabeza, A.J., Day, G.M. and Jones, W. (2009) Predicting inclusion behaviour and framework structures in organic crystals. *Chemistry: A European Journal*, **15** (47), 13033–13040.
[74] Habgood, M. (2013) Analysis of enantiospecific and diastereomeric cocrystal systems by crystal structure prediction. *Crystal Growth and Design*, **13**, 4549–4558.
[75] Braun, D.E., Karamertzanis, P.G. and Price, S.L. (2011) Which, if any, hydrates will crystallise? Predicting hydrate formation of two dihydroxybenzoic acids. *Chemical Communications*, **47** (19), 5443–5445.
[76] Issa, N., Barnett, S.A., Mohamed, S., *et al.* (2012) Screening for cocrystals of succinic acid and 4-aminobenzoic acid. *CrystEngComm*, **14** (7), 2454–2464.
[77] Bucar, D.K., Day, G.M., Halasz, I., *et al.* (2013) The curious case of (Caffeine).(Benzoic Acid): how heteronuclear seeding allowed the formation of an elusive cocrystal. *Chemical Science*, **4** (12), 4417–4425.
[78] Lancaster, R.W., Karamertzanis, P.G., Hulme, A.T., *et al.* (2006) Racemic progesterone: predicted in silico and produced in the solid state. *Chemical Communications* (**47**), 4921–4923.
[79] Schmidt, M.U., Paulus, E.F., Rademacher, N. and Day, G.M. (2010) Experimental and predicted crystal structures of pigment red 168 and other dihalogenated anthanthrones. *Acta Crystallographica Section B: Structural Science*, **66**, 515–526.

[80] Wu, H., Habgood, M., Parker, J.E., *et al.* (2013) Crystal structure determination by combined synchrotron powder x-ray diffraction and crystal structure prediction: 1: 1 L-Ephedrine D-Tartrate. *CrystEngComm*, **15** (10), 1853–1859.
[81] Braun, D.E., Tocher, D.A., Price, S.L. and Griesser, U.J. (2012) The complexity of hydration of phloroglucinol: a comprehensive structural and thermodynamic characterization. *Journal of Physical Chemistry B*, **116** (13), 3961–3972.
[82] Florence, A.J., Johnston, A., Price, S.L., *et al.* (2006) An automated parallel crystallisation search for predicted crystal structures and packing motifs of carbamazepine. *Journal of Pharmaceutical Sciences*, **95** (9), 1918–1930.
[83] Johnston, A., Johnston, B.F., Kennedy, A.R. and Florence, A.J. (2008) Targeted crystallisation of novel carbamazepine solvates based on a retrospective random forest classification. *CrystEngComm*, **10**, 23–25.
[84] Florence, A.J., Leech, C.K., Shankland, N., *et al.* (2006) Control and prediction of packing motifs: a rare occurrence of carbamazepine in a catemeric configuration. *CrystEngComm*, **8** (10), 746–747.
[85] Arlin, J.B., Price, L.S., Price, S.L. and Florence, A.J. (2011) A strategy for producing predicted polymorphs: catemeric carbamazepine form V. *Chemical Communications*, **47** (25), 7074–7076.

3

Solubilization of Poorly Soluble Drugs: Cyclodextrin-Based Formulations

Sachin S. Thakur[1], *Harendra S. Parekh*[1], *Carl H. Schwable*[2], *Yong Gan*[3], *and Defang Ouyang*[4]

[1] *School of Pharmacy, The University of Queensland, Australia*
[2] *School of Life and Health Sciences, Aston University, UK*
[3] *Shanghai Institute of Materia Medica, Chinese Academy of Sciences, China*
[4] *Institute of Chinese Medical Sciences, University of Macau, Macau*

3.1 Cyclodextrins in Pharmaceutical Formulations – Overview

The delivery of therapeutics preferentially to target tissue remains a considerable barrier to achieving effective clinical outcomes; this ultimately requires a molecule to display both optimal hydrophilic and lipophilic properties so it can be safely transported via the systemic circulation while readily partitioning and traversing biological membranes. While the majority of novel and established therapeutics demonstrate desirable permeation properties, poor water solubility remains a bottleneck to the successful administration of a broad range of drugs. Here, cyclodextrins (CDs) in particular have shown great promise, having been successfully applied as drug solubilizing agents and they are now routinely employed for this purpose in a range of pharmaceutical formulations.

CDs were first prepared in 1891 by French scientist A. Villiers, who sourced them through the bacterial digestion of starch [1]. The resulting product was named *cellulosine* as it represented a crude mixture of various dextrins with properties resembling cellulose,

---

Computational Pharmaceutics: Application of Molecular Modeling in Drug Delivery, First Edition.
Edited by Defang Ouyang and Sean C. Smith.

most notably remarkable resistance to acid hydrolysis. Over the next half century considerable effort was directed toward strategies for improving CD yield through better isolation techniques, analyzing their interaction with several organic compounds and attempting to prepare them via alternative, nonbacterial enzymatic methods [2–5]. Some 60 years after first being reported, Freudenberg, Cramer, and Plieninger obtained a patent which claimed the use of CDs as valuable molecules for drug complexation purposes [4, 6]. While research of these molecules was initially limited by their highly prohibitive preparation costs, the industrial production of CDs came to the fore in the 1970s, permitting their bulk preparation at commercially viable prices [4].

CDs represent a family of cyclic oligosaccharides comprising six or more (α-1,4-)-linked D-glucopyranose units [7]. The saccharide molecules are arranged in a chair conformation whereby the polar hydroxy groups orient themselves along the outer edges of the structure, while the center of the structure remains relatively apolar [7]. This conformation is responsible for the characteristic properties that CDs have become acclaimed for, that is a lipophilic inner cavity and a hydrophilic external surface. These properties enable CDs to accommodate poorly water-soluble drugs in their central (inner) cavity while retaining their inherent polar nature due to their exterior, hence forming water soluble drug-inclusion complexes. Of the many CDs available the relatively small six (αCD), seven (βCD), and eight (γCD) sugar-containing members of the family have been found to be most suitable in the field of formulation science, as shown in Table 3.1.

Although hydrophilic in nature, CDs have relatively low solubility in water when compared to acyclic saccharides. The reduced solubility can be attributed to the ring structure which provides a desirable orientation for strong intermolecular bonding facilitated by the numerous hydroxyl groups [7, 10–16]. Substitution of these hydroxyl groups can therefore help enhance their water solubility and several chemically modified CDs, including 2-hydroxypropyl-β-cyclodextrin (HPβCD), randomly methylated β-cyclodextrin (RMβCD),

Table 3.1 *Figures and properties of α, β, and γCD.*

Property	αCD	βCD	γCD
Molecular weight (Da)	972.84	1134.98	1297.12
Melting point (°C) [8]	250–260	255–265	240–245
Solubility in water (mg/ml) [9]	129.5	18.4	249.2
H donors	18	21	24
H acceptors	30	35	40
Height[a] (Å) [4]	7.8	7.8	7.8
Inner diameter[a] (Å) [4]	5.7	7.8	9.5
Outer diameter[a] (Å) [4]	13.7	15.3	16.9
Specific rotation $[\alpha]^{25}_{D}$ [8]	+150.5°	+162.0°	+177.4°

[a]Approximate values.

Table 3.2 *Examples of commercially available CD formulations and their administration routes [7].*

Drug/CD	Trade name	Formulation	Company (region)
Alprostadil/αCD	Caverject Dual®	Intravenous solution	Pfizer (Europe)
Aripiprazole/SBEβCD	Abilify®	Intramuscular solution	Bristol-Myers Squibb (USA); Otsuka America (USA)
Cisapride/HPβCD	Propulsid®	Suppository	Janssen (Europe)
Diclofenac sodium/ HPγCD	Voltaren Ophtha®	Eye drop solution	Novartis (Europe)
17β-Estradiol/RMβCD	Aerodiol®	Nasal spray	Servier (Europe)
Ethinylestradiol and drospirenone/βCD	Yaz®	Tablet	Bayer (Europe, USA)
Mitomycin	MitoExtra®, Mitozytrex®	Intravenous infusion	Novartis (Europe)
Nicotine/βCD	Nicorette®	Sublingual tablet	Pfizer (Europe)
PGE1/αCD	Prostavastin®	Parenteral solution	Ono (Japan); Schwarz (Europe)

sulfobutylether β-cyclodextrin (SBEβCD), and 2-hydroxypropyl-γ-cyclodextrin (HPγCD) have been successfully used in pharmaceutical formulations, as shown in Table 3.2 [17–25]. Other varieties of modified CDs include hydroxyethyl-β-cyclodextrin (HEβCD), glucosyl-β-cyclodextrin (GβCD), and 2-O-methyl-β-cyclodextrin (2OMβCD).

Numerous forms of CDs have been incorporated into no less than 35 commercially available drug formulations globally [7], and these products are administered via a variety of routes including the oral, sublingual, nasal, ocular, dermal, and parenteral routes. Due to their inherent hydrophilicity and comparatively high molecular weight, CDs are poorly absorbed across most biological membranes, which significantly limits their bioavailability when administered via the routes outlined above, leading to a very low potential for toxicity [17, 26]. Even when administered intravenously, CDs are rapidly excreted untransformed in the urine [27–29]. Given their relatively inert nature, CDs are considered virtually nontoxic to humans, with each of α, β, and γCD being classified by the United States Food and Drugs Administration as "generally regarded as safe" [27, 30–33].

In considering each route of administration, there exist certain route-specific advantages with the use of drug-CD inclusion complexes in a given formulation. With the oral route being the predominant route of drug administration, CD complexation has improved oral bioavailability of anti-inflammatory agents, anti-microbial agents, cardiac glycosides, anti-cancer agents, and peptides, to name just a few [34–43]. Aside from enhanced solubilization, CD complexation provides additional benefits to orally administered drugs such as enhanced stability through protection against the processes of oxidation, hydrolysis, and enzymatic degradation [10, 44–48]. Complexation also reduces the incidence and severity of drug-associated side effects such as gastric irritation common to NSAIDs, namely piroxicam, flurbiprofen, and naproxen [10, 49–52]. The bitter taste of drugs such as cetirizine can also be masked through complexation with CD, which is an important consideration for pediatric formulations [53]. Although not currently utilized for this purpose, CDs may be considered as active ingredients in oral hygiene products as they are known to complex with certain malodorous entities residing in the oral cavity [54].

Of particular interest here is the sublingual route as drugs entering the systemic circulation via this membrane avoid first-pass metabolism. Although the relatively low volume of saliva demands very high drug solubility in order to achieve effective dissolution, CDs can facilitate this [30]. To this end nitroglycerin, prostaglandin E2 (PGE2), testosterone, and 17β-oestradiol have all been successfully prepared as CD-based sublingual formulations [55–57].

Interest in nasal administration of drugs has received renewed attention in recent years and CD complexes can assist here in aiding drug solubility and so enhancing permeation and systemic absorption across the mucosal surface of the nasal cavity [30]. In particular, methylated CDs have shown the ability to enhance nasal absorption of various peptides including insulin in a rat model [58]. Other molecules including morphine and 17β-oestradiol have also demonstrated increased nasal bioavailability when administered in a CD complex [59, 60]. CDs are generally well tolerated by the nasal mucosa and may be utilized to mask the intrinsic toxicity of complexed molecules [44, 61]. CD concentration and exposure time are however both important considerations as higher doses and prolonged exposure of the carriers have shown potential to cause adverse effects, with a 20% w/v RMβCD solution causing severe damage to rat nasal mucosa [30, 61, 62].

Ocular delivery of CDs can be implemented in situations where rapid drug loss is encountered due to tear fluid drainage mechanisms. Topical ocular delivery is challenging as drugs need to be water soluble to penetrate the exterior surface of the eye (tear film and mucin) while also demonstrating adequate lipophilicity to penetrate the subsequent ocular barriers such as the cornea [30, 63]. CDs are able to efficiently solubilize hydrophobic drugs to allow permeation through these barriers [64] and drugs such as acetazolamide and pilocarpine have demonstrated enhanced corneal permeation when complexed with CDs while also enhancing drug stability and minimizing ocular toxicity [65, 66].

Dermal administration of drug-CD complexes is usually limited for use in aqueous-based topical formulations [30] with CDs unable to permeate through the outermost layer of the skin, the stratum corneum, to any appreciable extent rendering them ineffective as penetration enhancers [30, 67]. This drawback may however prove beneficial in slowing down the release of certain molecules which may otherwise permeate through the skin at a faster rate than desired. For example, the UV absorbing molecule oxybenzone present in sunscreens has its dermal permeation properties intentionally reduced through the addition of excess CD in the final formulation [68].

Intravenous administration of CD complexes is advantageous as it allows traditionally poorly water soluble molecules to be injected directly into the bloodstream without the need for potentially harmful organic co-solvents. However, CD-mediated hemolysis is a limiting factor here [7, 69, 70] with βCD and its methylated varieties banned from being administered via this route due to their high hemolytic potential and comparatively low water solubility of the parent CD [7, 69]. With the exception of these, most other CDs provide a safe and effective approach toward the intravenous administration of lipophilic drugs.

CDs are also being increasingly considered in the domain of gene delivery and for siRNA in particular given their increased potency and target selectivity (cf. antisense oligonucleotides) in correcting a variety of disease-specific cellular processes [71]. Genetic material however faces several significant barriers *in vivo* including interaction with plasma proteins, low stability, and low rates of permeation across biological membranes due to their high molecular volume and hydrophilic nature [72, 73]. A range of CD-based vectors continue to be investigated for siRNA delivery and their ability to protect genetic material from nucleases as well as enhance membrane absorption are particularly appealing [73]. To illustrate their potential, a self-assembling nanoparticulate system (CALAA-01) composed of a CD-containing polymer complexed with siRNA has progressed to phase I clinical safety studies [74, 75].

CD molecules have also been investigated in combination with various other vectors including liposomes, nanoparticles, and polymer hydrogels; this is extensively reviewed elsewhere [76].

3.2 Drug-CD Complexes – Preparation Methods

The literature boasts a range of methods for the formation of drug-CD complexes. The solvents and processing parameters used may vary, but the commonly used techniques can be broadly divided into spray drying, freeze drying, slurry method, kneading, solid phase complexation, co-precipitation, and neutralization. The two most commonly employed techniques for drug-CD complex formation are spray drying and freeze drying [27, 77]. Both techniques allow drug and CD to equilibrate in a polar solvent system prior to isolation of the solid complex by their respective techniques. In the case of spray drying, care must be taken to control the process of complex precipitation as too large a precipitate will obstruct the spray dryer nozzle [78]. The technique is also impractical when complexing volatile guest molecules as these may be lost during the drying process. This problem can be overcome with freeze drying, where volatile and heat labile molecules can be effectively complexed and subsequently dried at low temperatures [78]. Despite the ease with which the two methods can be adopted, both are restricted to a laboratory setting.

Industrially utilized techniques include the slurry method, kneading, and solid phase complexation, all of which are usually followed by subsequent extrusion [27, 78]. The slurry method and kneading both involve shear mixing of the drug and CD in minimal solvent. The primary difference here is the volume of solvent used, where kneading utilizes substantially less aqueous solvent (≈1 : 1 weight ratio of CD:water) than the slurry method [27, 77]. Solid phase complexation involves simple incorporation of drug and CD powders by the process of high shear mixing in the absence of solvent. Each of these techniques has lower reported complexation efficiencies than spray and freeze drying.

Less commonly used techniques for CD preparation include co-precipitation and neutralization. Co-precipitation involves yielding the complex precipitate by means of cooling the solvent whereas neutralization facilitates precipitation by modulating solvent to a pH where the drug becomes insoluble [27, 77]. Alternate complex formation techniques including melting, microwave irradiation, and supercritical anti-solvent processes have also been investigated [79–82].

3.3 Physicochemical Principles Underlying Drug-CD Complexes

3.3.1 Inclusion Drug-CD Complexes

A variety of CD inclusion complexes can form depending on the properties and size of the drug molecule as well as the drug-CD ratio and preparation method [83]. For example, some molecules will only associate with one CD molecule at a time, forming 1 : 1 inclusion complexes, whereas others may be able to interact with several CD molecules at a time (Figure 3.1, step A). The internal CD cavity dimensions also determine the extent of the interaction between the drug molecule and carrier. Apart from size, certain functional groups on the guest molecule may also facilitate better interaction with some CD varieties over others. Hence, determining the ideal drug-CD combination and ratio requires much preformulation work on a trial and error basis.

There exist several thermodynamic considerations to complex formation, a process which is generally performed in water or a similar polar solvent environment [30]. Possible forces driving complex formation include electrostatic interactions, van der Waals interactions, hydrophobic interactions, hydrogen bonding, release of conformational strain, exclusion of CD cavity-bound high-energy solvent, and charge-transfer interactions [87]. Optimal conditions are met when the solvent vacates the CD cavity and the net energetic driving force allows the drug to associate with the CD molecule in the most stable conformation.

Drug-CD complexes form in a reversible manner and the affinity of drug to the CD cavity is termed its stability constant (K) [11]. The stability constant can be derived from

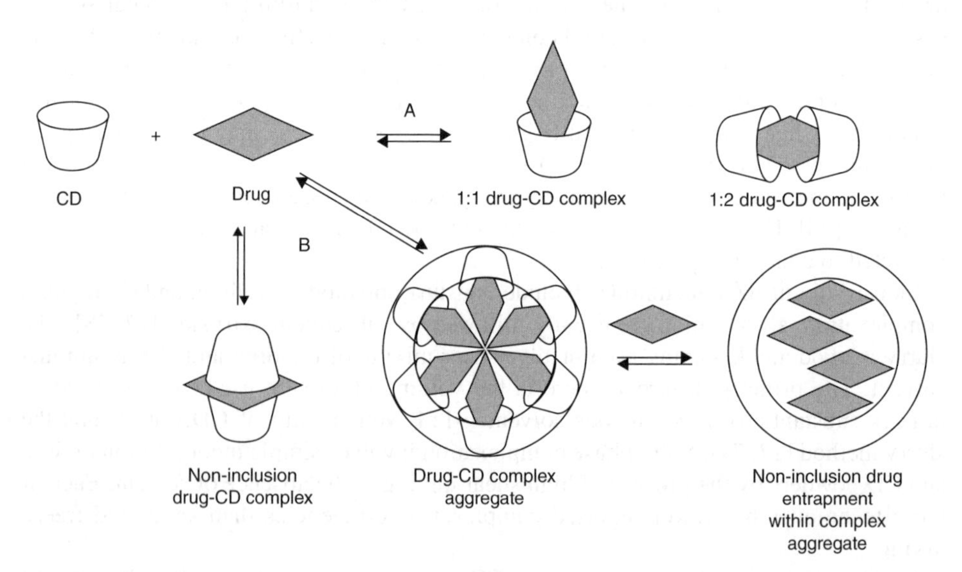

Figure 3.1 *Illustration of drug-CD complexes showing: path A – the ability to form 1 : 1 and 1 : 2 inclusion drug-CD complexes [83, 84]; B paths – possible mechanisms of non-inclusion drug-CD complex formation [85, 86] (see colour plate section). Path A reprinted by permission from Macmillan Publishers Ltd: Nature Drug Reviews [83], copyright 2004 and reprinted from [84] with permission from Elsevier. B paths reprinted with permission from [85], with kind permission from Springer Science and Business Media and [86] with permission from Elsevier.*

phase solubility isotherms [11, 27, 30]. When x drug molecules (D) associate with y CD molecules to form complex Dx/CDy, the following equilibrium results, with $Kx.y$ being the stability constant of the complex [7]:

$$x.\mathrm{D} + y.\mathrm{CD} \overset{K_{x.y}}{\rightleftharpoons} \mathrm{D}_x/\mathrm{CD}_y \tag{3.1}$$

Too low a stability constant will lead to difficulty in forming complexes whereas too high a stability constant will not allow timely dissociation of the drug from CD *in vivo*. Therefore, it is important to identify drug-CD affinity during the formulation optimization process.

An isotherm is obtained via the addition of drug in excess to a series of CD solutions of known concentration [77]. Following equilibration and removal of excess drug, the concentration of drug in solution can be plotted against the concentration of CD. In the case of 1 : 1 inclusion complexes, K is obtained using the following equation, where S_0 is the intrinsic solubility of drug in solution when $[\mathrm{CD}] = 0$ (y intercept):

$$K_{1:1} = \frac{\mathrm{slope}}{S_0(1-\mathrm{slope})} \tag{3.2}$$

In some cases however, drug may exist in dimeric, trimeric, or other aggregate conformations in solution. In such instances, an inability of CD to complex with the aggregates will lead to inaccurate values for S_0 and $K_{1:1}$. Due to this, the derivation of drug-CD complexation efficiency (CE) may serve as a more accurate representation of CD solubilization potential [11]. For a 1 : 1 drug-CD complex, CE is calculated using the following equation:

$$\mathrm{CE} = \frac{[\mathrm{D/CD}]}{[\mathrm{CD}]} = K_{1:1} \cdot S_0 = \frac{\mathrm{slope}}{1-\mathrm{slope}} \tag{3.3}$$

As a general rule, CD content within a formulation should be kept as low as possible. This is because increasing CD concentration decreases drug bioavailability, increases formulation bulk weight, and causes stability concerns due to intermolecular CD-CD associations resulting in aggregate formation [27]. As a result, it is important to consider means of enhancing the CE between drug and CD. CE can be enhanced via a variety of methods including changing drug crystallinity, altering a drug's ionization state, adding stability enhancers and employing co-solvent systems [27]. Considerations need to be given to a range of parameters; for example, a modification that increases the intrinsic solubility of drug will generally lead to the drug having a reduced affinity for the CD cavity due to a decrease in its lipophilicity. Thus, a careful balance between these parameters needs to be achieved to ensure optimum rather than maximal complexation.

3.3.2 Non-inclusion Drug-CD Complexes

Inclusion within the CD cavity is not absolutely required to enhance drug solubility. Other complexation phenomena may occur between the two species and these are the subject of much ongoing research [85]. For example, while ibuprofen and diflusinal are expected to form 2 : 1 complexes with HPβCD based on phase solubility studies; docking studies and

NMR investigations demonstrate that these substances only form 1 : 1 complexes [12]. As mentioned earlier, CD molecules may aggregate to form CD-CD complexes [12, 13, 16, 88, 89]. Further, hydrophobically modified CDs may behave as surfactants resulting in aggregation into micelle-like structures [14, 90, 91]. These and other mechanisms may allow the incorporation of hydrophobic drug within the aggregate and drive the formation of non-inclusion complexes (Figure 3.1, B paths). To demonstrate, the formation of large aggregates between hydrocortisone and HPβCD prevent the permeation of the former through cellophane membranes [14, 85, 92]. Other non-inclusion drug-CD interactions have also been identified such as the formation of "out of ring" hydrogen bonds between riboflavin and βCD/HPβCD, leading to enhanced solubility of the drug [93].

Using multiple types of CDs may also enable the formation of co-solvent systems. While dexamethasone forms micellar non-inclusion aggregates with HPγCD as well as with a γCD/HPγCD system, the latter system was able to more efficiently enhance both drug solubilization and drug release rates from the complexes [94]. Modulation of CD ratios could further control drug release by affecting solubility and particle size of the complexes.

Non-inclusion complexes have been further used to prolong the release of certain therapeutic agents. The innately hydrophilic drug meglumine antimoniate had its water solubility further improved by complexation with βCD, which allowed the administration of higher doses [95]. Spectroscopic characterization confirmed the formation of non-inclusion complexes which were able to slow down the release and absorption of drug, consequently leading to prolonged mean residence time of drug in serum [95].

CD/molecule inclusion complexes may also aggregate more readily than native CD molecules to allow subsequent non-inclusion complexation. For example, saturation of HPβCD cavities with cholesterol enhances the ability of this entity to solubilize cyclosporin A [86]. Drug-CD complexes have also been seen to form aggregates which may enable subsequent non-inclusion complexation (in a manner similar to that highlighted in Figure 3.1, B paths), with higher CE drugs generally leading to the formation of larger aggregates and thus poorer membrane permeation properties [15].

3.4 Characterization of Drug-CD Complexes

A variety of techniques have been employed to confirm the formation and utility of CD complexes. Selecting the most appropriate technique to use depends on the complexation process as well as the inherent properties of the drug, CD molecule, and their resultant complex. It is important for characterization to be able to distinguish between formed complexes and any "free" drug or CD. Drug-CD complexes may be characterized in the solid state or in solution. Both a simple physical mixture of the raw materials (drug and CD) as well as the raw materials independently subjected to processing conditions need to be simultaneously characterized in most instances in order to conclusively confirm complex formation.

3.4.1 Thermo-Analytical Methods

Thermo-analytical methods are commonly employed to characterize inclusion complexes [77, 96]. Differential scanning calorimetry (DSC) or differential thermal analysis (DTA) may be used to compare thermal profiles of drugs before and after CD complexation.

Molecular changes such as melting, oxidation, or decomposition are manifested as characteristic peaks or troughs on the derived thermograms [96]. Peak shifts and/or the gain and loss of features on the thermogram can confirm complex formation. Various therapeutics including salbutamol and famotidine have demonstrated the loss of characteristic peaks on thermograms when complexed with CD molecules [97, 98]. Thermograms may be further used to quantitatively determine the extent of drug complexation [77]. Such characterization requires a comparison of the area under the DSC curve of a physical mixture of drug-CD with that under the DSC curve of a formed complex. The proportional change in area can be correlated with the apparent degree of complex formation.

3.4.2 Microscopic Methods

Scanning electron microscopy (SEM) and transmission electron microscopy (TEM) can be used to image the crystalline state of the raw materials and the final product following complexation [16, 77]. The observed differences in structure can be used to indicate the formation of inclusion or non-inclusion complexes. Although the methods have been routinely used to characterize CD complexes, they are considered inconclusive in confirming complexation and should only be used as adjuncts to more robust characterization techniques.

3.4.3 Wettability/Solubility Studies

In wettability studies, powdered drug-CD complex is analyzed with regard to its contact angle and sedimentation and dissolution rates when exposed to water. While simple CD addition to a powder can enhance its cumulative wettability, the conduction of dissolution studies of the powder can give a better indication of whether drug-CD complexation has indeed occurred [77, 96]. Solid CD formulations may be pressed into tablets and subjected to a dissolution test. Following collection of media at preset intervals and subsequent analysis of drug content within the media, one can determine whether the CD complexation has improved a drug's dissolution property. Dissolution rates of complexes prepared via different methods can be compared to determine which method is most efficient at enhancing drug dissolution. Solubility studies using varying CD concentrations may also be assessed. If drug solubility is increased with increasing CD concentration, this can be taken as a clear indication of CDs interaction resulting in enhanced drug solubility.

3.4.4 Chromatographic Methods

Thin layer chromatography (TLC) and high pressure liquid chromatography (HPLC) can be used to distinguish drug, CD, and complexes due to differences in retention properties of the three samples. An issue, seen especially with TLC, is that complexation is a reversible process and the complexes may separate during chromatographic analysis. The technique may however be used to determine the affinity of drug and CD molecule [99]. Similar affinity studies may be designed using HPLC and as such imidazole/HPβCD and steroid/(βCD or γCD) complexes have been previously evaluated using this technique [100, 101].

3.4.5 Spectroscopic Methods

Spectroscopic characterization via ^{1}H-NMR is a robust technique for confirming the formation of inclusion complexes. The technique can also assist in determining the precise molecular groups of the drug and CD that are interacting during complex formation [102–104]. For example, spectra of hydrogen atoms located on the interior of the CD molecule show significant shifts when inclusion complexation occurs. The spectrum of the included molecule may also change, with relevant hydrogen peaks shifting to indicate their involvement in the interaction. When varying ratios of drug and CD are used, the extent of peak shift can help determine the ratio at which complexation is at a maximum. Salbutamol and indomethacin complexes with βCD have been confirmed using ^{1}H-NMR, helping to determine the precise extent of interaction between each molecule of drug and the CD [103, 104]. Although not routinely used, ^{13}C NMR may also be employed for complex identification, with the technique being most useful when complexation leads to conformational changes of the CD molecule [105]. CD interactions with polyaniline and piroxicam have previously been elucidated using ^{13}C NMR [106, 107].

Infrared (IR) spectroscopy enables the characterization of molecules through their structure-specific absorption of IR light. The technique is especially useful in cases where the drug contains functional groups which appear as characteristic bands on IR spectra, such as carbonyl or sulfonyl groups. If hydrogen bonds are formed between such characteristic groups and CD during complexation, increases in band intensity or widening of bands may be observed on the spectrum [96]. CDs complexed with various drugs, including clotrimazole, piroxicam, indomethacin, and naproxen, have been previously confirmed using IR [108–111].

Raman spectroscopy works in a complementary manner to IR spectroscopy and utilizes the inelastic scattering of monochromatic light to derive a characteristic molecular fingerprint. Drug-CD complexation can be analyzed via this technique by assessing the impact of intermolecular interactions on both band intensities and peak shifts on the spectra. Various drug-CD complexes have been confirmed using Raman spectroscopy, including those for carotenoids, ibuprofen, and piroxicam [106, 112, 113].

3.4.6 X-Ray Techniques

X-Ray diffraction (XRD) involves identifying complex formation by analyzing the crystal structure of a formed complex. Complex formation leads to changes in diffraction peaks based on the impact of formation on drug and CD crystallinity [96, 114, 115]. A decrease in peak sharpness on the diffractogram corresponds with a loss in crystallinity. Molecules including warfarin, quercetin, and celecoxib all demonstrate losses in crystalline peaks following complexation with various CDs, which indicates the amorphous nature of these complexes [114, 116, 117]. The crystal structures of some common CD molecules and the guest-CD complexes are presented in the Cambridge Structural Database (CSD, http://webcsd.cds.rsc.org/index.php).

An unrelated X-ray technique, termed X-ray scattering, has recently gained attention in the characterization of inclusion and non-inclusion complexes. The technique involves the derivation of the size and shape of macromolecular structures via analysis of elastic X-ray scattering patterns. Small angle X-ray scattering has been the most widely used variety of this technique, with intensity peaks able to yield information about lattice structures formed when CDs have been complexed with various molecules, including surfactants, polymers, and DNA [118–122].

3.4.7 Other Techniques

Several lesser used techniques for complex characterization are available and include UV-visible and fluorescence spectroscopy, electrochemistry, circular dichroism, drug degradation kinetics, phase distribution studies, and equilibrium dialysis. These techniques are not routinely used and the reader is referred to the literature where the aforementioned techniques have been employed and discussed [96, 102, 123–128].

3.5 Theoretical Progress of CD Studies

Molecular modeling is a characterization strategy which employs the use of computational methods to mimic the behavior of molecules. The distinguishing feature of the technique is its ability to characterize molecular systems at the atomistic level [129]. Currently there are hundreds of reports describing theoretical calculations of drug-CD complexes. Figure 3.2 and Table 3.3 classify recent research into five types: quantum mechanics (QM), molecular dynamics (MD), Monte Carlo (MC) simulations, docking, and quantitative structure-activity relationships (QSARs).

3.5.1 Quantum Mechanics

QM mathematically describes the spatial positions of all electrons and nuclei. It is able to predict the structural and electronic characteristics of molecules. Current QM methods for CD include semi-empirical methods, *ab initio* methods, and density functional theory (DFT) methods. Djemil *et al.* used different QM methods (e.g., PM6, HF, ONIOM) to

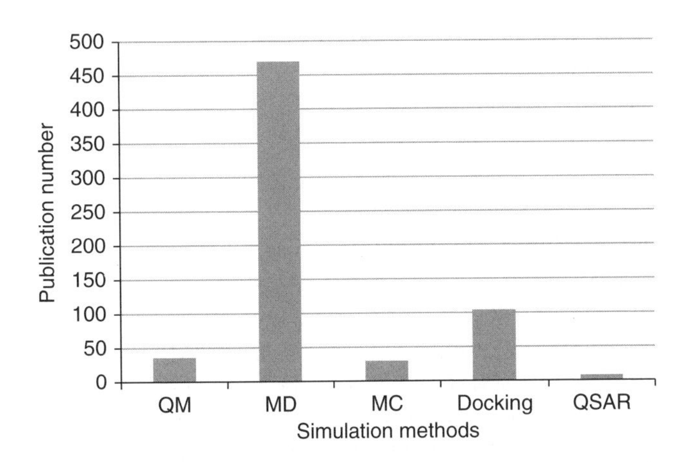

Figure 3.2 Publication numbers of different simulation methods for CDs until 2013.

Table 3.3 Examples of CD theoretical calculations.

Theoretical methods	Cyclodextrin	Guest molecule	References
Quantum mechanics	β-CD	Dopamine and epinephrine	[130]
	α-CD	Carboplatin, oxaliplatin, nedaplatin	[131]
	β-CD	Aflatoxin B1	[132]
Molecular dynamics	β-CD	57 Guest molecules	[133]
	β- and γ-CD	Amphotericin B	[134]
	α-, β-, and γ-CD	Cumene hydroperoxide	[135]
	α- and β-CD	1-Alkanols, substituted phenols, and substituted imidazoles	[136]
Monte Carlo simulations	β-CD	Praziquantel	[137]
	β-CD and methylated β-CD	Niobocene dichloride	[138]
	α-CD	Water	[139]
Docking	β-CD and four derivatives of β-CD	Luteolin	[140]
	HP-β-CD and methylated- β-CD	Voriconazole	[141]
	β-CD	4,4′-Dihydroxybiphenyl	[142]
Quantitative structure–activity relationship	α-CD	Benzene derivatives	[143]
	β-CD	Multiple compounds	[144]
	β-CD	233 Molecules	[145]

investigate the structures of dopamine and epinephrine/βCD complexes [130]. Their findings indicated that the catechol ring of the guest inserts into the hydrophobic cavity of βCD through intermolecular hydrogen bonding [130]. Another study compared the relative stability of the complexes between αCD and the platinum (II)-based drugs (e.g. carboplatin, oxaliplatin, nedaplatin) using the B3LYP/6-31G theory [131]. The relative stability studies demonstrated that, in this instance, hydrogen bonds were major contributors to complex formation and that the carboplatin/αCD complex was the most stable. However, the limitation of QM simulation to CD systems is that the large atom number of CD molecules requires extensive computing capabilities. Moreover, it is difficult to apply QM methods to solvated systems.

3.5.2 Molecular Dynamics Simulation

MD simulation mimics the physical motion of atoms and molecules under Newton's laws of physics [129]. It permits investigation of the structural, dynamic and energetic aspects of CD-guest complexes based on molecular mechanics (MM) and/or the empirical force field (e.g., AMBER, CHARMM, GROMOS, and CVFF) [146]. The binding free energy between CD and guest molecule can be calculated by the MM/PBSA, free energy perturbation, and thermodynamic integration methods. From Figure 3.2, we note there are over 400 publications describing CD-related MD simulations, which dominate research about

CD modeling. Previous MD studies have shown that amphotericin B (AmB) can form a stable complex with γCD via the insertion of the macrolide ring into the CD cavity, while the guest ring cannot enter the cavity of βCD due to space limitations [134]. van der Waals and electrostatic interactions are primary contributors to the formation of these particular CD inclusion complexes. The binding free energy between AmB and γCD is much higher than that of AmB and βCD, which is in agreement with the experimental observation of the bioavailability of AmB formulations [134]. MD simulations using the Amber force field have been applied to studying the inclusion complex formation between CD and organic molecules, for example, 1-alkanols, substituted phenols, and substituted imidazoles [136]. The calculated binding free energy by MM-PBSA method showed that van der Waals interactions were the dominant contributors of inclusion complex formation, and electrostatic interactions and the hydrophobic effect also contributed to the complex. Moreover, the flexibility of the guest molecule had a significant impact on complex stability [136].

3.5.3 Monte Carlo Simulation

MC simulation uses the same empirical force field as MD simulation. However, MC simulation features playing and recording the results in casino-like conditions by repeated random sampling [129]. After numerous results with low energies are achieved, the averaged energies and/or properties of the system will be provided [129]. Thus, unlike MD simulation, MC simulation cannot offer dynamic information with time evolution of the system in a form suitable for viewing. MC methods are especially useful for modeling systems with significant uncertainty and high degrees of freedom, such as polymer chains and protein membranes. With MC methods, a complex system is able to be sampled in numerous random configurations, and all data can be applied to simulate the whole system. MC simulations have been used to investigate the water solvation shells for praziquantel/CD complexes [137]. Water molecules were found to be useful in forming praziquantel/βCD complexes due to the hydrophilic effect imparted by the CD molecule [137]. MC simulations have also been used to study water molecules inside the hydrophobic cavity of αCD [139]. The results showed that about five water molecules are present in the CD cavity; however only 2.4 of these molecules on average interact with αCD by hydrogen bonds, which is fewer than the number interacting with the molecule outside the cavity [139].

3.5.4 Docking Studies

Docking is a simulation method for the prediction of one "ligand" molecule binding to a "receptor" molecule to form the "ligand–receptor" complex by scoring the functions with the association or binding affinity between two molecules. In rational drug design, docking is a popular approach for predicting the spatial orientation of drug candidates to their target protein or predicting the binding affinity between them. The inclusion complexes between luteolin and five varieties of CD molecules (βCD, RMβCD, HEβCD, HPβCD, GβCD) were previously studied using docking techniques [140]. Experimental results showed that the 1 : 1 luteolin/GβCD complex had the highest stability constant. Molecular docking results were in agreement with the experimental data [140]. Another study investigated the complexation between voriconazole with HPβCD and 2OMβCD [141]. Solubility studies suggested that 2OMβCD had a more efficient solubilization effect than HPβCD. However, results from docking simulation mismatched with these experimental results. The possible reason for the discrepancy between computer simulations and experimental results was the self-association and formation of multiple complexes between voriconazole and HPβCD molecules, rather than 1 : 1 complexes [141]. Therefore, we know that we cannot apply only the docking program to reliably characterize CD complexes.

3.5.5 Quantitative Structure–Activity Relationship

QSAR models are regression models with the summary of a relationship between chemical structures and biological activity within a given drug class, a technique widely used in drug design and development. Currently there are only a few QSAR studies for CD complexes. 3D-QSAR models were previously developed to predict the stability constants between benzene derivatives with αCD by comparative molecular field analysis region focusing (CoMFA-RF) and VolSurf methods [143]. A test set of 18 compounds was used for the predictive models. Results indicated that electrostatic interaction, hydrophobic effects, and shape were the three main contributors to these inclusion complexes while, separately, docking results also agreed with the combined 3D-QSAR models [143]. Another study developed a QSAR model for predicting complexation with βCD by the *top*ological *s*ubstructural *mo*lecular *de*sign (TOPS-MODE) approach [144]. Results revealed that hydrophobic effect and van der Waals interactions were the major driving forces for complexation [144].

3.6 Future Prospects of Cyclodextrin Formulation

Although numerous works have been published, the data arising from different physical methods compared to *in silico* approaches are often contradictory and conflicting. Moreover, existing *in silico* approaches are often convoluted and require significant specialist training both in underlying theory and associated software. It is therefore crucial to develop more universally acceptable, convenient, and reliable modeling methods applicable to CD formulations, which can serve to reduce the trial and error approach often adopted in a laboratory setting. In addition, the mechanism by which non-inclusion CD complexes for macromolecules form is still largely unclear and there is no widely agreeable principle or consensus on how best to develop non-inclusion CD formulations.

References

[1] Villiers, A., Sur la fermentation de la fécule par l'action du ferment butyrique. *Comptes Rendus de l'Academie des Sciences*, 1891. **112**: 536–538.
[2] Freudenberg, K., H. Boppel, and M. Meyer-Delius, Beobachtungen an der Stärke. *Naturwissenschaften*, 1938. **26**(8): 123–124.
[3] French, D., The schardinger dextrins. *Advances in Carbohydrate Chemistry*, 1957. **12**: 189–260.
[4] Szejtli, J., Introduction and general overview of cyclodextrin chemistry. *Chemical Reviews*, 1998. **98**(5): 1743–1753.
[5] Pringsheim, H., *A Comprehensive Survey of Starch Chemistry*. 1928: Chemical Catalogue Co.: New York.

[6] Freudenberg, K., Cramer, F., and Plieninger, H. Inclusion compounds of physiologically active organic compounds. 1953 Patent DE 895769 C.
[7] Loftsson, T. and M.E. Brewster, Pharmaceutical applications of cyclodextrins: basic science and product development. *Journal of Pharmacy and Pharmacology*, 2010. **62**(11): 1607–1621.
[8] Cook, W.G., Cyclodextrins, in *Handbook of Pharmaceutical Excipients*, R.C. Rowe, Sheskey, P.J; Cook, W.G, Fenton, M.E., eds, 2013, Medicines Complete: New York.
[9] Sabadini, E., T. Cosgrove, and F.C. Egídio, Solubility of cyclomaltooligosaccharides (cyclodextrins) in H2O and D2O: a comparative study. *Carbohydrate Research*, 2006. **341**(2): 270.
[10] Loftsson, T. and M.E. Brewster, Pharmaceutical applications of cyclodextrins. 1. Drug solubilization and stabilization. *Journal of Pharmaceutical Sciences*, 1996. **85**(10): 1017–1025.
[11] Kurkov, S.V. and T. Loftsson, Cyclodextrins. *International Journal of Pharmaceutics*, 2012, **436**, 359–378.
[12] Loftsson, T., Magnúsdóttir A., Másson M., Sigurjónsdóttir J.F., Self-association and cyclodextrin solubilization of drugs. *Journal of Pharmaceutical Sciences*, 2002. **91**(11): 2307–2316.
[13] Loftsson, T., M. Masson, and M.E. Brewster, Self-association of cyclodextrins and cyclodextrin complexes. *Journal of Pharmaceutical Sciences*, 2004. **93**(5): 1091–1099.
[14] Messner, M., Kurkov, S.V., Jansook, P., Loftsson, T., Self-assembled cyclodextrin aggregates and nanoparticles. *International Journal of Pharmaceutics*, 2010. **387**(1/2): 199.
[15] Messner, M., Kurkov, S.V., Flavia-Piera, R., Brewster, M.E., *et al.*, Self-assembly of cyclodextrins: the effect of the guest molecule. *International Journal of Pharmaceutics*, 2011. **408**(1): 235–247.
[16] Bonini, M., Rossi S., Karlsson G., Almgren M., *et al.*, Self-assembly of β-cyclodextrin in water. Part 1: Cryo-TEM and dynamic and static light scattering. *Langmuir*, 2006. **22**(4): 1478–1484.
[17] Loftsson, T. and M.E. Brewster, Pharmaceutical applications of cyclodextrins: effects on drug permeation through biological membranes. *Journal of Pharmacy and Pharmacology*, 2011. **63**(9): 1119–1135.
[18] Loftsson, T. and A.M. Sigurđardóttir, Improved acitretin delivery through hairless mouse skin by cyclodextrin complexation. *International Journal of Pharmaceutics*, 1995. **115**(2): 255–258.
[19] Zi, P., Yang X.H., Kuang H.F., Yang Y.S., *et al.*, Effect of HPβCD on solubility and transdermal delivery of capsaicin through rat skin. *International Journal of Pharmaceutics*, 2008. **358**(1): 151–158.
[20] Loftsson, T., Frioriksdottir H., Ingvarsdottir G., Jonsdottir B., *et al.*, The influence of 2-hydroxypropyl-β-cyclodextrin on diffusion rates and transdermal delivery of hydrocortisone. *Drug Development and Industrial Pharmacy*, 1994. **20**(9): 1699–1708.
[21] Doliwa, A., S. Santoyo, and P. Ygartua, Influence of piroxicam: hydroxypropyl-beta-cyclodextrin complexation on the in vitro permeation and skin retention of piroxicam. *Skin Pharmacology and Physiology*, 2001. **14**(2): 97–107.
[22] Järvinen, K., Jarvinen T., Thompson D.O., Stella V.J., *et al.*, The effect of a modified β-cyclodextrin, SBE4-β-CD, on the aqueous stability and ocular absorption of pilocarpine. *Current Eye Research*, 1994. **13**(12): 897–905.
[23] Loftsson, T., Gudmundsdottir H., Sigurjonsdottir J.F., *et al.*, Cyclodextrin solubilization of benzodiazepines: formulation of midazolam nasal spray. *International Journal of Pharmaceutics*, 2001. **212**(1): 29–40.
[24] Kalaiselvan, R., Mohanta G.P., Madhusudan S., Manna P.K., *et al.*, Enhancement of bioavailability and anthelmintic efficacy of albendazole by solid dispersion and cyclodextrin complexation techniques. *Die Pharmazie – An International Journal of Pharmaceutical Sciences*, 2007. **62**(8): 604–607.
[25] Le Corre, P., Dollo G., Chevanne F., Leverge R, Influence of hydroxypropyl-beta-cyclodextrin and dimethyl-beta-cyclodextrin on diphenhydramine intestinal absorption in a rat in situ model. *International Journal of Pharmaceutics*, 1998. **169**(2): 221–228.

[26] Arima, H., K. Motoyama, and T. Irie, Recent findings on safety profiles of cyclodextrins, cyclodextrin conjugates, and polypseudorotaxanes. *Cyclodextrins in Pharmaceutics, Cosmetics, and Biomedicine: Current and Future Industrial Applications*, 2011:Wiley-Blackwell: New York, pp. 91–122.

[27] Loftsson, T. and M.E. Brewster, Cyclodextrins as functional excipients: methods to enhance complexation efficiency. *Journal of Pharmaceutical Sciences*, 2012, **101**(9):3019–3032.

[28] Peeters, P., Passier, P., Smeets, J., Zwiers, A., *et al.*, Sugammadex is cleared rapidly and primarily unchanged via renal excretion. *Biopharmaceutics and Drug Disposition*, 2011. **32**(3): 159–167.

[29] Stella, V.J. and Q. He, Cyclodextrins. *Toxicologic Pathology*, 2008. **36**(1): 30–42.

[30] Loftsson, T., P. Jarho, M. Másson, and T. Järvinen, Cyclodextrins in drug delivery. *Expert Opinion on Drug Delivery*, 2005. **2**(2): 335–351.

[31] USFDA, GRAS Notice 000155: alpha-Cyclodextrin. 2004.

[32] USFDA, GRAS Notice 000074: beta-Cyclodextrin. 2001.

[33] USFDA, GRAS Notice 000046: gamma-Cyclodextrin. 2000.

[34] Muraoka, A., T. Tokumura, and Y. Machida, Evaluation of the bioavailability of flurbiprofen and its β-cyclodextrin inclusion complex in four different doses upon oral administration to rats. *European Journal of Pharmaceutics and Biopharmaceutics*, 2004. **58**(3): 667–671.

[35] Chow, D.D. and A.H. Karara, Characterization, dissolution and bioavailability in rats of ibuprofen-β-cyclodextrin complex system. *International Journal of Pharmaceutics*, 1986. **28**(2): 95–101.

[36] Hostetler, J.S., L. Hanson, and D. Stevens, Effect of cyclodextrin on the pharmacology of antifungal oral azoles. *Antimicrobial Agents and Chemotherapy*, 1992. **36**(2): 477–480.

[37] Luengo, J., Aranguiz T., Sepulveda J., *et al.*, Preliminary pharmacokinetic study of different preparations of acyclovir with β-cyclodextrin. *Journal of Pharmaceutical Sciences*, 2002. **91**(12): 2593–2598.

[38] McEwen, J., Clinical pharmacology of piroxicam-beta-cyclodextrin: implications for innovative patient care. *Clinical Drug Investigation*, 2000. **19**(S2): 27–31.

[39] Jambhekar, S., R. Casella, and T. Maher, The physicochemical characteristics and bioavailability of indomethacin from β-cyclodextrin, hydroxyethyl-β-cyclodextrin, and hydroxypropyl-β-cyclodextrin complexes. *International Journal of Pharmaceutics*, 2004. **270**(1): 149–166.

[40] Haeberlin, B., T. Gengenbacher, A. Meinzer, *et al.*, Cyclodextrins—useful excipients for oral peptide administration? *International Journal of Pharmaceutics*, 1996. **137**(1): 103–110.

[41] Uekama, K., Fujinaga T., Hirayama F., *et al.*, Improvement of the oral bioavailability of digitalis glycosides by cyclodextrin complexation. *Journal of Pharmaceutical Sciences*, 1983. **72**(11): 1338–1341.

[42] Pathak, S.M., P. Musmade, S. Dengle, *et al.*, Enhanced oral absorption of saquinavir with Methyl-Beta-Cyclodextrin—preparation and in vitro and in vivo evaluation. *European Journal of Pharmaceutical Sciences*, 2010. **41**(3): 440–451.

[43] Agüeros, M., V. Zabaleta, S. Espuelas, *et al.*, Increased oral bioavailability of paclitaxel by its encapsulation through complex formation with cyclodextrins in poly (anhydride) nanoparticles. *Journal of Controlled Release*, 2010. **145**(1): 2–8.

[44] Loftsson, T., Effects of cyclodextrins on the chemical stability of drugs in aqueous solutions. *Drug Stability*, 1995. **1**: 22–33.

[45] Garcia-Fuentes, M., A. Trapani, and M. Alonso, Protection of the peptide glutathione by complex formation with α-cyclodextrin: NMR spectroscopic analysis and stability study. *European Journal of Pharmaceutics and Biopharmaceutics*, 2006. **64**(2): 146–153.

[46] Loftsson, T., J. Baldvinsdóttir, and A.M. Sigurdardóttir, The effect of cyclodextrins on the solubility and stability of medroxyprogesterone acetate and megestrol acetate in aqueous solution. *International Journal of Pharmaceutics*, 1993. **98**(1): 225–230.

[47] Gorecka, B.A., Y.D. Sanzgiri, A.M. Sigurdardóttir, *et al.*, Effect of SBE4-beta-CD, a sulfobutyl ether beta-cyclodextrin, on the stability and solubility of O6-benzylguanine (NSC-637037) in aqueous solutions. *International Journal of Pharmaceutics*, 1995. **125**(1): 55–61.

[48] Uekama, K., T. Fujinaga, F. Hirayama, *et al.*, Effects of cyclodextrins on the acid hydrolysis of digoxin. *Journal of Pharmacy and Pharmacology*, 1982. **34**(10): 627–630.
[49] Loftsson, T., M.E. Brewster, and M. Másson, Role of cyclodextrins in improving oral drug delivery. *American Journal of Drug Delivery*, 2004. **2**(4): 261–275.
[50] Santucci, L., Fiorucci S, Chiucchiu S, *et al.*, Placebo-controlled comparison of piroxicam-β-cyclodextrin, piroxicam, and indomethacin on gastric potential difference and mucosal injury in humans. *Digestive Diseases and Sciences*, 1992. **37**(12): 1825–1832.
[51] Espinar, F., S. Anguiano Igea, J. Blanco Méndez, *et al.*, Reduction in the ulcerogenicity of naproxen by complexation with β-cyclodextrin. *International Journal of Pharmaceutics*, 1991. **70**(1): 35–41.
[52] Govindarajan, R. and M.S. Nagarsenker, Formulation studies and in vivo evaluation of a flurbiprofen-hydroxypropyl β-cyclodextrin system. *Pharmaceutical Development and Technology*, 2005. **10**(1): 105–114.
[53] Szejtli, J. and L. Szente, Elimination of bitter, disgusting tastes of drugs and foods by cyclodextrins. *European Journal of Pharmaceutics and Biopharmaceutics*, 2005. **61**(3): 115–125.
[54] Lantz, A.W., M.A. Rodriguez, S.M. Wetterer, and D.W. Armstrong, Estimation of association constants between oral malodor components and various native and derivatized cyclodextrins. *Analytica Chimica Acta*, 2006. **557**(1): 184–190.
[55] Fridriksdóttir, H., Loftsson T., Gudmundsson J.A., *et al.*, Design and in vivo testing of 17 beta-estradiol-HP beta CD sublingual tablets. *Pharmazie*, 1996. **51**(1): 39–42.
[56] Stuenkel, C., R. Dudley, and S. Yen, Sublingual administration of testosterone-hydroxypropyl-β-cyclodextrin inclusion complex simulates episodic androgen release in hypogonadal men. *Journal of Clinical Endocrinology and Metabolism*, 1991. **72**(5): 1054–1059.
[57] CycloLab, Approved Pharmaceutical Products Containing Cyclodextrins. 2013.
[58] Merkus, F.W., J.C. Verhoef, E. Marttin, *et al.*, Cyclodextrins in nasal drug delivery. *Advanced Drug Delivery Reviews*, 1999. **36**(1): 41–57.
[59] Kondo, T., K. Nishimura, T. Irie, K. Uekama, Cyclodextrin derivatives that modify nasal absorption of morphine and its entry into cerebrospinal fluid in the rat. *Pharmacy and Pharmacology Communications*, 1995. **1**(4): 163–166.
[60] Hermens, W.A., Deurloo, M.J.M., Romeyn S.G., *et al.*, Nasal absorption enhancement of 17β-estradiol by dimethyl-β-cyclodextrin in rabbits and rats. *Pharmaceutical Research*, 1990. **7**(5): 500–503.
[61] Laza-Knoerr, A., R. Gref, and P. Couvreur, Cyclodextrins for drug delivery. *Journal of Drug Targeting*, 2010. **18**(9): 645–656.
[62] Asai, K., M. Morishita, H. Hosoda, *et al.*, The effects of water-soluble cyclodextrins on the histological integrity of the rat nasal mucosa. *International Journal of Pharmaceutics*, 2002. **246**(1): 25–35.
[63] Ghate, D. and H.F. Edelhauser, Ocular drug delivery. *Expert Opinion on Drug Delivery*, 2006. **3**(2): 275–287.
[64] Loftsson, T. and T. Järvinen, Cyclodextrins in ophthalmic drug delivery. *Advanced Drug Delivery Reviews*, 1999. **36**(1): 59–79.
[65] Loftsson, T., Frioriksdottir H., Stefansson E., *et al.*, Topically effective ocular hypotensive acetazolamide and ethoxyzolamide formulations in rabbits. *Journal of Pharmacy and Pharmacology*, 1994. **46**(6): 503–504.
[66] Suhonen, P., T. Järvinen, K. Lehmussaari, *et al.*, Ocular absorption and irritation of pilocarpine prodrug is modified with buffer, polymer, and cyclodextrin in the eyedrop. *Pharmaceutical Research*, 1995. **12**(4): 529–533.
[67] Matsuda, H. and H. Arima, Cyclodextrins in transdermal and rectal delivery. *Advanced Drug Delivery Reviews*, 1999. **36**(1): 81–99.
[68] Felton, L.A., C.J. Wiley, and D.A. Godwin, Influence of hydroxypropyl-β-cyclodextrin on the transdermal permeation and skin accumulation of oxybenzone. *Drug Development and Industrial Pharmacy*, 2002. **28**(9): 1117–1124.
[69] Irie, T. and K. Uekama, Pharmaceutical applications of cyclodextrins. III. Toxicological issues and safety evaluation. *Journal of Pharmaceutical Sciences*, 1997. **86**(2): 147–162.

[70] Ohtani, Y., Irie T., Uekama K., *et al.*, Differential effects of alpha-, beta- and gamma-cyclodextrins on human erythrocytes. *European Journal of Biochemistry/FEBS*, 1989. **186**(1/2): 17.

[71] Aagaard, L. and J.J. Rossi, RNAi therapeutics: principles, prospects and challenges. *Advanced Drug Delivery Reviews*, 2007. **59**(2): 75–86.

[72] Li, W. and F.C. Szoka, Lipid-based nanoparticles for nucleic acid delivery. *Pharmaceutical Research*, 2007. **24**(3): 438–449.

[73] Chaturvedi, K., Ganguly K., Kulkarni AR, *et al.*, Cyclodextrin-based siRNA delivery nanocarriers: a state-of-the-art review. *Expert Opinion on Drug Delivery*, 2011. **8**(11): 1455–1468.

[74] Davis, M.E., The first targeted delivery of siRNA in humans via a self-assembling, cyclodextrin polymer-based nanoparticle: from concept to clinic. *Molecular Pharmaceutics*, 2009. **6**(3): 659–668.

[75] ClinicalTrials.gov, Safety Study of CALAA-01 to Treat Solid Tumor Cancers, 2008. http://www.clinicaltrials.gov/ct2/show/NCT00689065 (accessed 14 May 2013).

[76] Vyas, A., S. Saraf, and S. Saraf, Cyclodextrin based novel drug delivery systems. *Journal of Inclusion Phenomena and Macrocyclic Chemistry*, 2008. **62**(1): 23–42.

[77] ISP Pharmaceuticals, *CAVAMAX Cyclodextrins: Forming and Analyzing Drug Inclusion Complexes*, I.S. Products Editor, 2006.

[78] Del Valle, E.M., Cyclodextrins and their uses: a review. *Process Biochemistry*, 2004. **39**(9): 1033–1046.

[79] Jun, S.W., M.S. Kim, J.S. Kim, *et al.*, Preparation and characterization of simvastatin/hydroxypropyl-β-cyclodextrin inclusion complex using supercritical antisolvent (SAS) process. *European Journal of Pharmaceutics and Biopharmaceutics*, 2007. **66**(3): 413–421.

[80] Wulff, M. and M. Aldén, Solid state studies of drug–cyclodextrin inclusion complexes in PEG 6000 prepared by a new method. *European Journal of Pharmaceutical Sciences*, 1999. **8**(4): 269–281.

[81] Toropainen, T., S. Velaga, T. Heikkila, *et al.*, Preparation of budesonide/γ-cyclodextrin complexes in supercritical fluids with a novel SEDS method. *Journal of Pharmaceutical Sciences*, 2006. **95**(10): 2235–2245.

[82] Wen, X., F. Tan, Z. Jing, and Z. Liu, Preparation and study the 1: 2 inclusion complex of carvedilol with β-cyclodextrin. *Journal of Pharmaceutical and Biomedical Analysis*, 2004. **34**(3): 517–523.

[83] Davis, M.E. and M.E. Brewster, Cyclodextrin-based pharmaceutics: past, present and future. *Nature Reviews Drug Discovery*, 2004. **3**(12): 1023–1035.

[84] Atwood, J.L. and J.-M. Lehn, *Comprehensive Supramolecular Chemistry: Supramolecular Reactivity and Transport: Bioinorganic Systems*. Vol. **5**. 1996: Elsevier.

[85] Kurkov, S.V., E.V. Ukhatskaya, and T. Loftsson, Drug/cyclodextrin: beyond inclusion complexation. *Journal of Inclusion Phenomena and Macrocyclic Chemistry*, 2011. **69**(3/4): 297–301.

[86] Loftsson, T., K. Matthíasson, and M. Másson, The effects of organic salts on the cyclodextrin solubilization of drugs. *International Journal of Pharmaceutics*, 2003. **262**(1): 101–107.

[87] Liu, L. and Q.-X. Guo, The driving forces in the inclusion complexation of cyclodextrins. *Journal of Inclusion Phenomena and Macrocyclic Chemistry*, 2002. **42**(1): 1–14.

[88] Szente, L., J. Szejtli, and G.L. Kis, Spontaneous opalescence of aqueous γ-Cyclodextrin solutions: Complex formation or self-aggregation? *Journal of Pharmaceutical Sciences*, 1998. **87**(6): 778–781.

[89] Puskás, I., M. Schrott, M. Malanga, L. Szente, Characterization and control of the aggregation behavior of cyclodextrins. *Journal of Inclusion Phenomena and Macrocyclic Chemistry*, 2012: 1–8.

[90] Auzely-Velty, R., F. Djedaïni-Pilard, S. Désert, *et al.*, Micellization of hydrophobically modified cyclodextrins. 1. Micellar structure. *Langmuir*, 2000. **16**(8): 3727–3734.

[91] Witte, F. and H. Hoffmann, Aggregation behavior of hydrophobically modified β-cyclodextrins in aqueous solution. *Journal of Inclusion Phenomena and Molecular Recognition in Chemistry*, 1996. **25**(1/3): 25–28.

[92] Loftsson, T., M. Másson, and H.H. Sigurdsson, Cyclodextrins and drug permeability through semi-permeable cellophane membranes. *International Journal of Pharmaceutics*, 2002. **232**(1): 35–43.

[93] de Jesus, M.B., Fraceto, L.F., Florencia Martini, M., *et al.*, Non-inclusion complexes between riboflavin and cyclodextrins. *Journal of Pharmacy and Pharmacology*, 2012. **64**(6): 832–842.
[94] Jansook, P., G.C. Ritthidej, H. Ueda, *et al.*, yCD/HPyCD mixtures as solubilizer: solid-state characterization and sample dexamethasone eye drop suspension. *Journal of Pharmacy and Pharmaceutical Sciences*, 2010. **13**(3): 336–350.
[95] Ribeiro, R.R., Ferreria, W.A., Martins, P.S., *et al.*, Prolonged absorption of antimony (V) by the oral route from non-inclusion meglumine antimoniate-β-cyclodextrin conjugates. *Biopharmaceutics and Drug Disposition*, 2010. **31**(2/3): 109–119.
[96] Singh, R., N. Bharti, J. Madan, S.N. Hiremath, Characterization of cyclodextrin inclusion complexes–a review. *Journal of Pharmaceutical Science and Technology*, 2010. **2**(3): 171–183.
[97] Marques, H., J. Hadgraft, and I. Kellaway, Studies of cyclodextrin inclusion complexes. I. The salbutamol-cyclodextrin complex as studied by phase solubility and DSC. *International Journal of Pharmaceutics*, 1990. **63**(3): 259–266.
[98] Hassan, M.A., M.S. Suleiman, and N.M. Najib, Improvement of the in vitro dissolution characteristics of famotidine by inclusion in β-cyclodextrin. *International Journal of Pharmaceutics*, 1990. **58**(1): 19–24.
[99] Dąbrowska, M., J. Krzek, and E. Miękina, Stability analysis of cefaclor and its inclusion complexes of β-cyclodextrin by thin-layer chromatography and densitometry. *Journal of Planar Chromatography – Modern TLC*, 2012. **25**(2): 127–132.
[100] Morin, N., Cornet, S., Guinchard, C., *et al.*, HPLC retention and inclusion of imidazole derivatives using hydroxypropyl-β-cyclodextrin as a mobile phase additive, *Journal of Liquid Chromatography and Related Technologies*, 2000. **23**: 727–739.
[101] Flood, K.G., E.R. Reynolds, and N.H. Snow, Characterization of inclusion complexes of betamethasone-related steroids with cyclodextrins using high-performance liquid chromatography. *Journal of Chromatography A*, 2000. **903**(1): 49–65.
[102] Jadhav, G. and P. Vavia, Physicochemical, in silico and in vivo evaluation of a danazol-β-cyclodextrin complex. *International Journal of Pharmaceutics*, 2008. **352**(1): 5–16.
[103] Estrada, E., I. Perdomo-López, and J.J. Torres-Labandeira, Molecular modeling (MM2 and PM3) and experimental (NMR and thermal analysis) studies on the inclusion complex of salbutamol and β-cyclodextrin. *The Journal of Organic Chemistry*, 2000. **65**(25): 8510–8517.
[104] Fronza, G., A. Mele, E. Redenti, P. Ventura, 1H NMR and molecular modeling study on the inclusion complex β-cyclodextrin-indomethacin. *The Journal of Organic Chemistry*, 1996. **61**(3): 909–914.
[105] Schneider, H.-J., Hacket F., Rudiger V., *et al.*, NMR studies of cyclodextrins and cyclodextrin complexes. *Chemical Reviews*, 1998. **98**(5): 1755–1786.
[106] Redenti, E., Zanol M., Ventura P., *et al.*, Raman and solid state 13C-NMR investigation of the structure of the 1: 1 amorphous piroxicam: β-cyclodextrin inclusion compound. *Biospectroscopy*, 1999. **5**(4): 243–251.
[107] Hasegawa, Y., Inoue Y., Deguchi K., *et al.*, Molecular dynamics of a polyaniline/β-cyclodextrin complex investigated by 13C solid-state NMR. *The Journal of Physical Chemistry B*, 2012. **116**(6): 1758–1764.
[108] Van Hees, T., Geraldine, P., Brigitte E., *et al.*, Application of supercritical carbon dioxide for the preparation of a piroxicam-β-cyclodextrin inclusion compound. *Pharmaceutical Research*, 1999. **16**(12): 1864–1870.
[109] Lin, S.-Z., D. Wouessidjewe, M.C. Poelman, *et al.*, Indomethacin and cyclodextrin complexes. *International Journal of Pharmaceutics*, 1991. **69**(3): 211–219.
[110] Blanco, J., J.L. Vila-Jato, F. Otero, S. Anguiano, Influence of method of preparation on inclusion complexes of naproxen with different cyclodextrins. *Drug Development and Industrial Pharmacy*, 1991. **17**(7): 943–957.
[111] Bilensoy, E., M. Abdur Rouf, I., Vural, *et al.*, Mucoadhesive, thermosensitive, prolonged-release vaginal gel for clotrimazole: β-cyclodextrin complex. *AAPS PharmSciTech*, 2006. **7**(2): E54–E60.

[112] de Oliveira, V.E., Almeida E.W.C., Castro H.V., *et al.*, Carotenoids and β-cyclodextrin inclusion complexes: Raman spectroscopy and theoretical investigation. *The Journal of Physical Chemistry A*, 2011. **115**(30): 8511–8519.
[113] Rossi, B., P. Verrochio, G. Viliani, *et al.*, Vibrational properties of ibuprofen–cyclodextrin inclusion complexes investigated by Raman scattering and numerical simulation. *Journal of Raman Spectroscopy*, 2009. **40**(4): 453–458.
[114] Sinha, V., Anitha R., Ghosh S., *et al.*, Complexation of celecoxib with β-cyclodextrin: Characterization of the interaction in solution and in solid state. *Journal of Pharmaceutical Sciences*, 2005. **94**(3): 676–687.
[115] Lee, P.S., Han J.Y., Song T.W., *et al.*, Physicochemical characteristics and bioavailability of a novel intestinal metabolite of ginseng saponin (IH901) complexed with β-cyclodextrin. *International Journal of Pharmaceutics*, 2006. **316**(1): 29–36.
[116] Pralhad, T. and K. Rajendrakumar, Study of freeze-dried quercetin–cyclodextrin binary systems by DSC, FT-IR, X-ray diffraction and SEM analysis. *Journal of Pharmaceutical and Biomedical Analysis*, 2004. **34**(2): 333–339.
[117] Zingone, G. and F. Rubessa, Preformulation study of the inclusion complex warfarin-β-cyclodextrin. *International Journal of Pharmaceutics*, 2005. **291**(1): 3–10.
[118] Park, J.S., S. Jeong, B. Ahn, *et al.*, Selective response of cyclodextrin-dye hydrogel to metal ions. *Journal of Inclusion Phenomena and Macrocyclic Chemistry*, 2011. **71**(1/2): 79–86.
[119] Carlstedt, J., A. Bilalov, E. Krivtsova, *et al.*, Cyclodextrin–surfactant coassembly depends on the cyclodextrin ability to crystallize. *Langmuir*, 2012. **28**(5): 2387–2394.
[120] Tu, C.-W., S.-W. Kuo, and F.-C. Chang, Supramolecular self-assembly through inclusion complex formation between poly (ethylene oxide-bN-isopropylacrylamide) block copolymer and α-cyclodextrin. *Polymer*, 2009. **50**(13): 2958–2966.
[121] Polarz, S., Smarsly, B., Bronstein, L., *et al.*, From cyclodextrin assemblies to porous materials by silica templating. *Angewandte Chemie International Edition*, 2001. **40**(23): 4417–4421.
[122] Villari, V., Mazzaglia A., Darcy R., *et al.*, Nanostructures of cationic amphiphilic cyclodextrin complexes with DNA. *Biomacromolecules*, 2013. **14**(3): 811–817.
[123] Bekers, O., T. Loftsson, Matthias, K., *et al.*, Cyclodextrins in the pharmaceutical field. *Drug Development and Industrial Pharmacy*, 1991. **17**(11): 1503–1549.
[124] Aicart, E. and E. Junquera, Complex formation between purine derivatives and cyclodextrins: a fluorescence spectroscopy study. *Journal of Inclusion Phenomena and Macrocyclic Chemistry*, 2003. **47**(3–4): 161–165.
[125] Wei, Y.-L., Ding, L.H., Dong, C., *et al.*, Study on inclusion complex of cyclodextrin with methyl xanthine derivatives by fluorimetry. *Spectrochimica Acta, Part A: Molecular and Biomolecular Spectroscopy*, 2003. **59**(12): 2697–2703.
[126] Másson, M., T. Loftsson, S. Jónsdóttir, *et al.*, Stabilisation of ionic drugs through complexation with non-ionic and ionic cyclodextrins. *International Journal of Pharmaceutics*, 1998. **164**(1): 45–55.
[127] Másson, M., Sigurdardottir, B.V., Matthiasson, K., *et al.*, Investigation of drug-cyclodextrin complexes by a phase-distribution method: some theoretical and practical considerations. *Chemical and Pharmaceutical Bulletin*, 2005. **53**(8): 958–964.
[128] Ugwu, S.O., Alcala M.J., Bhardwaj R., Blanchard J., The application of equilibrium dialysis to the determination of drug-cyclodextrin stability constants. *Journal of Inclusion Phenomena and Molecular Recognition in Chemistry*, 1996. **25**(1/3): 173–176.
[129] Cai, W.S., Wang, T., Zhe Liu, Y., *et al.*, Free energy calculations for cyclodextrin inclusion complexes. *Current Organic Chemistry*, 2011. **15**(6): 839–847.
[130] Djemil, R. and D. Khatmi, Quantum mechanical study of complexation of dopamine and epinephrine with β-Cyclodextrin using PM6, ONIOM and NBO analysis. *Journal of Computational and Theoretical Nanoscience*, 2012. **9**(10): 1571–1576.
[131] Anconi, C.P.A., da Silva Delgado, L., Alves Dos Reis, J.B., *et al.*, Inclusion complexes of α-cyclodextrin and the cisplatin analogues oxaliplatin, carboplatin and nedaplatin: a theoretical approach. *Chemical Physics Letters*, 2011. **515**(1/3): 127–131.

[132] Ramírez-Galicia, G., R. Garduño-Juárez, and M.G. Vargas, Effect of water molecules on the fluorescence enhancement of Aflatoxin B1 mediated by Aflatoxin B1:β-cyclodextrin complexes. A theoretical study. *Photochemical and Photobiological Sciences*, 2007. **6**(1): 110–118.

[133] Wickstrom, L., He, P., Gallicchio, E., and Levy, R.M., Large scale affinity calculations of cyclodextrin host–guest complexes: Understanding the role of reorganization in the molecular recognition process. *Journal of Chemical Theory and Computation*, 2013. **9**(7): 3136–3150.

[134] He, J., Chipot, C., Shao, X., *et al.*, Cyclodextrin-mediated recruitment and delivery of amphotericin B. *Journal of Physical Chemistry C*, 2013. **117**(22): 11750–11756.

[135] Jiao, A., Zhou, X., Xu, X., and Jin, Z., Molecular dynamics simulations of cyclodextrin-cumene hydroperoxide complexes in water. *Computational and Theoretical Chemistry*, 2013. **1013**: 1–6.

[136] El-Barghouthi, M.I., C. Jaime, N.A. Al-Sakhen, *et al.*, Molecular dynamics simulations and MM-PBSA calculations of the cyclodextrin inclusion complexes with 1-alkanols, para-substituted phenols and substituted imidazoles. *Journal of Molecular Structure: THEOCHEM*, 2008. **853**(1/3): 45–52.

[137] Mota, G.V.S., C.X. Oliveira, A.M.J.C. Neto, *et al.*, Inclusion complexation of praziquantel and β-cyclodextrin, combined molecular mechanic and monte carlo simulation. *Journal of Computational and Theoretical Nanoscience*, 2012. **9**(8): 1090–1095.

[138] Pereira, C.C.L., Nolasco, M., Braga, S.S., *et al.*, A combined theoretical–experimental study of the inclusion of niobocene dichloride in native and permethylated β-cyclodextrins. *Organometallics*, 2007. **26**(17): 4220–4228.

[139] Georg, H.C., K. Coutinho, and S. Canuto, A look inside the cavity of hydrated α-cyclodextrin: a computer simulation study. *Chemical Physics Letters*, 2005. **413**(1/3): 16–21.

[140] Liu, B., Li, W., Zhao, J., *et al.*, Physicochemical characterisation of the supramolecular structure of luteolin/cyclodextrin inclusion complex. *Food Chemistry*, 2013. **141**(2): 900–906.

[141] Miletic, T., Kyriakos, K., Graovac, A., and Ibric, S., Spray-dried voriconazole-cyclodextrin complexes: solubility, dissolution rate and chemical stability. *Carbohydrate Polymers*, 2013. **98**(1): 122–131.

[142] Paramasivaganesh, K., Srinivasan, K., Manivel, A. *et al.*, Studies on inclusion complexation between 4,4'-dihydroxybiphenyl and β-cyclodextrin by experimental and theoretical approach. *Journal of Molecular Structure*, 2013. **1048**: 399–409.

[143] Ghasemi, J.B., M. Salahinejad, M.K. Rofouei, M.H. Mousazadeh, Docking and 3D-QSAR study of stability constants of benzene derivatives as environmental pollutants with α-cyclodextrin. *Journal of Inclusion Phenomena and Macrocyclic Chemistry*, 2012. **73**(1/4): 405–413.

[144] Pérez-Garrido, A., Morales Helguera, A., Corderio, M.N.D.S., and Escudero, GA., QSPR modelling with the topological substructural molecular design approach: β-cyclodextrin complexation. *Journal of Pharmaceutical Sciences*, 2009. **98**(12): 4557–4576.

[145] Pérez-Garrido, A., H.A. Morales, G.A. Abellán, *et al.*, Convenient QSAR model for predicting the complexation of structurally diverse compounds with β-cyclodextrins. *Bioorganic and Medicinal Chemistry*, 2009. **17**(2): 896–904.

[146] Wang, W., Donini, O., Reyes, C.M., and Kollman, P.A., Biomolecular simulations: recent developments in force fields, simulations of enzyme catalysis, protein–ligand, protein–protein, and protein–nucleic acid noncovalent interactions. *Annual Review of Biophysics and Biomolecular Structure*, 2001. **30**: 211–243.

4

Molecular Modeling of Block Copolymer Self-Assembly and Micellar Drug Delivery

Myungshim Kang[1], Dennis Lam[1], Dennis E. Discher[2], and Sharon M. Loverde[1]

[1] Department of Chemistry, College of Staten Island, City University of New York, USA
[2] Department of Chemical and Biomolecular Engineering, University of Pennsylvania, USA

4.1 Introduction

Amphiphilic diblock polymer molecules can self-assemble into a variety of morphologies in dilute solution – spherical micelles, cylindrical micelles, and bilayer (vesicle) morphologies. The hydrophobic blocks self-assemble in the core of the micelle, while the hydrophilic blocks maximize their contact with the solvent through the formation of a micellar corona. The free energy is a balance of the degree of stretching of the polymer chains, the polymer–polymer interactions, the polymer–solvent interactions, and the associated entropy, to give rise to an interfacial tension [1]. In a more general manner, the shape of the morphologies can be thought of in terms of a packing parameter, $P = v/al_c$ that is determined by molecular volume v of the packed hydrophobic segment, contour length l_c of the hydrophobic segment, and interfacial area a of the hydrophilic segment [2]. The range of P shifts the preferred assembly in dilute solution from spherical micelle ($P < 1/3$) to worm micelle ($1/3 < P < 1/2$) to vesicle ($P > 1/2$) or polymersome phase. In addition to the packing parameter, the copolymer concentration, the water content in solution, the nature of the

Computational Pharmaceutics: Application of Molecular Modeling in Drug Delivery, First Edition.
Edited by Defang Ouyang and Sean C. Smith.

cosolvent, as well as the kinetics of the aggregate formation can all affect the resulting morphology [3]. In general, diblock copolymers are characterized by the degree of immiscibility between the block components defined by the Flory–Huggins parameter, χ, the total number of monomers, N, and the hydrophilic mass fraction, f $(0<f<1)$, of the copolymer, with the degree of microphase separation determined by the product, χN [3]. N is the total length of the polymer and χ is defined by [4]:

$$\chi_{\mathrm{AB}} = \frac{Z}{kT}\left[\varepsilon_{\mathrm{AB}} - \frac{1}{2}(\varepsilon_{\mathrm{AA}} + \varepsilon_{\mathrm{BB}})\right] \tag{4.1}$$

Z is defined by the number of nearest neighbors, and ε_{AA}, ε_{BB}, and ε_{AB} are the interaction energies between A and B copolymer units. As outlined in Figure 4.1, as the hydrophilic mass fraction (f) increases, the preferred morphology changes from a bilayer vesicle to a worm-like micelle to a spherical micelle morphology. Applications of polymeric self-assemblies are in numerous fields – photonics, electronics, and biomedicine; additionally, multiblock polymers open up possibilities of new morphologies [5]. In particular, micelles are currently successfully used as pharmaceutical carriers for hydrophobic drugs and possess a number of attractive properties as drug carriers, including high stability and good biocompatibility [6].

The hydrophobic environment of the micellar core can serve as an environment for hydrophobic solutes, or drugs. As shown schematically in Figure 4.2, dependent on the interaction of drug with drug and drug with polymer, the hydrophobic drug can be solubilized homogeneously in the core, or the drug can locally segregate from the hydrophobic core of the micelle. The free energy for hydrophobic drug loading in the core of the micelle is a balance in terms of the enthalpy and the entropy between the solute and the hydrophobic diblock component, the interfacial energy, and the conformational energy of

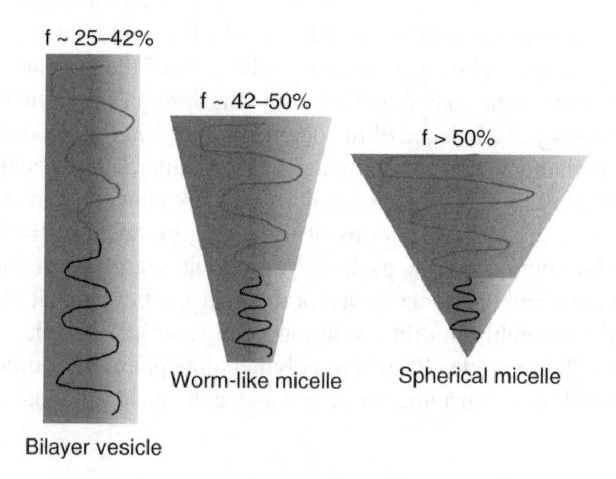

Figure 4.1 As the hydrophilic mass fraction, f, increases the preferred morphology changes from a bilayer vesicle to worm-like micelle to spherical micelle morphology (see colour plate section).

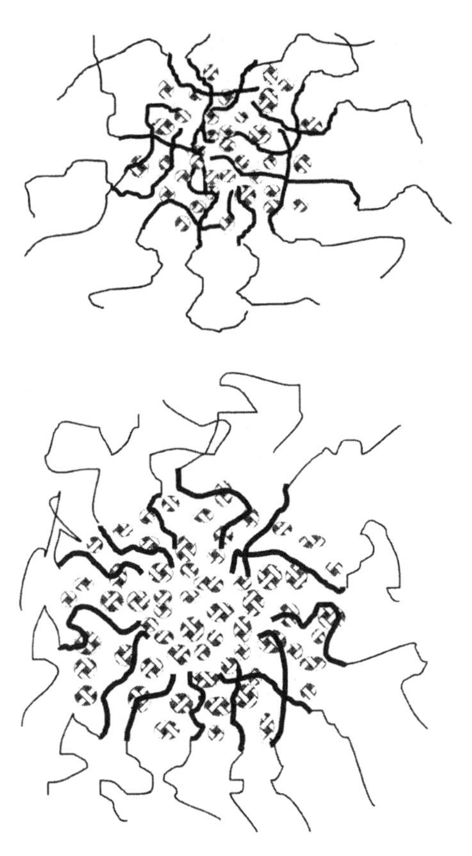

Figure 4.2 Two different schematic representations of spherical micelles containing solutes. The darker lines represent the hydrophobic block and the lighter lines represent the hydrophilic block. In the top sketch, the solute, or drug, is solubilized in the hydrophobic core of the micelle. In the bottom sketch the drug molecule segregates from the micellar core. Reprinted with permission from Ref. [7]. Copyright 2001 John Wiley & Sons, Ltd.

the polymer chains [8]. It is predicted that the solubility of the drug in the micelle is proportional to the χ_{dp} to some power, where χ_{dp} is defined as follows [7]:

$$\chi_{dp} = \frac{V_{drug}}{RT(\delta_{drug} - \delta_{polymer})^2} \tag{4.2}$$

V_{drug} is the volume of the drug, R is the ideal gas constant, T is the temperature, and δ_{drug} and $\delta_{polymer}$ are the Hildebrand solubility parameters of the drug and the polymer. Additionally, the inherent shape of the micelle is predicted to affect the ability to encapsulate different solutes [7]. Drug loading, and how it relates to micellar morphology, is a model problem to be addressed with multiscale molecular modeling that will be addressed within the scope of this chapter.

Upon entering the bloodstream, micelles are subjected to dilution, which can destabilize the assembly. There are various factors which affect the kinetic and thermodynamic stability of micelles, including the critical micelle concentration (CMC) of the polymer, the hydrophilic mass fraction (*f*), the T_g of the hydrophobic group, and also the drug content in the micelle [9]. The rate of disassembly of the micelle into unimers [10] depends on multiple factors, including the content of solvent in the core and the length of the hydrophilic and hydrophobic groups. Additionally, the copolymer may be degradable, for example, composed of the polyesters such as polylactic acid or polycaprolactone (PCL) connected with hydrophilic poly(ethylene oxide) (PEO) [11–13]. In particular, using copolymers that degrade in the pH range of the endosome will allow for the intracellular release of the encapsulated drugs. The rate of polyester hydrolysis is dependent on pH [14, 15], polymer molecular weight, and the degree of polymer crystallinity [16–18]. PEO stabilizes the micellar corona, inhibits the surface absorption of biological macromolecules, maximizing biocompatibility, and helps micelles escape the rapid reticuloendothelial system (RES) after intravenous administration, prolonging circulation time [6, 19]. Moreover, the properties of copolymers are tunable, with increased molecular weight imparting increased stability of the micelle [20]. Additionally, micelle size and shape have been shown to affect the circulation time in the bloodstream [21].

A tumor's newly developed vasculature is known to be more permeable than healthy blood vessels. High molecular weight species, such as polymers, can easily penetrate the vascular wall and seep into the tumor [22] – this concept has been termed the enhanced permeability and retention (EPR) effect. Many types of anticancer polymer-based self-assemblies have exploited this, including bioconjugates [23], liposomes [24], polymersomes [25], and polymeric micelles [26]. Micelles, particularly PEO-PCL micelles, are internalized via the mechanism of endocytosis [27]. It has been demonstrated by various groups that drugs or dyes inside the hydrophobic core of the micelle can be internalized inside the cell without the endocytosis of the micelle [28]. It is suggested that the plasma membrane mediates the internalization process through the uptake of hydrophobic dyes from the micellar core. Additionally, polymer micelles can increase drug uptake through other indirect mechanisms. For example, Pluronics have been shown to inhibit P-glycoprotein, enabling internalization of drug inside the cell [29].

Once a drug is delivered and released at the target site in the body, the next step is binding of the drug to the targeted protein. This leads to the desired modulation of the biochemical activities of cell physiology, where the inherent interactions between the bare drug itself and the protein take center stage. Figure 4.3 illustrates a thermodynamic cycle of a two-step ligand binding to a receptor. Primary concerns in drug discovery have focused on the optimization of the binding affinity for a more potent drug as well as the target selectivity of a drug to minimize the off-target related side effects. Therefore, the equilibrium dissociation constant (K_D) or the half-maximal inhibitory or effective concentration (IC_{50} or EC_{50}) have been commonly used as a measure of the binding affinity to the target or the off-target. More recently, there has been increased recognition of the kinetic aspects of drug–protein binding in addition to the equilibrium state, given the open, dynamic environments, and time-dependent factors *in vivo* [31, 32] and the concept of the drug residence time or, more specifically, the dissociation rate

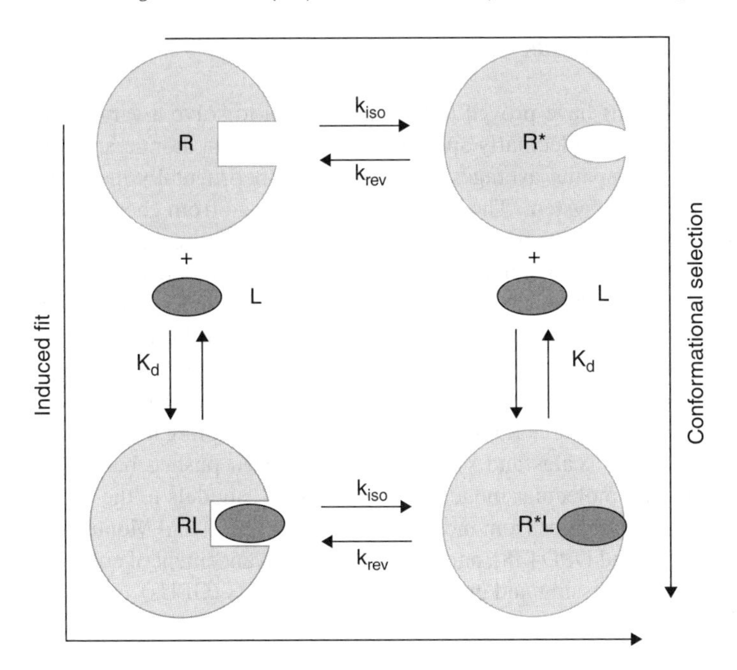

Figure 4.3 *Thermodynamic cycle for two-step ligand (L) binding to receptor (R). The reaction scheme for the conformational selection mechanism starts with the unliganded receptor state R (top left corner) and proceeds along the clockwise direction indicated by the arrow. The induced-fit mechanism also starts with unliganded receptor state R, but proceeds along the trajectory indicated by the counterclockwise arrow. In both models R and R* refer to distinct conformational states of the same receptor molecule. In the conformational selection model, the interconversion between states R and R* is slow and occurs prior to rapid ligand binding to state R*. In the induced-fit model, ligand binds rapidly to state R after which there is a slow conversion (i.e., receptor isomerization) to the bound state R*L. Reprinted with permission from Ref. [30]. Copyright Future-Science, 2011.*

constant (k_{off}), which has been suggested to be useful in preclinical lead optimization [31]. The dissociation rate or dissociative half-life has shown correlations with drug resistance [33], the *in vitro* cellular activity [34], the drug efficacy and duration *in vivo* [35], and the selectivity [36]. More active delivery systems may help improve the pharmacokinetics of the drug.

In the scope of this chapter, numerous aspects of multiscale molecular modeling for drug delivery will be discussed, including multiscale computational methods, micellar self-assembly and stability, the interaction of drugs with polymers and membranes, as well as the kinetics of drug binding. We will review atomistic, coarse-grained (CG) [37], and mesoscale simulation techniques, including dissipative particle dynamics (DPD) [38] and Brownian dynamics (BD) [39]. While the accuracy of force fields for proteins, nucleic acids, and polymers, continue to be improved, the length scales and time scales of micellar self-assembly, drug delivery, and the kinetics of binding in cellular environments entail the utilization of lower resolution or multiscale models.

4.2 Simulation Methods

Molecular simulations have proven to be a useful tool to solve a variety of problems by providing insight into molecularly specific interactions at an atomic and molecular level, whose temporal and spatial averages correspond to experimentally measurable collective properties of the given system. The application areas range from chemistry and physics to biophysics, chemical engineering, and pharmaceutics, and are growing with no visible limits. The quality of the simulation results fundamentally depends on the given force fields and sampling methods: as a bottom-up approach, the macroscopic properties of the simulated system are entirely based on the force fields used. Furthermore, efficient sampling methods are also critical to achieve statistically significant results. Therefore tremendous efforts from both the theoretical and experimental sides have concentrated on the development of accurate force fields, as well as efficient sampling methods. The limitations in the applicable time scales and system sizes have been pushed back in various ways, including multiscale molecular modeling from atomistic models to the CG level [40, 41], different dynamics strategies from molecular dynamics (MD) and Monte Carlo (MC) simulations to BD [39] and DPD [38], and, of course, the advancement of computational power including parallel processing and graphics processing units (GPUs).

4.2.1 All-Atom Models

Classical mechanics governs the motion of particles in MD simulations. In classical atomistic models, every atom is considered as a bead and the chemical bond between them can be represented as a connecting spring. Force fields for atomistic molecular models define how to calculate the intra- and intermolecular interactions. The properties of each atom include the mass, the charge, and the van der Waals radius. The total energy per atom includes the potentials for bond, angle, and dihedral, as well as the nonbonded potentials for electrostatics and van der Waals interactions:

$$U = U_{\text{bond}} + U_{\text{angle}} + U_{\text{dihedral}} + U_{\text{elec}} + U_{\text{vdw}}. \tag{4.3}$$

Many groups have developed more accurate force fields through both theoretical and experimental efforts [42–46]. These classical atomistic models allow us to expand our investigations to systems many orders of magnitude larger than quantum mechanics can reach by introducing the concept of the point charge instead of the electron cloud. More recently, polarizable force fields have been developed to catch the instantaneously induced charge due to the changes of the electrostatic environment, at the expense of computational cost [47–49]. Smith and colleagues demonstrated that the optimization of partial charges based on Kirkwood–Buff theory of liquids leads to better performance than the scaled-up charges based on gaseous state calculations without increasing the computational cost [42].

4.2.2 Coarse-Grained Models

In order to reach larger systems and time scales of interest, CG molecular models have been introduced at many levels where the so-called bead (interacting center) can represent several neighboring atoms, a residue, or an entire protein domain [50–55]. The number of the particles in the system dramatically drops and therefore the computational cost decreases accordingly. The beads are connected by a virtual bond and parameterized in many ways, including re-mapping from available atomistic simulations [56, 57], such as Boltzmann inversion [40, 57], as well as parameterization via experimental data for bulk liquid and surface properties. For example, the Shinoda–DeVane–Klein (SDK) CG model is briefly described in this section as an example of a CG model.

Micellar self-assembly and stability, the interaction of drugs with polymers and membranes, as well as the kinetics of drug binding are all examples of systems that can be approached with molecular modeling at multiple length scales – atomistic, CG, mesoscopic, in complement with advanced sampling techniques, as reviewed in this chapter. CG molecular models have been applied to study the behavior of numerous phenomena in science and engineering, including polymers, proteins, and membranes [58–63]. Atomistically detailed molecular models are inherently limited to the smaller length and time scales; however, CG models allow sampling of much longer length and time scales due to decreased modeling resolution. Mesoscale models, such as DPD and BD, to be described in more detail, allow for simulation of even longer time and length scales, while sacrificing the atomic level detail. Mesoscale methods have captured the phase behavior of block copolymers, surfactants, and membranes [64], in addition to the kinetics of protein–ligand binding in crowded environments [65]. The method of approach to go from the atomistic level to a higher resolution model has been rigorously examined [66]. Structural and thermodynamic approaches aim for close agreement with measured properties from simulations at the all-atom level, or experimental properties, such as radial distribution functions with a force-matching approach [52], and/or thermodynamic properties [50] of chemical groups, such as surface tension or bulk liquid density. Moreover, a combined structural and thermodynamic approach has been shown to be fairly transferrable over a range of temperatures close to room temperature. CG MD, including mesoscopic simulation techniques, have shown a potential to capture the self-assembly of PEO-based surfactants and polymers [50, 67]. Several examples of CG MD, as applied to drug delivery, will be outlined in the next section of this chapter.

As an example, the SDK CG force field approach is briefly outlined here. For the SDK methodology, the intramolecular interactions are modeled via harmonic potentials given by $V(r)_{\text{bond}} = K_{\text{b}}\left(r - r_{\text{o}}\right)^{2}$ and $V(r)_{\text{angle}} = K_{\text{a}}\left(\theta - \theta_{\text{o}}\right)^{2}$. Here, K_{b} and r_{o} represent the equilibrium force constant and distance for bonds, and K_{a} and θ_{o} represent the equilibrium force constant and equilibrium angle. These constants are obtained from the respective all-atomistic simulations using an inverse Boltzmann technique, such that $U_{\text{bond}}(r)\alpha - k_{\text{B}}Tln\dfrac{P(r)}{r^{2}}$ and $U_{\text{angle}\alpha} - k_{\text{B}}Tln\dfrac{P(\theta)}{\sin\theta}$, while the nonbonded interaction is set by a pairwise additive potential based on the Lennard–Jones (LJ) potential: $U_{\text{LJ9-6}} = \dfrac{27}{4}\varepsilon\left\{\left(\dfrac{\sigma}{r}\right)^{9} - \left(\dfrac{\sigma}{r}\right)^{6}\right\}$ or $U_{\text{LJ12-4}} = \dfrac{3\sqrt{3}}{2}\varepsilon\left\{\left(\dfrac{\sigma}{r}\right)^{12} - \left(\dfrac{\sigma}{r}\right)^{4}\right\}$ [50]. The SDK CG approach [50] has been systematically applied to study the self-assembly of materials properties of lipids [68], proteins [69], polyesters [70], and surfactants [37]. Likewise, the Martini force field [71], a similar methodology, has been applied to numerous soft matter systems – lipids, proteins, buckyballs, as well as polymers, such as polystyrene [72]. The SDK approach, at long sampling times and high concentrations, when extrapolated to low concentrations, has been shown to exhibit good agreement with experimental CMCs of short PEO-based surfactants at room temperature [73–75]. Additionally the SDK approach, combined with enhanced sampling techniques, has been applied to PCL [70]. This will be discussed in the next section of this chapter in comparison with experimental results.

4.2.3 Mesoscale Methods: BD and DPD

Another way to extend the time scales and length scales of the system is to adopt a more mesoscopic approach to simulation. Here we briefly introduce BD and DPD as examples. These methods have also been coupled with CG molecular descriptions, such as proteins or polymers. BD [39] is a stochastic approach to simulate the dynamics of molecular systems. Traditional MD simulations calculate particle displacements over a specified time interval by integrating Newton's equations of motion. BD calculations start with the Langevin equation of internal motion, which contains frictional and random forces representing the collision with solvent. It is assumed that the momentum relaxation time is faster than the position relaxation, which allows us to consider time intervals longer than the momentum relaxation time scale. The BD trajectories are obtained by solving the Langevin equation assuming no average acceleration during the simulation [39, 76]: $\Delta r_i = \frac{D_i}{k_B T} f_i \Delta t + R_i(\Delta t)$, where the systemic force on the particle i, f_i, is obtained the usual way as a negative gradient of the potential energy. D_i is the pre-computed 6×6 diffusion tensor of particle i, k_B is the Boltzmann constant, and T is the absolute temperature. $R_i(\Delta t)$ is the random displacement vector due to the collision of surrounding solvent with a Gaussian distribution with zero mean and the variance–covariance of $R_i(\Delta t) R_j(\Delta t) = 2D_i \delta_{ij} \Delta t$. Here δ_{ij} is the Kronecker delta. The Cholesky factorization of D can give the random displacement. For simplicity [77], D_i can be replaced with the diffusion coefficient of particle i, and is given by the Stocks–Einstein equation $D_i = \frac{k_B T}{6\pi\eta a}$, where η is the viscosity of solvent, and a is the radius of the molecule. In order to include the hydrodynamic interactions [39, 78] and make BD more realistic, the $6N \times 6N$ diffusion tensor for the entire system of the total N spherical molecules is added in the propagation scheme: $\Delta r = (\nabla D)\Delta t + \frac{D}{k_B T} f \Delta t + R(\Delta t)$. Here the diffusion tensor D depends on the positions of particles, meaning that each BD step needs to accompany the Cholesky decomposition of D to compute the random displacement with considerable computational cost [79]. In BD simulations, solvents are not treated explicitly: BD takes into consideration the viscosity of solvent, but does not reproduce the electrostatic screening effect of solvent and the hydrophobic effect like implicit solvent models do. Such a way to treat solvent implicitly makes it possible to simulate large systems for a very long time along with larger time intervals to calculate the particle displacements.

DPD [80] method is another approximate, CG scheme designed for the simulations of liquids at mesoscopic scales beyond which conventional MD can reach. The full atomistic details of the solvent are not necessary and inefficient in the study of Brownian motion of colloids, where only its viscosity, density, and temperature matter. As a coarse-graining of MD, DPD provides an efficient model to predict the complex hydrodynamic behavior: particles in DPD represent clusters of molecules with pairwise interactions. In addition to the conservative force used in the conventional MD, a dissipative force and a random force compose the DPD algorithm: $F_i = \sum_{j \neq i} \left[f^C(r_{ij}) + f^D(r_{ij}, v_{ij}) + f^R(r_{ij}) \right]$, where the conservative force on the particle i, f^C, is system dependent and is derived from a pair potential between particles i and j. The dissipative force f^D is the frictional force and represents the viscous resistance within the real fluid: $f^D(r_{ij}, v_{ij}) = -\gamma\omega^D(r_{ij})(v_{ij} \cdot e_{ij})e_{ij}$, where γ is a friction coefficient with a distance variation of $\omega^D(r_{ij})$, $r_{ij} = r_i - r_j$, and e_{ij} is the unit vector in the direction of r_{ij}, and $v_{ij} = v_i - v_j$. The random force f^R is given in a similar form to make up for the degrees of freedom eliminated after the coarse graining: $f^R(r_{ij}) = \sigma\omega^R(r_{ij})\xi_{ij}e_{ij}$, where σ is a coefficient controlling the magnitude of the random pair force between the DPD particles, whose distance variation is described in $\omega^R(r_{ij})$. The random variable ξ_{ij} has a Gaussian distribution and unit variance, with $\xi_{ij} = \xi_{ji}$. The fluctuation and dissipation are related in such a way that they satisfy $\omega^D(r_{ij}) = [\omega^R(r_{ij})]^2$ and $\sigma^2 = 2k_B T\gamma$ for the thermal equilibrium of the system. The DPD method takes a very similar approach to that of BD in terms of the employment of the frictional and random forces. The difference is that this combination of the frictional and the random forces in the BD does not conserve the momentum, while the frictional and the random forces in DPD are formulated to conserve momentum and therefore reproduce hydrodynamic behavior on long time scales and large length scales.

4.2.4 Free Energy Methods

It is essential to determine the free energy profile in order to understand most chemical and biological processes. However, the needs of thorough sampling of the configuration space make it extremely costly to calculate the free energy of complex systems. Various free energy calculation methods have been developed. Here we briefly introduce three of them: steered molecular dynamics (SMD), adaptive biasing force (ABF) method, and metadynamics. SMD [81, 82] is based on nonequilibrium simulation using Jarzynski's equality [83]. In SMD simulations, an external force is applied to the system to induce the structural changes or process of interest with minimized sampling cost. A guiding potential $h(r;\lambda) = \frac{k}{2}[\xi(r) - \lambda]^2$ is introduced in such a way that an external parameter λ is correlated with the reaction coordinate ξ, which acts as a spring that constrains ξ to be near λ with a force constant k. Typically, either velocity or force is held constant during the process. With a constant velocity v, the work done on the system w is calculated as $w_{0\to t} = -kv\int_0^t dt'(\xi(r_{t'}) - \lambda_0 - vt')$. According to Jarzynski's equality, the free energy difference ΔF is equal to the average work (W) done on the system through nonequilibrium processes regardless of the speed of the process. In principle: $e^{-\beta\Delta F} = e^{-\beta W}$, where β is $1/(k_B T)$ with the Boltzmann constant k_B and the absolute temperature T. The potential of mean force (PMF), that is, the free energy profile as a function of a coordinate, can be calculated by averaging the work. Due to the difficulty in sampling rare trajectories, the application of SMD has a practical limit to slow processes with the fluctuation of the work comparable to the temperature [82]. SMD has been used to investigate numerous problems such as structural transitions like unfolding or stretching of structures [82], protein–ligand binding and unbinding processes [84], and penetration of solutes or ions across channels and membranes [85].

The ABF method [86, 87] takes a different and very efficient approach to enhance the sampling for free energy calculations. ABF involves estimating the mean force along the collective

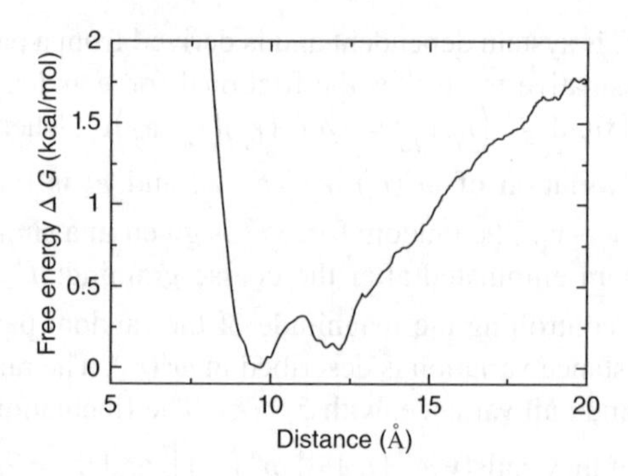

Figure 4.4 *The potential of mean force between two Taxol molecules in water as a function of the distance calculated using the ABF method.* G_{ref} *is taken as the free energy minimum at approximately 12 Å.*

variables (CVs) ξ from a running average in the appropriate bin during unconstrained simulations, $F_\xi \mid \xi^* = -dA\left(\xi^*\right)/d\xi$, where $A(\xi)$ is the free energy as a function of an order parameter ξ and the bracket $\mid \xi^*$ denotes the conditional average, removing this force by applying a opposing external force of $-F_\xi \mid \xi^* \nabla \xi$. The resulting zero mean force on ξ leads to a random walk with no barrier along ξ, meaning uniform sampling along ξ. ABF method has several advantages, including a very small statistical error and excellent convergence. An example of PMF calculation using ABF is shown in Figure 4.4. This simple and efficient sampling method has found many successful applications in chemical and biological systems studying conformational equilibrium [88], selective uptake of a small molecule by protein [89], and dissociation and association free energy profiles of a wide range of complexes [90, 91].

Metadynamics [92, 93] also uses an external potential to accelerate the reaction process to improve the sampling of rare events. In this algorithm, the history-dependent potential is applied to recursively reconstruct the free energy from the bottom of the well through the exhaustive exploration on a predefined region in the CV space. A small repulsive Gaussian potential with the height ω and width δ_{ξ_i} is added at every τ_G steps during the MD run, centered at previously explored configurations and tending to accumulate in regions of lower effective free energy as the system evolves. The sum of these potential constructs the external metadynamics potential on the CVs $\xi = (\xi_1, \xi_2, \ldots, \xi_{nc})$ at time t as follows:

$$V_G(\xi) = \omega \sum_{\substack{t'=\tau_G, 2\tau_G, \ldots \\ t'<t}} \exp\left\{ -\sum_{i=1}^{nc} \frac{\left[\xi_i - \xi_i(t')\right]^2}{2\delta_{\xi_i}^2} \right\}. \tag{4.4}$$

The sum of the underlying PMF $A(\xi)$ and $V_G(\xi)$ forms the effective free energy $\bar{A}(\xi) = A(\xi) + V_G(\xi)$. After a sufficiently long time, the underlying free energy is estimated as $\lim_{t\to\infty} V_G(\xi) \sim -A(\xi)$. While both ABF and metadynamics utilize external forces, metadynamics requires three parameters to define the external potentials: the Gaussian width δ_{ξ_i}, the height ω, and the frequency τ_G at which the Gaussians are updated. These parameters have a strong influence on the quality of the reconstructed free energy and the sampling efficiency: the ratio ω/τ_G determines the accuracy of the reconstructed free energy profile. A much smaller ratio of ω/τ_G than $\kappa_B T/\tau\xi$, where $\tau\xi$ is the longest correlation time of ξ values, is usually recommended for good statistical convergence. Clearly, a finite number of chosen CVs should fully describe the process of interest for the reliability of metadynamics: they should distinguish the initial, intermediate, and the final states clearly and describe all the slow relevant "events" of interest. Metadynamics has been applied and proven useful in a multitude of areas, including phase transitions [94], crystal structure prediction [95], chemical reactions [96], and biophysics [97, 98].

4.3 Simulations of Micellar Drug Delivery

As the hydrophilic mass fraction, f, increases the preferred morphology changes from a bilayer vesicle to worm-like micelle to spherical micelle morphology. For example, in systems composed of neutral polymers such as the diblock copolymer poly(ethylene oxide)-polybutadiene (PEO-PBD), the phase behavior has been thoroughly characterized [99–101]. The first characterization of the dilute solution phase behavior of neutral amphiphilic diblock copolymers established that as the weight fraction of the hydrophilic PEO block ($f_{hydrophilic}$) is increased, the preferred assemblies shifts from lower to higher curvatures–vesicles to worms to spheres [100]. In this study, worm micelles were also formed within a very narrow range of $f_{hydrophilic}$ ($0.5 < f_{hydrophilic} < 0.55$), while vesicles and spherical micelles are formed at $f_{hydrophilic}$ values above and below, respectively. As shown in Figure 4.5, the preferred morphology of assemblies of PEO-PCL was studied for block copolymers containing different hydrophilic/hydrophobic ratios and mapped into a plot of core block hydrophobicity (M_{CH_2}) versus hydrophilic mass fraction, $f_{hydrophilic}$ (f) [102]. In the phase diagram, there is a characteristic shift from bilayer to worm-like micelle to spherical micelle morphology, where the formation of worm-like micelles is limited to a narrow range of molecular weights, with a critical molecular weight of tail necessary for worm-like micelle formation. Figure 4.5a shows representative fluorescent micrographs of spherical micelles, worm micelles, and vesicles labeled with a lipophilic dye. The inset shows localization of hydrophilic dye within PEO-PCL vesicles. Figure 4.5b shows the phase behavior of PEO-PCL assemblies in dilute solution. Filled symbols in gray represent corresponding morphologies: triangles for spherical micelles, squares for worm micelles, and circles for vesicles. The gray lines indicate the approximate phase boundaries for regions of phase coexistence. In comparison with the PEO-PBD worm micelle phase, the PEO-PCL worm micelle phase coexists with spheres and vesicles. Additionally, as previously mentioned, below a critical length of the hydrophobic block, worm micelles are observed, and increasing $f_{hydrophilic}$ shifts the preferred morphology directly from vesicles to spheres.

4.3.1 Characterization of PCL Micelles with Simulation

Poly(ethelyne glycol) (PEG)-PCL copolymers with different hydrophilic fractions ($f_{hydrophilic}$) and different molecular weights (Table 4.1) were simulated with CG MD techniques and compared to the previously described experimental phase diagram [102]. Simulations were performed utilizing the SDK parameterization methodology with the LAMMPS MD code [103]. While the experimental polymer concentrations are 100 μM, the simulation

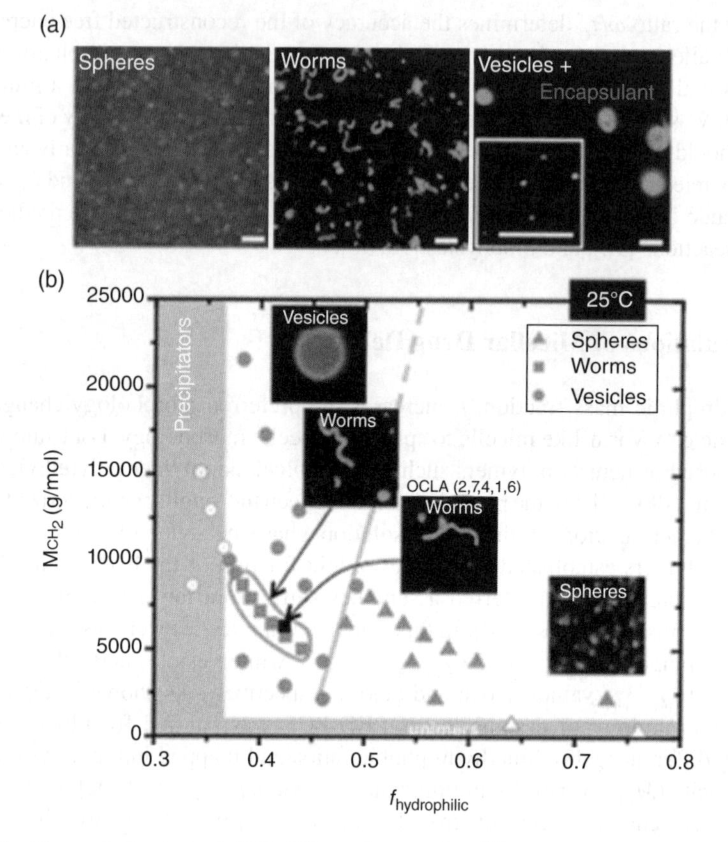

Figure 4.5 *Phase behavior of PEO-PCL polymers in dilute solution. (a) Representative fluorescent micrographs of spherical micelles, worm micelles and vesicles, labeled with a lipophilic dye. Inset shows localization of hydrophilic dye within PEO-PCL vesicles. (b) Phase behavior of PEO-PCL assemblies in dilute solution. Filled symbols in gray represent corresponding morphologies: triangles for spherical micelles, squares for worm micelles, and circles for vesicles. Gray lines indicate the approximate phase boundaries. Scale bar is 10 μm. Reprinted with permission from Ref. [102]. Copyright 2010 American Chemical Society.*

concentrations are limited by the explicit CG water in the simulation box. Each box contains up to 0.6 million CG water beads (which represent three molecular waters each), with polymer concentrations of approximately 15 mM (CMC is submicromolar). To begin with, PEG_{23}-PCL_4 and PEG_x-PCL_y, where x and y represent the number of monomers per polymer, formed loose aggregates in simulation, but the experimental morphology was undetermined.

Moreover, it is found that the stable morphology in solution ranges from polymer vesicle bilayers to worm micelles to spherical micelles with increasing $f_{hydrophilic}$, as shown in Figure 4.6. Homogeneous dispersion of PEG_{1000}-PCL_{3000} assembles into a bilayer (Figure 4.6a). Here x and y represent the molecular weight of the polymer. As shown, there are random fluctuations in a homogeneous mixture of copolymer and water that are soon followed by segregation with both hydrophobic and hydrophilic entanglements throughout the periodic box.

Table 4.1 Comparison of experimental and simulation morphologies for PEO-PCL self-assemblies.

Diblock	Diblock (DA)	Experimental morphology	Simulation morphology	Aggregate core dimension (simulation)
PEO_{23}-PCL_4	PEO1000-PCL500	Undetermined	Loose aggregates	N/A
PEO_{23}-PCL_9	PEO1000-PCL1000	Spherical (dominant)	Spherical	~
PEO_{23}-PCL_{26}	PEO1000-PCL3000	Vesicle (dominant) + spherical	Bilayer/spherical	42 Å (bilayer)
PEO_{45}-PCL_{44}	PEO2000-PCL5000	Worm (dominant by wt%) + spherical	Worm/spherical	60 Å (worm)
PEO_{46}-PCL_{67}	PEO2000-PCL7700	Vesicle	Vesicle	50 Å (bilayer)
PEO_{111}-PCL_{66}	PEO4800-PCL7500	Worm	Worm	71 Å (worm)
PEO_{148}-PCL_{88}	PEO6500PCL10000	Worm	Worm	76 Å (worm)

Reproduced from Ref. [70].

Following there is complete segregation of water and polymer into a bicelle (30 ns), that over a long time reveals a bilayer (100 ns). Figure 4.6b shows such a PEG_{2000}-PCL_{7700} bilayer patch of a "polymersome" composed of PEG_{2000}-PCL_{7700} and Figure 4.6c shows a stable worm micelle of PEG_{2000}-PCL_{5000}, which has a much lower $f_{hydrophilic}$. Orange chains of PEG in the sectioned bilayer and worm micelle (Figure 4.6b,c) are indeed seen to interact with the PCL core in addition to generating the expected corona hydrated with light blue waters. This is confirmed by the interfacial density profile. PEG and PCL are weakly immiscible polymers, so this is to be expected. Additionally, PEG_{2000}-PCL_{1000} generates a spherical form as shown in the schematic phase diagram in Figure 4.6d, in comparison with the experimental phase diagram of Figure 4.5b. We see that the two phase diagrams agree qualitatively.

4.3.2 Advantages of Worm-Like Micelles, Breakup of Micelles

It has been established that worm micelles are functionally better than spherical micelles [21] for numerous reasons. For example, for delivery applications, flexible wormlike micelles possess a long circulation time of up to a week. Additionally, they are also more efficient carriers of hydrophobic drugs such as paclitaxel (Taxol) as compared with spherical micelles [104]. On the other hand, polymer vesicles display a shorter circulation and delivery time, but can also deliver hydrophilic drugs [105–107]. Worm micelles can stretch and align with flow due to predictably strong hydrodynamic effects for micelles much greater than 1 μm in length, as shown in Figure 4.7. Here, we see a flow chamber with immobilized phagocytes, with long worm-like micelles flowing past the cells, aligned with the flow due to hydrodynamic forces. Occasionally, the worm-like micelles leave a fragment, while smaller micelles and vesicles are captured by the phagocyte.

Common mechanisms of micellar instability and drug release are due to the degradation of the copolymer – for polyester-based copolymers the polymer shortens through the gradual degradation of the hydrophobic caprolactone or lactic acid tail [14]. It has also been shown specifically for the polymersome case, or bilayer case, that a blend of inert

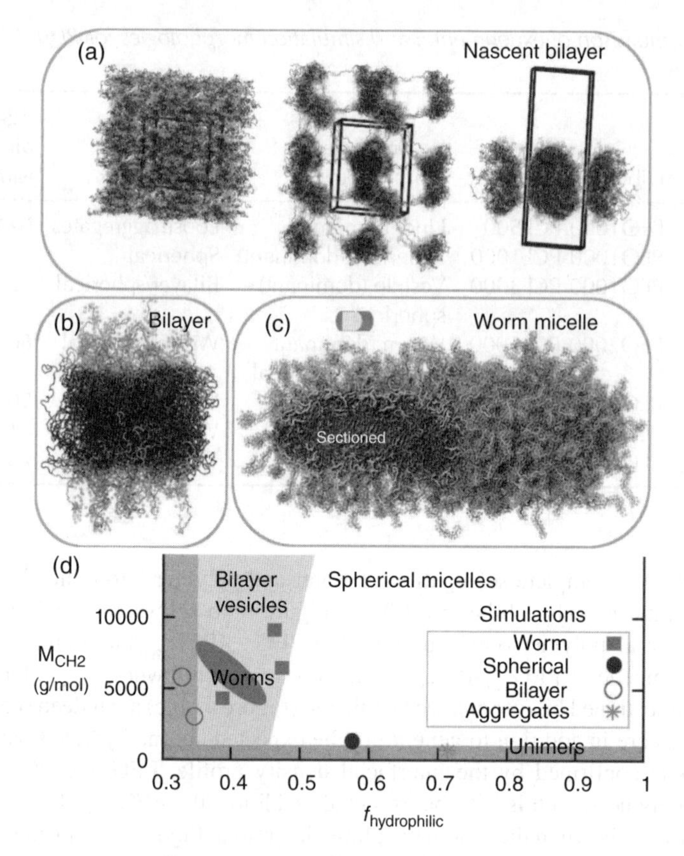

Figure 4.6 Coarse grain self-assembly of PEG-PCL with comparison to experimental phase diagram. (a) Homogeneous dispersion of PEG_{1000}-PCL_{3000} that assembles into a bilayer. CG water forms a hydration layer (≤5Å) around PEG (sectioned). Fluctuations in the random mixture are followed by segregation with extensive polymer entanglements in the periodic box, and then formation of a bicelle-like frustrated bilayer (at 30 ns). (b,c) Bilayer of PEG_{2000}-PCL_{7700}, and worm of PEG_{2000}-PCL_{5000} are both stable morphologies. (d) Phase diagram for CG simulations fit within a recent experimental phase diagram of the dominant phases. Reprinted with permission from Ref. [70]. Copyright 2012 WILEY-VCH Verlag GmbH & Co. KGaA, Weinheim.

copolymers can also be mixed in the membrane [108] to control the degradation rate. With worm-like micelles it is confirmed that the degradation proceeds through a "chain end cleavage" rather than a "random scission" process [109] and promotes a morphology transition from worms to spheres.

Polydispersity may have an inherent effect on the breakup mechanism–additional physical mechanisms in worm-like assemblies that could contribute to this transition include phase segregation of the polydisperse components on the surface and the resulting preferred curvature of segregated phases. For example, segregation on the surface of lipid vesicles coupled with line tension has been considered to be one of the driving forces for curvature and budding [110]. Using mesoscale simulation techniques such as DPD, we have explored the nonequilibrium breakup of heterogeneous worm-like micelles [111],

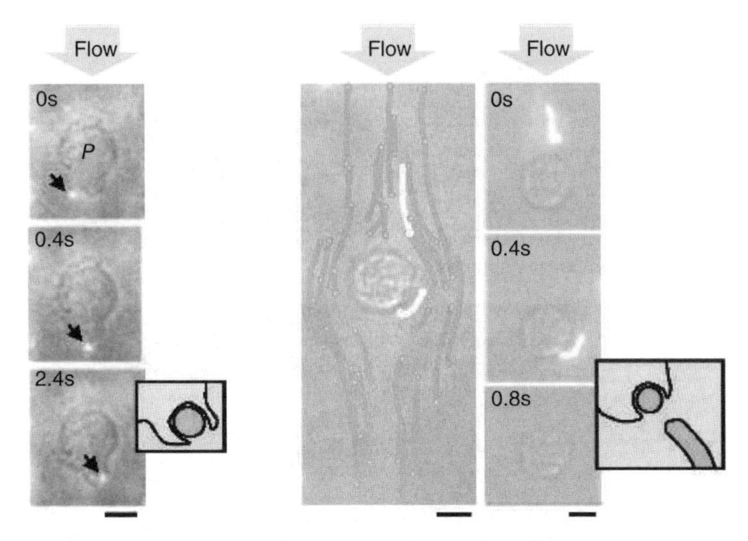

Figure 4.7 *In a flow chamber with immobilized phagocytes, long worm-like micelles flow past the cells, occasionally leaving a fragment, while smaller micelles and vesicles are captured (left, arrows point to small micelles and vesicles). Flow velocity is 25 mm/s. The scale bars represent 5 mm. Reprinted with permission from Macmillan Publishers Ltd: Nature Nanotechnology, Ref. [21], copyright 2007.*

as shown in Figure 4.8. The worm-like micelles are composed of two copolymers, with two different hydrophilic fractions, $f_{EO}=0.82$ and $f_{E0}=0.50$. Using mixed length hydrophobic tails, the polymer core develops radial undulations, leading to pinching, budding, and breakup after several microseconds. The undulating shape of the polymeric worm micelle core during breakup as shown in Figure 4.8 is analogous to Rayleigh-type instabilities and mixtures of PEO-PBD polymers [101]. Understanding the micellar instability in relation to drug delivery copolymers remains a relatively unexplored topic. It is shown that the concentration is linearly correlated to the mean curvature of the undulating shape of the worm micelle during the nonequilibrium breakup process, as seen in the simulation snapshots in Figure 4.8b–f. The molecular interfacial concentration is thus related to the mesoscale shape of the assembly and may drive the process of micellar breakup.

Additionally, polymers such as PCL are semi-crystalline. Polylactide, for example, can display different curvatures in confinement thin films depending on the chirality of the molecule [112]. Crystallinity is not a desirable property for the core of micelles, because it impedes the fluidity of the core and hydrodynamic interaction in the blood stream. For example, worm-like micelles composed of PEO-PCL have been shown to crystallize in nanodomains along the worm-like micelle length. They demonstrate partial rigidity over time, but incorporation of amorphous monomer along the PCL backbone disrupts and suppresses this crystallization, which is amenable for an effective delivery system [102]. The flexibility and persistent length of worm-like micelles can be probed experimentally [113], while simulations of CG worm-like micelles assembled from amphiphilic components suggest that, upon compression, a characteristic buckling occurs [114]. Studies of the bending rigidity, and correlations to the interfacial properties, of large worm-like micelles will add

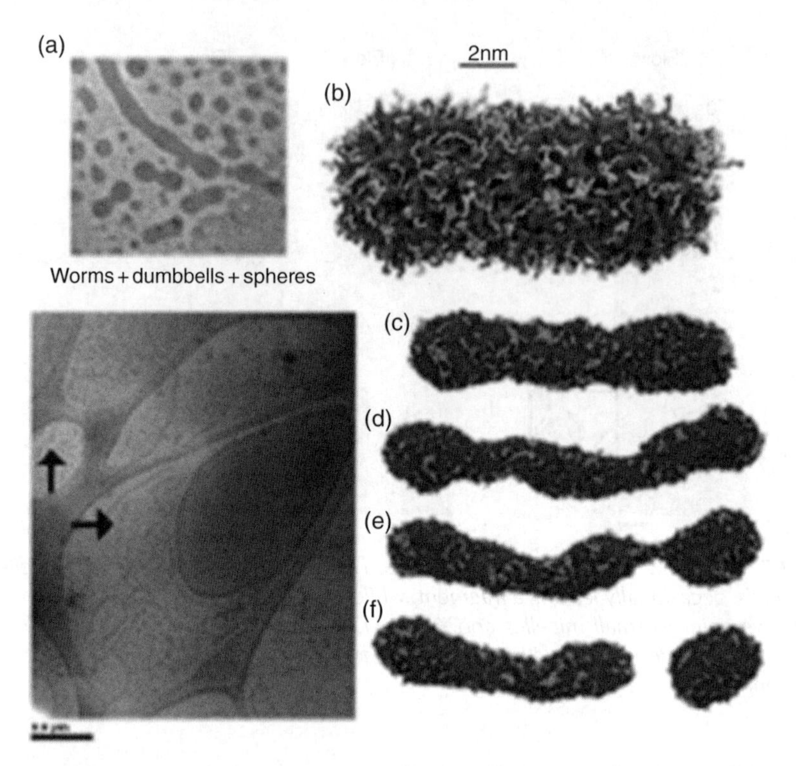

Figure 4.8 *Mechanism of worm to sphere transition. (a) Cryo-TEM images show the breakup of worm-like micelles of PEO–PCL (M_n = 11 500 kg/mol), with undulation and budding. Arrows in the bottom image highlight breakup events. (b–f) Simulation results for a binary copolymer micelle, which details the transition dynamics of the breakup of the worm to spheres. Reprinted with permission from Ref. [111]. Copyright 2010 Royal Society of Chemistry.*

increased insight into rheological properties of biopolymer networks. This is an additional example of how interfacial organization at the nanoscale is intrinsically related to the mesoscopic properties and shape of micelles, in addition to the breakup process described above.

4.4 Taxol

Paclitaxel (or Taxol) is one of the first natural anticancer drugs discovered by Arthur S. Barclay in the bark of the Pacific Yew tree in 1962. Taxol acts to impede mitosis through promotion of the polymerization of tubulin into mictrotubules [115], both of which are key components of the cytoskeleton. A molecular model of Taxol is shown in Figure 4.9. The main group of Taxol, the taxane portion of the molecule – one eight- and two six-membered rings – is rigid, but slightly more hydrophilic than the three phenyl groups as side chains. The crystal structure in water is an alternating head to tail packing, as described by Mastropaolo *et al.* [116]. However, the structure in hydrophobic environments such as chloroform is proposed such that Taxol stacks in a head to head manner [117]. Due to its

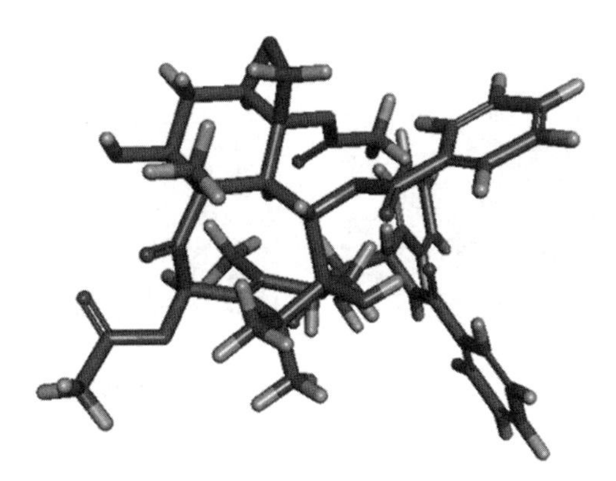

Figure 4.9 *Structure of paclitaxel (Taxol).*

low degree of solubility in water, Taxol partitions into the hydrophobic cores of membranes, as well as hydrophobic micelles. Taxol has a partition coefficient into 1-palmitoyl-2-oleoyl-sn-glycero-3-phosphocholine (POPC) lipid membranes of $K \approx 10^4$ at 22 °C and $K \approx 10^3$ at 37 °C [118]. The corresponding oil–water partition coefficient is known to be log $P_{o/w} = 4.4$ [119]. Moreover, Taxol also displays a significantly higher partitioning in a polymer membrane than in a lipid membrane [15]. The relative difference in the partitioning between lipid and polymer phases can be estimated from the difference of taxol solubility parameters respectively [120]. Worm-like micelles of PEO-PCL have been shown to solubilize a higher percentage weight (wt%) of Taxol (~10 mol%) in their hydrophobic cores than spherical micelles [21]. Thus, solubility can be modified with multiple strategies. The ability for PEO-PCL micelles to load drugs is as a balance of the size of the drug and the associated entropy and interfacial tension between the component copolymer blocks, as well as the effective internal pressure of the micellar core assembly [121]. Additionally, it has been found that the intrinsic shape of the micelle loading the drug has an effect on the ability to encapsulate different "guest" drugs [7]. Experimentally, the loading of Taxol is indicated to be dependent on the geometry of the micelle itself as well as the length of the polymer [122]. This suggests significant interfacial effects on loading, consistent with previous theoretical arguments.

How Taxol interactions change with membrane shape and size is an ideal question to address with multiscale modeling. All-atomistic simulations of a melt of PCL 7 polymer chains are used to obtain the flexibility parameters of a CG model of PCL, while all-atomistic simulations of Taxol in octanol and water are utilized to obtain parameters for the drug. The CG models break each molecule into two to five heavy atoms per bead, depending on the specific chemistry and symmetry of the underlying chemical structure, thus decomposing the intermolecular potentials into chemical components [123] and optimizing intramolecular potentials based an all-atomistic simulations [66, 124]. To test the interaction between Taxol molecules in a hydrophobic environment, a CG model of Taxol was loaded into worm-like PEO_{2000}-PCL_{5000} micelles at 3 and 9 wt% drug loading. The 9 wt% case is shown in Figure 4.10. After microsecond simulations, drug molecules appear dispersed and the micellar structure remains stable, with a slight shift in the interfacial density concentration of the PEO and the drug concentration respectively. Within the micelle,

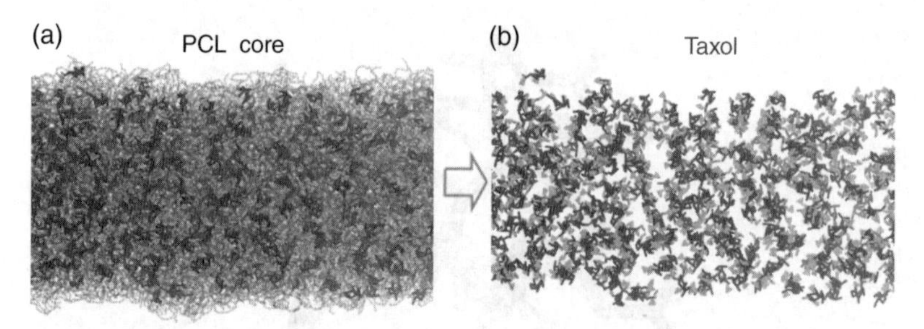

Figure 4.10 *Taxol dispersion at realistic drug concentration within a worm micelle: (a) CG Taxol at 9 wt% in a* PEG_{2000}*-*PCL_{5000} *worm micelle core, (b) Taxol dispersion with PCL chains omitted for clarity. Reprinted with permission from Ref. [70]. Copyright 2012 WILEY-VCH Verlag GmbH & Co. KGaA, Weinheim.*

Taxol moves towards the PEO-PCL interface and interacts strongly with interfacial PEO. Additionally a Taxol–Taxol network is formed, the periodicity of which is denoted by the peak in the $g(r)$ at 20 Å and appears consistent with locally correlated head to head stacking of the drug. Using mean squared displacement, the diffusion constant of the Taxol in the worm is estimated. At 300 ns, 3 wt% Taxol gave $D_{\mathrm{Tax/PCL}} \sim 3\,\mu m^2/s$ whereas 9 wt% gave slightly lower $D_{\mathrm{Tax/PCL}} \sim 1\,\mu m^2/s$. Loading of the drug near the interface is consistent with a molecular explanation of the phenomenon of "burst release" [125], whereby the drug localizes toward the inner corona of the micelle and is likewise attributed with faster diffusion upon release from the micelle.

An ideal way to capture the complexity for advanced delivery systems, while maintaining a strong connection with the underlying polymer and drug chemistry is to use a multiscale CG MD approach. Micelles of PEG_{2000}-PCL_{5000} form in water in both spherical and worm micelle morphologies. We calculate the free energy of the drug across the block copolymer interface, for the two micellar shapes: worms and spheres. Close to the interface (50 Å), the Taxol orients with aromatic portions of the drug facing the PCL core. We calculate the free energy profile of the drug across the interface by utilizing thermodynamic integration along a radial path from the center of the micelle mass. To begin with, Taxol's free energy displays a minimum close to the interface between the core and the corona of the worm micelle, while the minimum is in the center of the spherical micelle.

Accounting for morphology dependent drug loading requires accurate modeling of the subtly distinct contributions to the aggregate free energy – namely the entropies of hydrophilic and hydrophobic blocks, the various interaction energies near and far from the interface, the solvent, and the drug. Additionally, there is a greater mean change in free energy from the minimum to the interface for the case of the worm (21.1 ± 6.6 kcal/mol) as compared for the case of the sphere (20.9 ± 8.1 kcal/mol). Normalizing by the interfacial area of the worm and sphere respectively, the relative difference in magnitude is approximately 1.65 (±0.5), which is within the factor of two as seen in experiments [126]. Thus, a morphology change should allow drug release; and likewise, drug loading could change the morphology due to the corresponding change in interfacial tension.

In this chapter we have highlighted several multiscale simulations in consideration of micellar drug delivery, the loading and shape of the nanoparticle/micelle, as well as micellar stability. Additional investigations could help clarify the interactions of micelles with cell membranes, as well as the interaction of hydrophobic drugs themselves with the cell membrane. Furthermore, the diffusion and binding of the drug itself with its target (Taxol and microtubule) is within the realm of multiscale simulation. In the next sections we will address multiscale simulations of Taxol with model cellular membranes, as well multiscale modeling as it applies to the kinetics of ligand–protein association.

4.4.1 Taxol Behavior in Membranes

It is known that Taxol penetrates and accumulates in the membranes and has effects on the membrane's physical and electrical properties, including the lipid order parameters, elasticity, phospholipid phase transition, lipid curvature profiles, and fluidity [117, 127]. Taxol has a free energy of binding of $\Delta G = -7.9$ kcal/mol in a POPC lipid membrane [118]. The calculated binding free energy using the ABF method of atomistic models for a single molecule of Taxol and POPC bilayer shows close to the experimental value: ~−6.5 kcal/mol, with a minimum around the demarcation of the hydrophobic tail and the hydrophilic head of POPC and an increase toward the core of the POPC (authors' unpublished work). It is supported by differential scanning calorimetry data that suggests some access of the acyl group of Taxol into the bilayer via alignment with the acyl chain of the phosphatidylcholine component [128]. This kind of initial positioning of small molecules in the interfacial region of a membrane at a low concentration was observed in amphipathic molecules like curcumin and some helical antimicrobial peptides with a membrane thinning effect, although the interfacial region in the amphipathic drug case was between water and lipid. After a threshold concentration, only the concentration in the core of the membrane increased. While antimicrobial peptides that display a concentration dependent threshold to oligomeric states form pores, in contrast curcumin with a monomeric state does not form pores [129]. A CG representation of the system shown in Figure 4.11 can be used to investigate if Taxol shows a similar concentration dependence in its free energy profile along the axis of the lipid bilayer; possibly the free energy in the hydrophobic core of the bilayer will lower to form a second minimum as the concentration goes up. Moreover, Tuszynski and colleagues observed that, at high concentrations, Taxol permeabilized lipid bilayer

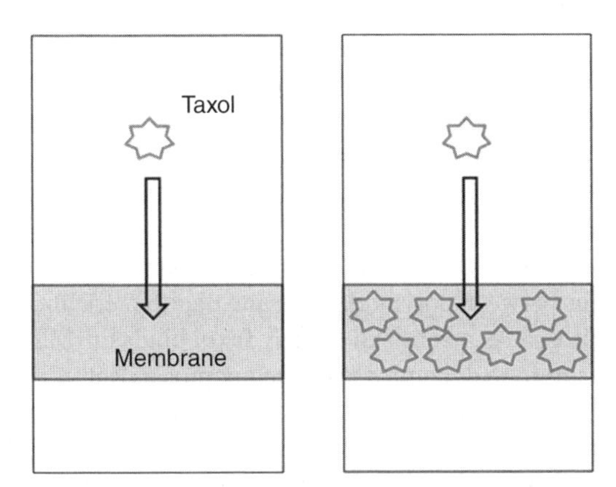

Figure 4.11 *A schematic figure of Taxol insertion into the membrane.*

membranes by forming ion-flowing pores which featured triangular conductance events unlike the rectangular ones of ion channels in electrophysiology recordings. Additionally, they performed atomistic MD simulation of each Taxol–lipid complex to study the drug–lipid physical interactions [130]. In order to study this concentration-dependent behavior of Taxol in the membrane, especially in the high concentration where the shift of the free energy profile is expected, the scale of the system of interest tends to increase. Rational coarse grain (rCG) models of Taxol and the membrane can be used to extend the length and time scales of the system further.

4.4.2 Ligand-Protein Binding

However, drug delivery to the target site is not the final step. The released drug needs to diffuse and bind to the active site of the target protein. Drug action is based on the binding of the drug to its targets and the subsequent physiological modulation of the protein function in most cases. The binding affinity of a drug to a targeted protein has provided a useful measure to guide drug developments. The binding affinity is usually described by the dissociation constant K_D, or a ratio of the relevant concentrations at equilibrium. For a simple drug–receptor (D-R) binding reaction to form a binary complex (DR): $D + R \underset{k_{off}}{\overset{k_{on}}{\rightleftarrows}} DR$, $K_D = [D][R]/[DR]$. K_D is also described as a ratio of the association rate constant, k_{on}, and the dissociation rate constant, k_{off}: $K_D = k_{on} / k_{off}$, indicating the relation between the thermodynamics and the kinetics of the binding reaction. K_D depends on the free energy difference between the unbound and bound states, $\Delta G = RT \ln[K_D]$, while the rate constants k_{on} and k_{off} are related to the free energy barriers of the transition states. This is a simple case of a static lock and key model of drug–receptor binding. Figure 4.3 shows a thermodynamic cycle for two-step ligand (L) binding to a receptor (R), where the conformational changes of the receptor are involved in the drug–receptor binding. The conformational changes are induced by the binding of the drug or ligand in the induced fit model, or the receptor may have a number of conformations regardless of the binding of the drug in the conformational selection model. If a third player, an effector, comes into the game, and it binds to the receptor in a site other than the ligand binding site, inducing conformational changes of the receptor which are good or bad for the ligand binding, the allosteric effect has to be considered. The implication of the kinetics of drug–protein binding on drug efficacy and drug safety is increasingly appreciated and thus the determinants of binding kinetics are being sought for a new, temporal dimension during lead optimization in drug discovery [30, 131].

4.4.3 Taxol–Tubulin Binding

Microtubules (MTs) are key components of the cytoskeleton in eukaryotic cells; the characteristic dynamics between polymerization and depolymerization processes is critical in cell division and mitosis. As mentioned earlier, Taxol binds the MT to promote polymerization and disturb the mitotic dynamics, causing apoptosis. But, it is not fully understood exactly how Taxol and other MT-stabilizing agents induce this mitotic arrest and the consequent cell death. α and β tubulin form a heterodimer as a building block and polymerize to protofilaments and then to MTs. The phosphorylation state of the nucleotide bound to β tubulin in the exposed end determines the direction of the polymerization process: guanosine triphosphate (GTP) bound to β tubulin stabilizes the straight conformation of the MT, making it grow, while guanosine diphosphate (GDP) bound to β tubulin favors disassembly [132]. The binding of Taxol makes polymerization favored even with GDP on the β subunit. Taxol binding at the luminal site in β-tubulin is shown in Figure 4.12. Upon the binding of Taxol, the M-loop protrudes from the lateral surface, enhancing the interprotofilament interactions with the H1′-S2 loops, as well as the H2–S3 loops in the adjacent protofilaments and therefore stabilizing the MT [133]. Snyder and colleagues constructed a model of Taxol in the bound conformation of Taxol using docking based on crystallographic density analysis and confirmed that the Taxol binding pocket lies within a "deep hydrophobic cleft" [134]. But this luminal taxoid site could not explain the experimental observations indicating the fast kinetic features of the Taxol binding and therefore an unhindered access to the binding site, including the rapid staining of MTs with the fluorescently labeled Taxol analog Flutax [135] and the rapid flexibility induction of Taxol in MTs *in vitro* [136]. The search for openings on the MT surface available for the diffusion of Taxol to its luminal binding site showed the kinetic inaccessibility of the hidden taxoid binding site through the high resolution model of the MTs. To explain this fast binding, the existence of an external binding site has been hypothesized: the inter-subunit space from adjacent αβ heterodimers is described as pore type I. This binding model has two mutually exclusive binding sites with a shared "switching component." The observed 1 : 1 stoichiometry of Taxol supports this. Through computational analysis, a two-step binding mechanism is suggested, involving a first step of the fast binding of Taxol on the external pore type I and a slow second step of the internalization of Taxol from the external binding site to the luminal taxoid binding site through a conformational change of the flexible H6-H7 loop in β tubulin [137–140]. Freedman and colleagues used computational molecular modeling to confirm the role of the H6-H7 loop as a hinge in this hypothesis [141].

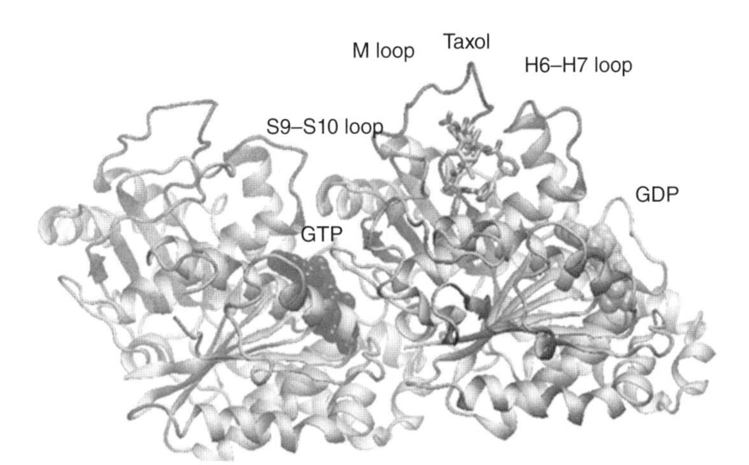

Figure 4.12 *The luminal taxoid binding site of Taxol in the αβ tubulin heterodimer. The dimer is displayed in a cartoon representation, and important loops are highlighted. GTP and GDP are displayed with a van der Waals representation. (PDB ID: 1JFF). In the two-step hypothesis, Taxol first binds the external type I pore near H6-H7, and the H6-H7 loop in the β-tubulin acts as a hinge. Conformational changes lead to a slow shift of Taxol to the internal binding site.*

Although there are arguments against this based on MD and modeling work [142], claiming that the external pore type I is just an entry site, not a binding site, more evidence has supported this two-step binding hypothesis. In the atomistic structure of a tubulin tetramer built from electron microscopy (EM) and atomic structures with the help of contour-based fitting [138, 143], the free energy profile of this proposed Taxol internalization process from the external binding site to the luminal taxoid binding site of MTs was studied by multiply targeted molecular dynamics (MTMD) and the following refinement procedure and computational results were supported by kinetics studies with fluorescent Taxol derivatives [144]. Since the scale of MTs with an outer diameter of about 25 nm and a length of up to tens of microns is way beyond the realm of atomistic simulations, most of the atomistic simulations have involved only their building block of un-assembled tubulin heterodimer, which turned out to be curved even after the Taxol binding in contrast to the Taxol-bound straight form in MTs. Recently a tetramer constructed from EM and atomic structures using contour-based fitting [138, 144] was used to include the proposed external binding site in the inter-dimer space. And the applications of CG techniques provided valuable insights in the study of MTs [145, 146].

4.5 Summary and Conclusions

High molecular weight polymer-based micellar vehicles have been designed for the delivery of both hydrophilic and hydrophobic drugs. The continued improvement of force fields and sampling methods have made possible detailed insight into intermolecular interactions using atomistic simulations. Lower resolution or multiscale models have been essential to bridge the gap between simulation results and experimentally accessible properties in the study of micellar self-assembly, drug delivery and drug–protein binding in cellular environments. In this chapter, numerous aspects of multiscale molecular modeling for micellar drug delivery are discussed, including multiscale computational methods, micellar self-assembly and stability, the interaction of drugs with polymers and membranes, and drug binding as the next step after drug delivery. We highlighted several multiscale simulations in consideration of micellar drug delivery, the loading and shape of the nanoparticle/micelle, and micellar stability. The advantages of worm-like micelles were reviewed. As a well-studied application in the micellar drug delivery, we addressed multiscale simulations of Taxol with model cellular membranes, as well as multiscale modeling of Taxol–tubulin association. Furthermore, in complement with experimental investigations, we will possess more control in the rational design of drug delivery systems.

References

[1] Noolandi, J.; Hong, K. M. (1983) Theory of block copolymer micelles. *Macromolecules*, **16** 1443–1448.
[2] Israelachvili, J. N. *Intermolecular and Surface Forces*; 2nd edn; Academic: London, 1991.
[3] Mai, Y. Y.; Eisenberg, A. (2012) Self-assembly of block copolymers. *Chemical Society Reviews*, **41** (18) 5969–5985.
[4] Bates, F. S. (1991) Polymer-polymer phase-behavior. *Science*, **251** (4996) 898–905.
[5] Bates, F. S.; Hillmyer, M. A.; Lodge, T. P.; *et al.* (2012) Multiblock polymers: panacea or pandora's box? *Science*, **336** (6080) 434–440.

[6] Torchilin, V. P. (2007) Micellar nanocarriers: pharmaceutical perspectives. *Pharmaceutical Research*, **24** (1) 1–16.
[7] Nagarajan, R. (2001) Solubilization of "Guest" molecules into polymeric aggregates. *Polymers for Advanced Technologies*, **12** 23–43.
[8] Lebens, P. J. M.; Keurentjes, J. T. F. (1996) Temperature-induced solubilization of hydrocarbons in aqueous block copolymer solutions. *Industrial and Engineering Chemistry Research*, **35** (10) 3415–3421.
[9] Allen, C.; Maysinger, D.; Eisenberg, A. (1999) Nano-engineering block copolymer aggregates for drug delivery. *Colloids and Surfaces B: Biointerfaces*, **16** (1-4) 3–27.
[10] Aniansson, E. A. G.; Wall, S. N. (1974) Kinetics of step-wise micelle association. *Journal of Physical Chemistry*, **78** (10) 1024–1030.
[11] Soo, P. L.; Luo, L. B.; Maysinger, D.; Eisenberg, A. (2002) Incorporation and release of hydrophobic probes in biocompatible polycaprolactone-block-poly(ethylene oxide) micelles: implications for drug delivery. *Langmuir*, **18** (25) 9996–10004.
[12] Piskin, E.; Kaitian, X.; Denkbas, E. B.; Kucukyavuz, Z. (1995) Novel pdlla/peg copolymer micelles as drug carriers. *Journal of Biomaterials Science Polymer Edition*, **7** (4) 359–373.
[13] Yokoyama, M.; Fukushima, S.; Uehara, R.; *et al.* (1998) Characterization of physical entrapment and chemical conjugation of adriamycin in polymeric micelles and their design for in vivo delivery to a solid tumor. *Journal of Controlled Release*, **50** (1/3) 79–92.
[14] Ahmed, F.; Discher, D. E. (2004) Self-porating polymersomes of PEG-PLA and PEG-PCL: hydrolysis-triggered controlled release vesicles. *Journal of Controlled Release*, **96** (1) 37–53.
[15] Ahmed, F.; Pakunlu, R. I.; Srinivas, G.; *et al.* (2006) Shrinkage of a rapidly growing tumor by drug-loaded polymersomes: Ph-triggered release through copolymer degradation. *Molecular Pharmaceutics*, **3** (3) 340–350.
[16] Park, T. G. (1995) Degradation of poly(lactic-co-glycolic acid) microspheres: effect of copolymer composition. *Biomaterials*, **16** (15) 1123–1130.
[17] Belbella, A.; Vauthier, G.; Fessi, H.; *et al.* (1996) In vitro degradation of nanospheres from poly(D, L-lactides) of different molecular weights and polydispersities. *International Journal of Pharmaceutics*, **129** 95–102.
[18] Husmann, M.; Schenderlein, S.; Luck, M.; *et al.* (2002) Polymer erosion in PLGA microparticles produced by phase separation method. *International Journal of Pharmaceutics*, **242** (1/2) 277–280.
[19] Klibanov, A. L.; Maruyama, K.; Torchilin, V. P.; Huang, L. (1990) Amphipathic polyethyleneglycols effectively prolong the circulation time of liposomes. *FEBS Letters*, **268** (1) 235–237.
[20] Bermudez, H.; Brannan, A. K.; Hammer, D. A.; *et al.* (2002) Molecular weight dependence of polymersome membrane structure, elasticity, and stability. *Macromolecules*, **35** (21) 8203–8208.
[21] Geng, Y.; Dalhaimer, P.; Cai, S. S.; *et al.* (2007) Shape effects of filaments versus spherical particles in flow and drug delivery. *Nature Nanotechnology*, **2** (4) 249–255.
[22] Maeda, H.; Wu, J.; Sawa, T.; *et al.* (2000) Tumor vascular permeability and the epr effect in macromolecular therapeutics: a review. *Journal of Controlled Release*, **65** (1/2) 271–284.
[23] Luo, Y.; Prestwich, G. D. (1999) Synthesis and selective cytotoxicity of a hyaluronic acid-antitumor bioconjugate. *Bioconjugate Chemistry*, **10** (5) 755–763.
[24] Torchilin, V. P. (2005) Recent advances with liposomes as pharmaceutical carriers. *Nature Reviews Drug Discovery*, **4** (2) 145–160.
[25] Xu, J. P.; Ji, J.; Chen, W. D.; Shen, J. C. (2005) Novel biomimetic polymersomes as polymer therapeutics for drug delivery. *Journal of Controlled Release*, **107** (3) 502–512.
[26] Kataoka, K.; Matsumoto, T.; Yokoyama, M.; *et al.* (2000) Doxorubicin-loaded poly(ethylene glycol)-poly(beta-benzyl-l-aspartate) copolymer micelles: their pharmaceutical characteristics and biological significance. *Journal of Controlled Release*, **64** (1/3) 143–153.
[27] Allen, C.; Yu, Y. S.; Eisenberg, A.; Maysinger, D. (1999) Cellular internalization of PCL20-b-PEO44 block copolymer micelles. *Biochimica et Biophysica Acta, Biomembranes*, **1421** (1) 32–38.
[28] Chen, H.; Kim, S.; Li, L.; *et al.* (2008) Release of hydrophobic molecules from polymer micelles into cell membranes revealed by forster resonance energy transfer imaging. *Proceedings of the National Academy of Sciences of the United States of America*, **105** 6596.

[29] Zastre, J. A.; Jackson, J. K.; Wong, W.; Burt, H. M. (2008) P-glycoprotein efflux inhibition by amphiphilic diblock copolymers: relationship between copolymer concentration and substrate hydrophobicity. *Molecular Pharmaceutics*, **5** (4) 643–653.
[30] Copeland, R. A. (2011) Conformational adaptation in drug-target interactions and residence time. *Future Medicinal Chemistry*, **3** (12) 1491–1501.
[31] Copeland, R. A.; Pompliano, D. L.; Meek, T. D. (2007) Drug-target residence time and its implications for lead optimization (vol 5, pg 730, 2006). *Nature Reviews Drug Discovery*, **6** (3) 249.
[32] Swinney, D. C. (2004) Biochemical mechanisms of drug action: what does it take for success? *Nature Reviews Drug Discovery*, **3** (9) 801–808.
[33] Shuman, C. F.; Markgren, P. O.; Hamalainen, M.; Danielson, U. H. (2003) Elucidation of hiv-1 protease resistance by characterization of interaction kinetics between inhibitors and enzyme variants. *Antiviral Research*, **58** (3) 235–242.
[34] Berezov, A.; Zhang, H.-T.; Greene, M. I.; Murali, R. (2001) Disabling erbB receptors with rationally designed exocyclic mimetics of antibodies: structure-function analysis. *Journal of Medicinal Chemistry*, **44** (16) 2565–2574.
[35] Vanderheyden, P. M. L.; Fierens, F. L. P.; Vauquelin, G. (2000) Angiotensin II type 1 receptor antagonists: why do some of them produce insurmountable inhibition? *Biochemical Pharmacology*, **60** (11) 1557–1563.
[36] So, O.-Y.; Scarafia, L. E.; Mak, A. Y.; *et al.* (1998) The dynamics of prostaglandin H synthases. Studies with prostaglandin h synthase 2 Y355F unmask mechanisms of time-dependent inhibition and allosteric activation. *Journal of Biological Chemistry*, **273** (10) 5801–5807.
[37] Shinoda, W.; DeVane, R.; Klein, M. L. (1998) Coarse-graining of Condensed Phase and Biomolecular Systems, G. A. Voth, ed. CRC: London, pp. 329–342.
[38] Groot, R. D.; Warren, P. B. (1997) Dissipative particle dynamics: bridging the gap between atomistic and mesoscopic simulation. *Journal of Chemical Physics*, **107** (11) 4423–4435.
[39] Ermak, D. L.; McCammon, J. A. (1978) Brownian dynamics with hydrodynamic interactions. *Journal of Chemical Physics*, **69** (4) 1352–1360.
[40] Carbone, P.; Varzaneh, H. A. K.; Chen, X. Y.; Muller-Plathe, F. (2008) Transferability of coarse-grained force fields: the polymer case. *Journal of Chemical Physics*, **128** (6) 064904.
[41] Zhang, Z. Y.; Pfaendtner, J.; Grafmuller, A.; Voth, G. A. (2009) Defining coarse-grained representations of large biomolecules and biomolecular complexes from elastic network models. *Biophysical Journal*, **97** (8) 2327–2337.
[42] Kang, M.; Smith, P. E. (2006) A kirkwood-buff derived force field for amides. *Journal of Computational Chemistry*, **27** (13) 1477–1485.
[43] Jorgensen, W. L.; Tirado-Rives, J. (1988) The opls potential functions for proteins energy minimizations for crystals of cyclic peptides and crambin. *Journal of the American Chemical Society*, **110** (6) 1657–1666.
[44] van Gunsteren, W. F.; Billeter, S. R.; Eising, A. A.; *et al. Biomolecular Simulation: The Gromos96 Manual and User Guide* ETH: Zurich, 1996.
[45] Case, D.A., Darden, T.A., Cheatham, I.T.E.; *et al.* (2010) Coarse-graining of Condensed Phase and Biomolecular Systems, University of California, San Francisco.
[46] A. D. MacKerell, Jr.; D. Bashford, M. Bellott; *et al.* (1998) All-atom empirical potential for molecular modeling and dynamics studies of proteins. *Journal of Physical Chemistry B*, **102** (18) 3586–3616.
[47] Leontyev, I. V.; Stuchebrukhov, A. A. (2012) Polarizable mean-field model of water for biological simulations with amber and charmm force fields. *Journal of Chemical Theory and Computation*, **8** (9) 3207–3216.
[48] He, X.; Lopes, P. E. M.; Mackerell, A. D., Jr. (2013) Polarizable empirical force field for acyclic polyalcohols based on the classical drude oscillator. *Biopolymers*, **99** (10) 724–738.
[49] Schyman, P.; Jorgensen, W. L. (2013) Exploring adsorption of water and ions on carbon surfaces using a polarizable force field. *Journal of Physical Chemistry Letters*, **4** (3) 468–474.
[50] Shinoda, W.; DeVane, R.; Klein, M. L. (2007) Multi-property fitting and parameterization of a coarse grained model for aqueous surfactants *Molecular Simulation*, **33** (1/2) 27–36.
[51] Marrink, S. J.; Risselada, H. J.; Yefimov, S.; *et al.* (2007) The martini force field: coarse grained model for biomolecular simulations. *Journal of Physical Chemistry B*, **111** (27) 7812–7824.

[52] Izvekov, S.; Voth, G. A. (2005) A multiscale coarse-graining method for biomolecular systems. *Journal of Physical Chemistry B*, **109** (7) 2469–2473.
[53] Pickholz, M.; Saiz, L.; Klein, M. L. (2005) Concentration effects of volatile anesthetics on the properties of model membranes: a coarse-grain approach. *Biophysical Journal*, **88** (3) 1524–1534.
[54] Chaudhri, A., Kamerzell, T., and Voth, G. A. (2010) Coarse-grained modeling of antibody self-association. Abstracts of Papers of the American Chemical Society, vol. 240.
[55] Head-Gordon, T.; Brown, S. (2003) Minimalist models for protein folding and design. *Current Opinion in Structural Biology*, **13** (2) 160–167.
[56] Tozzini, V.; Rocchia, W.; McCammon, J. A. (2006) Mapping all-atom models onto one-bead coarse-grained models: general properties and applications to a minimal polypeptide model. *Journal of Chemical Theory and Computation*, **2** (3) 667–673.
[57] Muller-Plathe, F. (2002) Coarse-graining in polymer simulation: from the atomistic to the mesoscopic scale and back. *Chemphyschem*, **3** (9) 754–769.
[58] Shinoda, W.; DeVane, R.; Klein, M. L. (2012) Computer simulation studies of self-assembling macromolecules. *Current Opinion in Structural Biology*, **22** (2) 175–186.
[59] Marrink, S. J.; de Vries, A. H.; Tieleman, D. P. (2009) Lipids on the move: simulations of membrane pores, domains, stalks and curves. *Biochimica Et Biophysica Acta, Biomembranes*, **1788** (1) 149–168.
[60] Kamerlin, S. C. L.; Vicatos, S.; Dryga, A.; Warshel, A. (2011) Coarse-grained (multiscale) simulations in studies of biophysical and chemical systems. *Annual Review of Physical Chemistry*, **62** 41–64.
[61] Allen, E. C.; Rutledge, G. C. (2008) A novel algorithm for creating coarse-grained, density dependent implicit solvent models. *Journal of Chemical Physics*, **128** (15) 154115.
[62] Percec, V.; Wilson, D. A.; Leowanawat, P.; *et al.* (2010) Self-assembly of janus dendrimers into uniform dendrimersomes and other complex architectures. *Science*, **328** (5981) 1009–1014.
[63] Reynwar, B. J.; Illya, G.; Harmandaris, V. A.; *et al.* (2007) Aggregation and vesiculation of membrane proteins by curvature-mediated interactions. *Nature*, **447** (7143) 461–464.
[64] Groot, R. D. (2004) Applications of dissipative particle dynamics. *Lecture Notes in Physics*, **640** 5–38.
[65] Kang, M.; Roberts, C.; Cheng, Y. H.; Chang, C. E. A. (2011) Gating and intermolecular interactions in ligand-protein association: coarse-grained modeling of HIV-1 protease. *Journal of Chemical Theory and Computation*, **7** (10) 3438–3446.
[66] Noid, W. G.; Chu, J.; Ayton, G. S.; *et al.* (2008) The multiscale coarse-graining method. 1. A rigorous bridge between atomistic and coarse-grained models. *Journal of Chemical Physics*, **128** 244114.
[67] Srinivas, G.; Discher, D. E.; Klein, M. L. (2004) Self-assembly and properties of diblock copolymers by coarse-grain molecular dynamics. *Nature Materials*, **3** (9) 638–644.
[68] Shinoda, W.; DeVane, R.; Klein, M. L. (2010) Zwitterionic lipid assemblies: molecular dynamics studies of monolayers, bilayers, and vesicles using a new coarse grain force field. *Journal of Physical Chemistry B*, **114** (20) 6836–6849.
[69] DeVane, R.; Shinoda, W.; Moore, P. B.; Klein, M. L. (2009) A transferable coarse grain non-bonded interaction model for amino acids. *Journal of Chemical Theory and Computation*, **5** (8) 2115–2124.
[70] Loverde, S. M.; Klein, M. L.; Discher, D. E. (2012) Nanoparticle shape improves delivery: rational coarse grain molecular dynamics (rCG-MD) of taxol in worm-like PEG-PCL micelles. *Advanced Materials*, **24** (28) 3823–3830.
[71] Periole, X.; Marrink, S. In *Biomolecular Simulations: Methods and Protocols*; L.M.A.E. Salonen, ed. Springer: New York, 2013; vol. **294**, pp. 533–565.
[72] Rossi, G.; Monticelli, L.; Puisto, S. R.; *et al.* (2011) Coarse-graining polymers with the martini force-field: polystyrene as a benchmark case. *Soft Matter*, **7** (2) 698–708.
[73] Levine, B. G.; LeBard, D. N.; DeVane, R.; *et al.* (2011) Micellization studied by GPU-accelerated coarse-grained molecular dynamics. *Journal of Chemical Theory and Computation*, **7** (12) 4135–4145.

[74] LeBard, D. N.; Levine, B. G.; Mertmann, P.; *et al.* (2011) Self-assembly of coarse-grained ionic surfactants accelerated by graphics processing units. *Soft Matter*, **8** (8) 2385–2397.
[75] Jusufi, A.; LeBard, D. N.; Levine, B. G.; Klein, M. L. (2012) Surfactant concentration effects on micellar properties. *Journal of Physical Chemistry B*, **116** (3) 987–991.
[76] Fernandes, M. X.; de la Torre, J. G. (2002) Brownian dynamics simulation of rigid particles of arbitrary shape in external fields. *Biophysical Journal*, **83** (6) 3039–3048.
[77] Gabdoulline, R. R.; Wade, R. C. (1998) Brownian dynamics simulation of protein-protein diffusional encounter. *Methods*, **14** (3) 329–341.
[78] Dickinson, E.; Allison, S. A.; McCammon, J. A. (1985) Brownian dynamics with rotation-translation coupling. *Journal of the Chemical Society, Faraday Transactions 2*, **81** 591–601.
[79] Geyer, T.; Winter, U. (2009) An O(N-2) approximation for hydrodynamic interactions in brownian dynamics simulations. *Journal of Chemical Physics*, **130** (11) 114905.
[80] Hoogerbrugge, P. J.; Koelman, J. M. V. A. (1992) Simulating microscopic hydrodynamic phenomena with dissipative particle dynamics. *Europhysics Letters*, **19** (2) 155–160.
[81] Isralewitz, B.; Gao, M.; Schulten, K. (2001) Steered molecular dynamics and mechanical functions of proteins. *Current Opinion in Structural Biology*, **11** (2) 224–230.
[82] Park, S.; Khalili-Araghi, F.; Tajkhorshid, E.; Schulten, K. (2003) Free energy calculation from steered molecular dynamics simulations using Jarzynski's equality. *Journal of Chemical Physics*, **119** (6) 3559–3566.
[83] Jarzynski, C. (1997) Equilibrium free-energy differences from nonequilibrium measurements: a master-equation approach. *Physical Review E*, **56** 5018–5035.
[84] Huang, D.; Caflisch, A. (2011) The free energy landscape of small molecule unbinding. *PLoS Computational Biology*, **7** (2) e1002002.
[85] Chen, H.; Ilan, B.; Wu, Y.; *et al.* (2007) Charge delocalization in proton channels, I: the aquaporin channels and proton blockage. *Biophysical Journal*, **92** (1) 46–60.
[86] Darve, E.; Rodriguez-Gomez, D.; Pohorille, A. (2008) Adaptive biasing force method for scalar and vector free energy calculations. *Journal of Chemical Physics*, **128** (14) 144120.
[87] Chipot, C.; Lelievre, T. (2011) Enhanced sampling of multidimensional free-energy landscapes using adaptive biasing forces. *SIAM Journal on Applied Mathematics*, **71** (5) 1673–1695.
[88] Henin, J.; Schulten, K.; Chipot, C. (2006) Conformational equilibrium in alanine-rich peptides probed by reversible stretching simulations. *Journal of Physical Chemistry B*, **110** (33) 16718–16723.
[89] Henin, J.; Tajkhorshid, E.; Schulten, K.; Chipot, C. (2008) Diffusion of glycerol through Escherichia coli aquaglyceroporin GlpF. *Biophysical Journal*, **94** (3) 832–839.
[90] Yonetani, Y.; Kono, H. (2013) Dissociation free-energy profiles of specific and nonspecific DNA-protein complexes. *Journal of Physical Chemistry B*, **117** (25) 7535–7545.
[91] Caballero, J.; Poblete, H.; Navarro, C.; Alzate-Morales, J. H. (2013) Association of nicotinic acid with a poly(amidoamine) dendrimer studied by molecular dynamics simulations. *Journal of Molecular Graphics and Modelling*, **39** 71–78.
[92] Laio, A.; Parrinello, M. (2002) Escaping free-energy minima. *Proceedings of the National Academy of Sciences of the United States of America*, **99** (20) 12562–12566.
[93] Laio, A.; Gervasio, F. L. (2008) Metadynamics: a method to simulate rare events and reconstruct the free energy in biophysics, chemistry and material science. *Reports on Progress in Physics*, **71** (12) 126601.
[94] Yao, Y.; Klug, D. D. (2012) Reconstructive structural phase transitions in dense mg. *Journal of Physics: Condensed Matter*, **24** (26) 265401.
[95] Oganov, A. R.; Lyakhov, A. O.; Valle, M. (2011) How evolutionary crystal structure prediction works – and why. *Accounts of Chemical Research*, **44** (3) 227–237.
[96] Ensing, B.; De Vivo, M.; Liu, Z. W.; *et al.* (2006) Metadynamics as a tool for exploring free energy landscapes of chemical reactions. *Accounts of Chemical Research*, **39** (2) 73–81.
[97] Pietrucci, F.; Marinelli, F.; Carloni, P.; Laio, A. (2009) Substrate binding mechanism of hiv-1 protease from explicit-solvent atomistic simulations. *Journal of the American Chemical Society*, **131** (33) 11811–11818.
[98] Leone, V.; Marinelli, F.; Carloni, P.; Parrinello, M. (2010) Targeting biomolecular flexibility with metadynamics. *Current Opinion in Structural Biology*, **20** (2) 148–154.
[99] Won, Y. Y.; Davis, H. T.; Bates, F. S. (1999) Giant wormlike rubber micelles. *Science*, **283** (5404) 960–963.

[100] Jain, S.; Bates, F. S. (2003) On the origins of morphological complexity in block copolymer surfactants. *Science*, **300** (5618) 460–464.
[101] Jain, S.; Bates, F. S. (2004) Consequences of nonergodicity in aqueous binary PEO-PB micellar dispersions. *Macromolecules*, **37** (4) 1511–1523.
[102] Rajagopal K.; Mahmud, A.; Christian D.A.; *et al.* (2010) Curvature-coupled hydration of semicrystalline polymer amphiphiles yields flexible worm micelles but favors rigid vesicles: polycaprolactone-based block copolymers. *Macromolecules*, **2010** (43) 9736–9746.
[103] Plimpton, S. (1995) Fast parallel algorithms for short-range molecular dynamics. *Journal of Computational Physics*, **117** 1–19.
[104] Kim, Y.; Dalhaimer, P.; Christian, D. A.; Discher, D. E. (2005) Polymeric worm micelles as nano-carriers for drug delivery. *Nanotechnology*, **16** (7) S484–S491.
[105] Discher, D. E. (2007) Emerging applications of polymersomes in delivery: from molecular dynamics to shrinkage of tumors. *Progress in Polymer Science*, **32** 838–857.
[106] Kim, Y.; Tewari, M.; Pajerowski, J. D.; *et al.* (2009) Polymersome delivery of sirna and antisense oligonucleotides. *Journal of Controlled Release*, **134** 132–240.
[107] Christian, D. A.; Cai, S.; Garbuzenko, O. B.; *et al.* (2009) Flexible filaments for in vivo imaging and delivery: persistent circulation of filomicelles opens the dosage window for sustained tumor shrinkage. *Molecular Pharmaceutics*, **6** 1343–1352.
[108] Ahmed, F.; Hategan, A.; Discher, D. E.; Discher, B. M. (2003) Block copolymer assemblies with cross-link stabilization: from single-component monolayers to bilayer blends with PEO-PLA. *Langmuir*, **19** (16) 6505–6511.
[109] Geng, Y.; Discher, D. E. (2005) Hydrolytic degradation of poly(ethylene oxide)-block-polycaprolactone worm micelles. *Journal of the American Chemical Society*, **127** (37) 12780–12781.
[110] Baumgart, T.; Hess, S. T.; Webb, W. W. (2003) Imaging coexisting fluid domains in biomembrane models coupling curvature and line tension. *Nature*, **425** (6960) 821–824.
[111] Loverde, S. M.; Ortiz, V.; Kamien, R. D.; *et al.* (2010) Curvature-driven molecular demixing in the budding and breakup of mixed component worm-like micelles. *Soft Matter*, **6** 1419–1425.
[112] Maillard, D.; Prud'homme, R. E. (2006) Chirality information transfer in polylactides: from main-chain chirality to lamella curvature. *Macromolecules*, **39** 4272–4275.
[113] Dalhaimer, P.; Wagner, O. I.; Leterrier, J. F.; *et al.* (2005) Flexibility transitions and looped adsorption of wormlike chains. *Journal of Polymer Science Part B: Polymer Physics*, **43** (3) 280–286.
[114] den Otter, W. K.; Shkulipa, S. A.; Briels, W. J. (2003) Buckling and persistence length of an amphiphilic worm from molecular dynamics simulations. *Journal of Chemical Physics*, **119** (4) 2363–2368.
[115] De Brabander, M.; Geuens, G.; Nuydens, R.; *et al.* (1981) Taxol induces the assembly of free microtubules in living cells and blocks the organizing capacity of the centrosomes and kinetochores. *Proceedings of the National Academy of Sciences of the United States of America*, **78** (9) 5608–5612.
[116] Mastropaolo (1995) Crystal and molecular structure of paclitaxel. *Proceedings of the National Academy of Sciences of the United States of America*, **92** 6920–6924.
[117] Balasubramanian, S.; Alderfer, J. L.; Staubinger, R. M. (1994) Solvent- and concentration-dependent molecular interactions of taxol. *Journal of Pharmaceutical Sciences*, **83** 1470–1476.
[118] Wenk, M. R.; Fahr, A.; Reszka, R.; Seelig, J. (1996) Paclitaxel partitioning into lipid bilayers. *Journal of Pharmaceutical Sciences*, **85** (2) 228–231.
[119] Forrest, M. L.; Yanez, J. A.; Remsberg, C. M.; *et al.* (2008) Paclitaxel prodrugs with sustained release and high solubility in poly(ethylene glycol)-b-poly(epsilon-caprolactone) micelle nanocarriers: pharmacokinetic disposition, tolerability, and cytotoxicity. *Pharmaceutical Research*, **25** (1) 194–206.
[120] Nagarajan, R. and Wang, C.C. (1995) Estimation of surfactant tail transfer free energies from polar solvents to micelle core. *Langmuir*, **11** 4673–4677.
[121] Kumar, V.; Prudhomme, R. K. (2007) Thermodynamic limits on drug loading in nanoparticle cores. *Journal of Pharmaceutical Sciences*, **97** 4904.
[122] Park, E. K.; Lee, S. B.; Lee, Y. M. (2005) Preparation and characterization of methoxy poly(ethylene glycol)/poly(epsilon-caprolactone) amphiphilic block copolymeric nanospheres for tumor-specific folate-mediated targeting of anticancer drugs. *Biomaterials*, **26** (9) 1053–1061.

[123] Siepmann, J. I.; Karaborni, S.; Smit, B. (1993) Simulating the critical behavior of complex fluids. *Nature*, **365** 330–333.
[124] Shelley, J. C.; Shelley, M. Y.; Reeder, R. C.; *et al.* (2001) A coarse grain model for phospholipid simulations. *Journal of Physical Chemistry B*, **105** 4464–4470.
[125] Teng, Y.; Morrison, M. E.; Munk, P.; *et al.* (1998) Release kinetics studies of aromatic molecules into water from block copolymer micelles. *Macromolecules*, **31** 3578–3587.
[126] Cai, S.; Vijayan, K.; Cheng, D.; *et al.* (2007) Micelles of different morphologies–advantages of worm-like filomicelles of peo-pcl in paclitaxel delivery. *Pharmaceutical Research*, **24** (11) 2099–2109.
[127] Sonee, M.; Barron, E.; Yarber, F. A.; Hamm-Alvarez, S. F. (1998) Taxol inhibits endosomal-lysosomal membrane trafficking at two distinct steps in cv-1 cells. *American Journal of Physiology*, **275** (6, part 1) C1630–C1639.
[128] Ali, S.; Minchey, S.; Janoff, A.; Mayhew, E. (2000) A differential scanning calorimetry study of phosphocholines mixed with paclitaxel and its bromoacylated taxanes. *Biophysical Journal*, **78** (1) 246–256.
[129] Huang, H. W. (2009) Free energies of molecular bound states in lipid bilayers: lethal concentrations of antimicrobial peptides. *Biophysical Journal*, **96** (8) 3263–3272.
[130] Ashrafuzzaman, M.; Tseng, C. Y.; Duszyk, M.; Tuszynski, J. A. (2012) Chemotherapy drugs form ion pores in membranes due to physical interactions with lipids. *Chemical Biology and Drug Design*, **80** (6) 992–1002.
[131] Pan, A. C.; Borhani, D. W.; Dror, R. O.; Shaw, D. E. (2013) Molecular determinants of drug-receptor binding kinetics. *Drug Discovery Today*, **18** (13–14) 667–673.
[132] Akhmanova, A.; Steinmetz, M. O. (2008) Tracking the ends: a dynamic protein network controls the fate of microtubule tips. *Nature Reviews Molecular Cell Biology*, **9** (4) 309–322.
[133] Sui, H.; Downing, K. H. (2010) Structural basis of interprotofilament interaction and lateral deformation of microtubules. *Structure*, **18** (8) 1022–1031.
[134] Snyder, J. P.; Nettles, J. H.; Cornett, B.; *et al.* (2001) The binding conformation of taxol in beta-tubulin: a model based on electron crystallographic density. *Proceedings of the National Academy of Sciences of the United States of America*, **98** (9) 5312–5316.
[135] Evangelio, J. A.; Abal, M.; Barasoain, I.; *et al.* (1998) Fluorescent taxoids as probes of the microtubule cytoskeleton. *Cell Motility and the Cytoskeleton*, **39** (1) 73–90.
[136] Dye, R. B.; Fink, S. P.; Williams, R. C., Jr. (1993) Taxol-induced flexibility of microtubules and its reversal by MAP-2 and Tau. *Journal of Biological Chemistry*, **268** (10) 6847–6850.
[137] Buey, R. M.; Calvo, E.; Barasoain, I.; *et al.* (2007) Cyclostreptin binds covalently to microtubule pores and lumenal taxoid binding sites. *Nature Chemical Biology*, **3** (2) 117–125.
[138] Magnani, M.; Maccari, G.; Andreu, J. M.; *et al.* (2009) Possible binding site for paclitaxel at microtubule pores. *FEBS Journal*, **276** (10) 2701–2712.
[139] Diaz, J. F.; Strobe, R.; Engelborghs, Y.; *et al.* (2000) Molecular recognition of taxol by microtubules. Kinetics and thermodynamics of binding of fluorescent taxol derivatives to an exposed site. *The Journal of Biological Chemistry*, **275** (34) 26265–26276.
[140] Diaz, J. F.; Barasoain, I.; Souto, A. A.; *et al.* (2005) Macromolecular accessibility of fluorescent taxoids bound at a paclitaxel binding site in the microtubule surface. *Journal of Biological Chemistry*, **280** (5) 3928–3937.
[141] Freedman, H.; Huzil, J. T.; Luchko, T.; *et al.* (2009) Identification and characterization of an intermediate taxol binding site within microtubule nanopores and a mechanism for tubulin isotype binding selectivity. *Journal of Chemical Information and Modeling*, **49** (2) 424–436.
[142] Prussia, A. J.; Yang, Y.; Geballe, M. T.; Snyder, J. P. (2010) Cyclostreptin and microtubules: is a low-affinity binding site required? *Chembiochem*, **11** (1) 101–109.
[143] Chacon, P.; Wriggers, W. (2002) Multi-resolution contour-based fitting of macromolecular structures. *Journal of Molecular Biology*, **317** (3) 375–384.
[144] Maccari, G.; Mori, M.; Rodriguez-Salarichs, J.; *et al.* (2013) Free energy profile and kinetics studies of paclitaxel internalization from the outer to the inner wall of microtubules. *Journal of Chemical Theory and Computation*, **9** (1) 698–706.
[145] Sept, D.; MacKintosh, F. C. (2010) Microtubule elasticity: connecting all-atom simulations with continuum mechanics. *Physical Reviews Letters*, **104** (1) 018101.
[146] Grafmueller, A.; Noya, E. G.; Voth, G. A. (2013) Nucleotide-dependent lateral and longitudinal interactions in microtubules. *Journal of Molecular Biology*, **425** (12) 2232–2246.

5

Solid Dispersion – a Pragmatic Method to Improve the Bioavailability of Poorly Soluble Drugs

Peng Ke[1], Sheng Qi[2], Gabriele Sadowski[3], and Defang Ouyang[4]

[1] *Merck & Co, Inc., UK*
[2] *School of Pharmacy, University of East Anglia, UK*
[3] *Department of Biochemical and Chemical Engineering, TU Dortmund University, Germany*
[4] *Institute of Chinese Medical Sciences, University of Macau, Macau*

5.1 Introduction of Solid Dispersion

Oral drug delivery is the most convenient route for drug administrations, with solid oral dosage forms being preferred over other types of dosage forms because of their smaller bulk size, better stability, accurate dose, and ease of preparation [1]. Drugs contained in solid oral dosage forms need to have good solubility in order to result in good bioavailability provided they also have good permeability. However, based on some recent estimates, about 40% of present drugs are poorly soluble in water, and even up to 60% of compounds coming directly from synthesis encounter the same problem [1]. Therefore, finding an efficient approach to increase the dissolution rate and solubility of the drugs is a major challenge for pharmaceutical researchers.

Solid dispersions, first developed and prepared by Sekiguchi and Obi [2], are a useful method which can overcome the above-mentioned limitations [2]. They are currently one of the most promising strategies for the enhancement of the oral bioavailability of poorly water-soluble drugs. Chiou and Riegelman [3] defined a solid dispersion as a solid matrix consisting

Computational Pharmaceutics: Application of Molecular Modeling in Drug Delivery, First Edition.
Edited by Defang Ouyang and Sean C. Smith.

of at least two components: generally a hydrophilic carrier and a hydrophobic drug [3]. The carrier can be either crystalline or amorphous and the drug can be dispersed molecularly, in amorphous particles (clusters), or in crystalline particles. The particle size of the drugs in a solid dispersion can reach as low as molecular level in favor of rapid dissolution [4]. In addition, the drugs can exist in their amorphous forms which in theory represent the most energetic solid state of a material, and hence they should have the biggest advantage in terms of apparent solubility [5]. There are a few marketed products available using solid dispersion strategy and the number has been increasing in the recent years, as shown in Table 5.1.

After nearly 50 years of continuing research and development, many solid dispersion formulations have been prepared. According to the carrier types, molecular arrangements and addition of surfactants, solid dispersions have been classified into three generations [6]. The first generation carriers of solid dispersions are crystalline carriers, such as urea and sugars. The best example of the first generation solid dispersion is the eutectic mixture prepared by Sekiguchi and Obi [2], the first solid dispersion prepared in the literature [2]. In that study, sulfathiazole and urea were melted together at a temperature above the eutectic point and then cooled in an ice bath. The dispersed drugs were trapped within the carrier matrix in the resultant solid eutectic, which was then milled to reduce the particle size. A eutectic mixture usually contains two crystalline materials and has a melting point lower than that for any other ratio of mixtures of the same materials. Later, some other solid dispersion systems were developed by preparing solid solutions through molecular dispersions of active pharmaceutical ingredient (API) into crystalline carriers (e.g., mannitol) instead of eutectic mixtures [7, 8]. The disadvantage of forming crystalline solid dispersions

Table 5.1 *Marketed products using solid dispersion approach.*

Product	Indication	Carrier system	Company
Certican (Everolimus)	Immunosuppressant	HPMC	Novartis
Cesamet (Nabilone)	Antiemeticum	PVP	Valeant Pharma
Fenoglide (Fenofibrate)	Hyperlipidemia	PEG 6000, Poloxamer 188	Santarus
Gris-PEG (Griseofulvin)	Antimycoticum	PEG 400, PEG 8000	Pedinol Pharma
Incivek (Telepravir)	HCV	HPMCAS	Vertex
Intelence (Etravirine)	HIV	HPMC	Tibotec/J&J
Isoptin SRE (Verapamil)	Hypertension	HPC/HPMC (sustained release)	Abbott
Kaletra (Lopinavir/ Ritonavir)	HIV	Copovidone	Abbott
Kalydeco (Ivacaftor)	Cystic fibrosis	HPMCAS	Vertex
Nimotop (Nimodipine)	Calcium channel blocker	PEG	Bayer
Norvir (Ritonavir)	HIV	Copovidone	Abbott
Prograf (Tacrolimus)	Immunosuppressant	HPMC	Fujisawa
Rezulin (Troglitazone)	Diabetes	PVP	Wyeth
Sporanox (Itraconazole)	Anti-fungal	HPMC	Janssen/Ortho McNeil
Zelboraf (Vemurafenib)	Melanoma	HPMCAS	Roche

is that it makes the dissolution of drug slower than when it is in the amorphous state. Therefore, the second generation of solid dispersions was developed, containing amorphous drugs and carriers. These kinds of solid dispersions are the most commonly used nowadays. The carriers used here are mostly polymeric, such as pvp, copovidone, HPMC, HPMCAS, and so forth. Depending on the molecular arrangement between the drug and carriers, the second generation solid dispersions can be divided into solid solution, solid suspension, or a mixture of both. Compared to the first generation of solid dispersion, the second generation provides a better dissolution of drug since the drug particle size can be reduced to nearly a molecular level and the drug can exist in its amorphous form. A third generation of solid dispersions has been developed by using a carrier which has surface activity or self-emulsifying properties. The solid dispersions contain either a surfactant carrier or both polymeric and surfactant carriers. It was found that these types of systems could help to significantly improve the bioavailability of a poorly-soluble drug and at the same time help to improve the stability of the drug for a long period of time [9]. Recently, more and more controlled release systems utilize solid dispersion technology to achieve an extended release profile of poorly water-soluble drugs with a short biological half-life, which represents the fourth generation of solid dispersions [10]. In these systems, a poorly water-soluble drug is molecularly dispersed in solubilizing carriers or swellable polymers which can prolong drug release from the solid dispersion system. The polymers commonly used in controlled release solid dispersions include ethyl cellulose, polyethylene oxide (PEO), HPC, Eudragit RS, RL, Kollicoat SR, Kollidon SR, HPMC, HPMCAS, and so on. The drug is usually released from the system through mechanisms of diffusion and erosion.

5.2 Preparation Methods for Solid Dispersions

There are generally two main methods for the preparation of solid dispersions: one is through the use of a liquid phase such as melting and solvent methods [11], and the other one is through a solid phase such as mechanical methods, for example, milling [12]. It should be noted that the method used to prepare solid dispersions can play an important role in the physical stability of the systems [1], therefore formulators would need to take into account what is the best manufacture method for the amorphous system when developing a solid dispersion formulation.

5.2.1 Melting Method

The melting method is widely used for preparing solid dispersions and some of the first developed solid dispersions for pharmaceutical applications were prepared by this method. Generally, in a melting process, drug and carrier are melted together and the high molecular mobility of the components allows them to be incorporated into each other. Cooling usually follows to solidify the melted mixtures and the processes include ice bath agitation, immersion in liquid nitrogen, stainless steel thin layer spreading followed by a cold draught, spreading on plates with dry ice, and so forth. Pulverization is often needed to break the cake into small particles after cooling.

However, there are some significant drawbacks which might limit the application of the melting method. First, degradation can happen to some drugs and carriers during heating to temperatures needed to melt them. Second, incomplete miscibility between drug and carrier may occur because of the high viscosity of some carriers in the molten state. Therefore, some modified melting methods, such as hot melt extrusion, have been developed in order to avoid these problems [13]. In this approach, the carrier and drug used are first heated together to the melting temperature or a temperature lower than that for a short period of time, and then extrusion at high rotational speed follows. Finally, the resulting materials are cooled at room temperature and milled. In some cases, the use of carbon dioxide as a plasticizer can greatly reduce the heating temperature, which makes this method even more useful to thermosensitive compounds. Hot melt extrusion has been shown to be a versatile technique for preparation of different dosage forms, such as powders, granules, spheres, films and patches. Recently, there is an emerging trend of combining hot melt extrusion and injection moulding as a pharmaceutical continuous process (direct shaping) for preparing tablets containing drugs in polymeric matrices for controlled drug release as well as immediate release purposes [14]. Egalet is a novel delivery system using such a technique to prepare controlled release products based on modulating the erosion of the matrix [15]. Apart from making tablets, injection moulding has also been used for preparation of serval implants. Similar to injection moulding, Abbott developed a hot melt extrusion related technology called "Calendaring" which was the first directly shaped hot melt extrusion product on the market [16]. Thus far, there are a few commercial products available on the market containing solid dispersions prepared by hot melt extrusion technique and some examples include: Kaletra (developed by Abbott), a protease-inhibitor combination product used for treatment of human immunodeficiency virus; Implannon and Nuvaring, used for contraception and both marketed by Merck & Co.; and Zoladex, marketed by AstraZeneca and it is a goserelin acetate implant which is used for the treatment of prostate cancer.

5.2.2 Solvent Method

In the solvent method, drug and carrier are dissolved in a solvent (usually volatile organic solvents) and then the solvent is removed, leading to the production of solid dispersions. Solvent methods are very useful for thermosensitive compounds because the solvent evaporation usually occurs at a low temperature. There are many different types of solvent methods which are determined by the solvent removal process, such as vacuum drying, heating on a hot plate, slow evaporation of solvent at low temperature, use of a rotary evaporator, use of nitrogen, spray drying, freeze drying, use of supercritical fluids, and so on. Among these methods, spray drying and freeze drying are probably the most commonly used. In a spray drying process, the carrier and drug are dissolved or suspended in a certain solvent and then the solution is sprayed through an atomizer, giving very small droplets which can be dried easily using a heated air flow. During freeze drying, the drug and carrier are firstly dissolved in a common solvent and then the solution would go through three successive steps: freezing, primary drying and secondary drying. It should be mentioned that spray freeze drying is a combination of spray drying and freeze drying techniques. This method provides a faster freezing rate compared to freeze drying, reducing the risk of phase separation of the amorphous system and allows the generation of reduced particle size of drug particles, providing rapid dissolution. The use of super critical fluids, usually carbon dioxide, has also been receiving increasing attention recently [17]. Its advantages include reduced particle size and residual solvent content, disappearance of any degradation and high product yield. The co-precipitation method and spin-coated films are also two very useful solvent methods for the preparation of solid dispersions. Electrostatic spinning has also gained interest in recent years for the preparation of solid dispersions and the advantages of this technique are the generation of fibers with micron or submicron diameters which provide extremely high surface area per unit mass, facilitating fast and efficient removal of solvent and leading to better amorphization; and the high surface area also provides a significant improvement of dissolution rate compared to other solvent drying techniques [81]. The limitation regarding the solvent methods is the use of organic solvents which increases the preparation cost and leads to a great risk to the environment.

5.2.3 Ball Milling

Ball milling differs from the melting and solvent methods in that solid dispersions of drug and carrier are prepared in the solid state. This method has been widely used to make the amorphous phase [18]. During milling, strong mechanical forces occur which can help to facilitate the incorporation of drug and carrier [19]. Heat might be generated in the local collision region between the balls and the wall of the pot, which could help to dissolve the drug into a carrier with good miscibility [12]. Studies have shown that solid solutions can be made by using high-energy milling approaches such as planetary ball milling. Advantages of ball milling to prepare solid dispersions include its suitability for thermosensitive materials and its ease in handling, avoiding the use of large amount of solvents. However, compared to melting and solvent methods, ball milling might be difficult to scale up. Prolonged milling might also lead to potential degradation of the materials.

5.3 Thermodynamics of Solid Dispersions

Solid dispersions usually consist of at least and amorphous polymer and an API. As for all other mixtures, the thermodynamic behavior of such a solid dispersion depends on temperature and concentration, which is shown in typical phase diagrams in Figure 5.1 [20].

Above the solubility line, the mixture of polymer an API forms a homogeneous liquid. The polymer acts as solvent for the API. From thermodynamic point of view, this is quite the same as an API solution in a common solvent. The only difference is that the viscosity of the API/polymer system is quite high compared to a common API solution and the dispersion therefore might look like "solid" wherefore API/polymer dispersions are often called solid dispersions

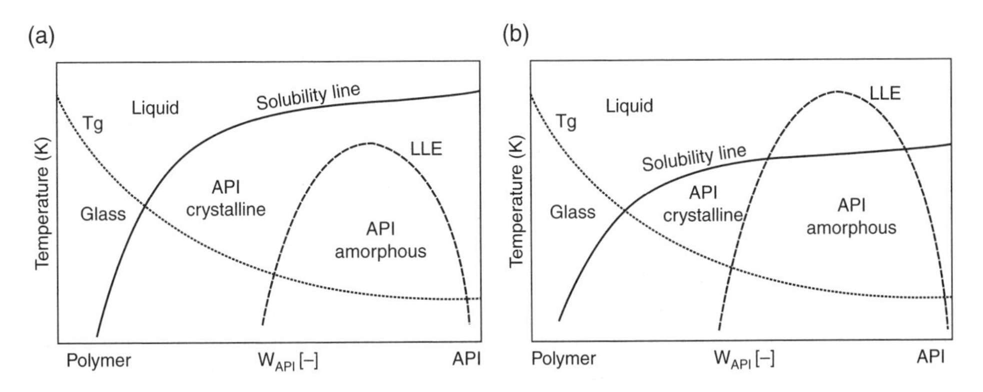

Figure 5.1 *Thermodynamic phase diagrams of a solid dispersions consisting of a polymer and an API. (a) Amorphous two-phase region below the solubility line. (b) Amorphous two-phase region exceeding the solubility line.*

(although from thermodynamic point of view above the solubility line there are liquids). In this region, no API crystals are formed. For that reason, this is the preferred region for a stable, amorphous dispersion. However, as seen in Figure 5.1, this region only exists at high temperatures (usually far above room temperature) or at very low API concentrations.

Solid dispersions as used for pharmaceutical applications are usually below the solubility line. In this region, the API tends to crystallize. At a certain temperature, the solubility line gives the concentration of the non-crystalline API in the polymer which is in thermodynamic equilibrium with API crystals. This, however, does not mean that crystals are necessarily formed when decreasing the temperature from a temperature above the solubility line to a temperature below. Although crystal formation is thermodynamically preferred, it is kinetically hindered due to the often high viscosity of the system. After very long times (e.g., during storage of API solid dispersions), however, crystal formation might occur as this is the thermodynamically stable state.

Besides this, very often another region can be found in the temperature–concentration space, where the API/polymer mixture demixes into two amorphous phases (an API-rich one and an API-poor one) which from thermodynamic point of view is a liquid–liquid equilibrium (LLE). This amorphous two-phase region might be either completely located below the solubility line (as shown in Figure 5.1a) or exist both below and above the latter (see Figure 5.1b). In both cases, demixing might occur for temperatures/concentrations below the LLE line.

In the first case (Figure 5.1a), as this region is located completely below the solubility line, crystallization is preferred thermodynamically, but (metastable) liquid–liquid demixing is preferred kinetically. Thus, mixtures below the LLE line tend first to demix into two amorphous phases and afterwards tend to crystallize. Due to the usually high viscosity both demixing and crystallization might take quite a long time but there definitely is a thermodynamic driving force for that.

In temperature/concentration regions of the LLE which are located above the solubility line (Figure 5.1b), amorphous demixing is thermodynamically preferred compared to crystallization. Here, no crystals are formed but liquid–liquid demixing is likely (only kinetically hindered at high viscosities).

It becomes obvious that, at low temperatures, solid dispersions are only stable (without demixing and/or crystal formation) at very low API concentrations. As the therapeutic API concentration of most pharmaceutical formulations is higher than that, these formulations are usually metastable and tend to demix or crystallize. High viscosities however dramatically increase the time for the thermodynamically stable state to be reached. This especially applies to API/polymer formulations in the glassy state (a glass transition temperature higher than room temperature). For that reason, Figures 5.1a and 5.1b also contain a line that gives the glass transition temperature of an API/polymer system as a function of composition. All metastable states (below solubility line and/or LLE line) that can be found also below the glass transition line are stabilized by the extremely high viscosity of the mixture and can thus be kept for a certain time (weeks to years) without demixing or crystallization.

Another factor which dramatically influences the thermodynamic behavior of API/polymer systems is water. Depending on the temperature and relative humidity of the surroundings, a certain amount of water dissolves in the API/polymer solid dispersion and thus influences the phase behavior (solubility line and liquid–liquid demixing) as well as the glass transition temperature. The magnitude of the water influence of course depends on the particular API and polymer. A qualitative picture of this influence is shown in Figure 5.2.

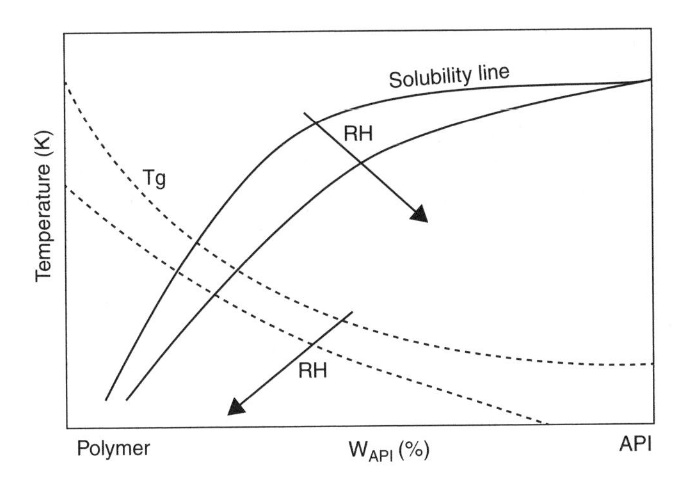

Figure 5.2 *Influence of relative humidity RH on the phase behavior of a polymer/API solid dispersion.*

Very often, the solubility of the API in the polymer decreases in the presence of water, which means that the tendency of API crystallization in the solid dispersion of a given API concentration increases. At the same time, the glass transition temperature of the API/polymer solid dispersion dramatically decreases with increasing water content, which also favors API crystallization.

The phase behavior schematically shown in Figures 5.1 and 5.2 can be (at least partly) experimentally determined. Solubility curves as well as glass transition temperatures of the solid dispersions are accessible, for example, via DSC. The amorphous demixing, however, can hardly be determined experimentally since it is often located in the metastable region where crystallization is preferred.

However, both the solubility line of the crystalline API as well as the region of amorphous demixing are assessable via thermodynamic modeling. The solubility line is determined by the thermodynamic equilibrium of crystalline API on the one hand and saturated liquid mixture of API and polymer on the other hand. At a certain temperature, the concentration of the API in the liquid mixture is the API solubility, which is shown along the solubility line in Figure 5.1. The phase equilibrium condition for determining this solubility reads as:

$$x_{\text{API}} = \frac{1}{\gamma_{\text{API}}} \exp\left\{ -\frac{\Delta h_{\text{API}}}{RT} \left(1 - \frac{T}{T_{\text{API}}^{\text{SL}}} \right) \right\} \tag{5.1}$$

whereas x_{API} is the mole fraction of the API dissolved in the polymer, Δh_{API} and $T_{\text{API}}^{\text{SL}}$ are the melting enthalpy and the melting temperature of the pure API, respectively. γ_{API} is the so-called activity coefficient of the API in the polymer. It depends on the kind of API and polymer, on temperature, and on concentration (in the case of water-containing solid dispersions it also depends on the water content). Thus, determining the activity coefficient is crucial for calculating the API solubility. For the solubility in polymers it is usually far from unity and cannot be neglected. It can be determined either from excess Gibbs energy (g^{E}) models or from equations of state.

Although by far not the best-suited model, very often the g^E model by Flory–Huggins [1] is used for that purpose. It accounts for the remarkable difference in size of API and polymer molecules. It is however weak as far as the energetic interactions and in particular the formation of hydrogen bonds are concerned. This usually leads to an only qualitative description of the temperature difference of API solubility using this model. In particular during the last two decades, a whole series of equations of state have been developed which explicitly account for hydrogen-bond formation (association) and which are therefore much better suited for describing API solubilities in polymers in the absence as well as in the presence of water. Examples for those models are, for example, the statistical associating fluid theory (SAFT) [2] or, more recently, the perturbed-chain statistical associating fluid theory (PC-SAFT) [3], whereas the latter is in particular suitable for polymer systems.

The phase equilibrium conditions for modeling the liquid–liquid demixing are different from those for the API solubility and read as:

$$x_{\text{API}}^{\text{L1}} \cdot \gamma_{\text{API}}^{\text{L1}} = x_{\text{API}}^{\text{L2}} \cdot \gamma_{\text{API}}^{\text{L2}}; \quad x_{\text{Polymer}}^{\text{L1}} \cdot \gamma_{\text{Polymer}}^{\text{L1}} = x_{\text{Polymer}}^{\text{L2}} \cdot \gamma_{\text{Polymer}}^{\text{L2}} \tag{5.2}$$

Because the API and the polymer distribute between the two amorphous phases L1 and L2, phase equilibrium conditions for both API and polymer are required. Equation 5.2 contains activity coefficients, this time for API as well as for polymer. As before, both of them can be obtained in the simplest case from the Flory–Huggins model or from more sophisticated models, like PC-SAFT.

Figure 5.3 shows as an example the solubility of two sulfonamides as measured and predicted from Flory–Huggins theory as well as from PC-SAFT. For both models, calculations are pure predictions, that is, binary parameters were set to zero in both cases and solubility was predicted based on information on the pure API and the pure polymer only.

It becomes obvious that PC-SAFT predictions are much more reliable than those from the Flory–Huggins model, in particular at low temperatures which are usually of practical interest.

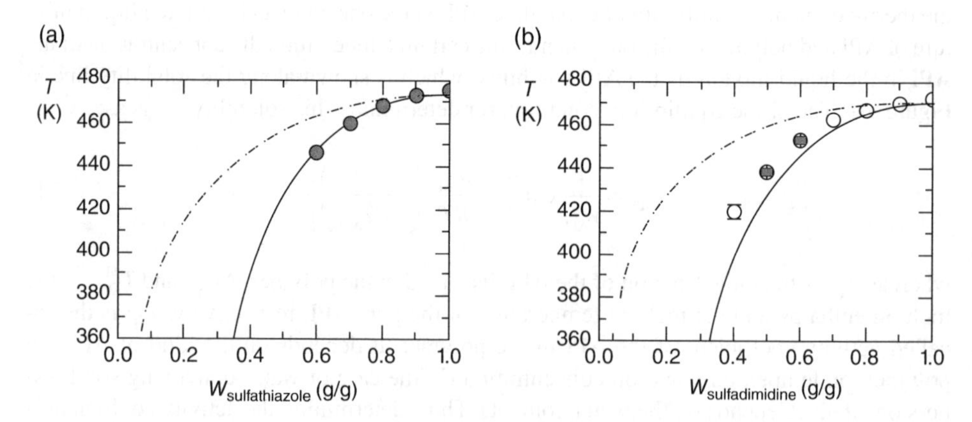

Figure 5.3 Solubilities of (a) sulfathiazole and (b) sulfadimidine in PVP. Lines are predictions using PC-SAFT (full lines) and Flory–Huggins theory (dashed lines) [4]. Symbols are experimental data. Data taken from Ref. [5].

5.4 Molecular Structure of Amorphous Solid Dispersions

Currently there are two main molecular models for amorphous solid dispersions: a conventional model and a new model. Usually polymer carriers are more likely to form amorphous solid solutions, while low molecular weight carriers tend to form simple eutectic mixture or solid solutions [21]. In the conventional model of amorphous solid dispersions, the drug molecules are irregularly trapped within the polymeric network in the solid dispersions, as shown in Figure 5.4a [21].

Molecular modeling is a specialized form of computer modeling used to mimic the behavior of molecules [23]. The distinguishing characteristic of the technique is to describe molecular systems at an atomistic level. Molecular modeling techniques have great potential for the study of the structure of solid dispersions at the molecular level. Table 5.2 summarizes recent molecular modeling studies describing solid dispersions. Most studies with simple models [26–29, 31] or simple simulation protocols [24, 30, 32] were not enough to simulate real experimental conditions of solid dispersions, and therefore provided very limited information about the molecular structure of solid dispersions. However, a new molecular structure of amorphous solid dispersions was developed by computer modeling methods [22]. In this research, the molecular structure of solid dispersions with a hot melt preparation method was investigated by the simulated annealing method. Modeling results indicated that linear polymer chains form random coils under heat and the drug molecules stick on the surface of the polymer coils, as shown in Figure 5.4b [22]. These results suggested that the amorphous state of drug molecules in solid dispersions improves its dissolution properties. At the same time, the drug molecules at the surface of the polymer coils are able to easily move and aggregate together, which may explain the physical instability of solid dispersions. If the drug molecules are trapped within the cavity of a polymer chain network (as shown in Figure 5.4a), according to conventional theory, the drug molecules should be more difficult to move and recrystallize due to the very high energy barrier of polymeric networks, which contradicts numerous experiment results with physical aging of solid dispersions. The new model of solid dispersions may present more reasonable molecular images of solid dispersions than the conventional theory [22].

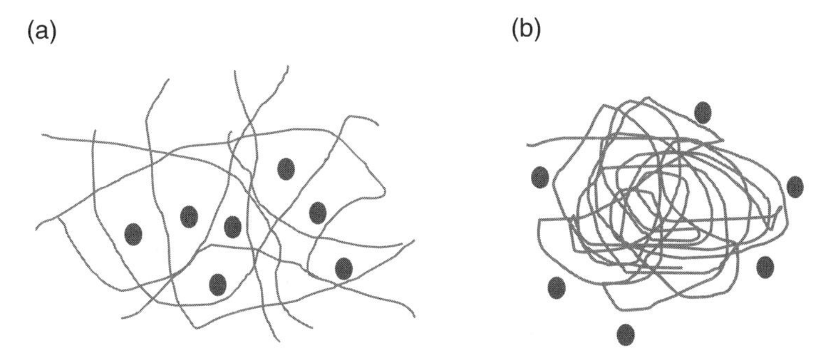

Figure 5.4 *Schematic representation of amorphous solid dispersions with polymer carriers: (a) the conventional model of amorphous solid solution [21] and (b) a new model of amorphous solid dispersions [22] (see colour plate section).*

Table 5.2 Molecular modeling studies of solid dispersions.

Materials	Simulation programme	Simulation details	References
Isomalt and other drug molecules	Molecular dynamics (MDs) simulation GROMACS package and GROMACS all-atom force field	40 isomalt molecules and one drug molecule in the systems	[24]
Poly(3-hydroxybutyrate) and poly(ethylene oxide)	Molecular dynamics (MD) simulations Cerius2 software Dreiding 2.21 force field	100 repeated units for each polymer	[25]
Ibuprofen, naproxen, and PVP	Molecular modeling program SYBYL Molecular mechanics methods and semiempirical quantum chemistry method	PVP monomer, trimer structure of PVP with ibuprofen, or naproxen	[26]
Celecoxib and PVP	Molecular modeling program SYBYL MMFF94 force field	Nine celecoxib molecules and 18 molecules of monomer PVP	[27, 28]
Felodipine, PVP, and PEG	Density functional theory (DFT) Gaussian 98 program	Felodipine and PVP or PEG monomer	[29]
Aryl propionic acid derivatives	Molecular docking Docking and discover module of Insight II software	Single drug molecule and PVP with 30 repeating units	[30]
Imidazolidine derivatives and PEG/PVP polymers	Density functional theory (DFT) Gaussian 03W program	Monomeric model: Imidazolidine derivatives and PEG/PVP monomer	[31]
Indomethacin, carriers (polyethylene oxide (PEO), glucose, and sucrose)	MD simulations Material Studio 4.0 COMPASS force field	Ten *in silico* amorphous models with multiple drug molecules and carrier molecules	[32]
Ibuprofen, PEG, PVP, Poloxamer	MD simulations AMBER12 program GAFF force field	20 repeated units for each polymer and different molar ratios between drug and polymers	[22]

5.5 Physical Stability of Solid Dispersions

The physical stability of a solid dispersion is highly dependent on the state of the drug in the dispersion. Here solid dispersions were assigned to two broad classes based on the physical state of the drug in the systems: crystalline drugs in solid dispersions and amorphous solid dispersions (solid solution) whereby drugs completely disperse into the polymer chain at the molecular level. The solid dispersions containing crystalline drugs normally have little physical instability risk, and in some cases the instabilities are more associated with the physical instability of the carrier material. Crystalline carrier materials, such as PEG and lipids, often pose more risks of instability on aging [33]. The discussion of how to assess the instability of these systems is discussed elsewhere, and here we mainly focus on the discussion of testing the physical instability of amorphous solid dispersions. For amorphous solid dispersions, although they have shown a high potential for improving the dissolution rate of poorly water-soluble drugs, the commercial application of such solid dispersions is very limited. There are many hurdles for commercializing amorphous solid dispersions, and in many cases the physical stability of the dispersions is the major concern [11, 34–37].

An amorphous drug–polymer solid dispersion is a thermodynamically instable delivery system since molecularly dispersed drugs (high energy level) in the system will tend to convert back to the more stable crystalline form (low energy level), leading to the physical stability issue. Freshly prepared amorphous solid dispersions may have the drug molecularly dispersed in the polymeric carriers. On aging the physical instability of amorphous solid dispersions could occur in the form of phase separation and recrystallization due to the relaxation of high energy-level drug molecules and the molecular mobility of drugs (can be accelerated by storage conditions, i.e., stressed humidity or temperature). Phase separation refers to the procedure in amorphous solid solutions where molecularly dispersed drug molecules migrate together to form a drug-rich phase eventually containing a higher drug concentration than the average bulk drug concentration. Recrystallization could then occur within the drug-rich phase whereby a high concentration of amorphous drug recrystallizes out to form the more stable crystalline state.

Drug recrystallization in aged solid dispersions could also take place if nuclei are present. Nuclei can be a residual crystalline drug in the solid dispersion, or foreign particles obtained during processing, or certain insoluble excipients added to the formulation [38]. With the occurrence of phase separation or recrystallization, the attempt to improve the dissolution rate of poorly water-soluble drugs using amorphous solid dispersions will fail as the recrystallized drugs do not have as high a dissolution rate as the amorphous form. Therefore, detection of the physical instability is essential for amorphous solid dispersion-based formulation development.

5.5.1 Detection of Physical Instability of Amorphous Solid Dispersions

Factors which can potentially affect the physical stability of amorphous solid dispersions have been investigated widely, and glass transition temperatures (T_g), molecular mobility, physical stability of amorphous drugs alone, miscibility between drugs and polymers, and solid solubility of drugs in polymers have been considered as key factors influencing the physical stability of solid dispersions [2, 8–11]. In terms of detecting the physical instabilities of amorphous solid dispersions, the commonly used key indicators are T_g, molecular mobility, drug–polymer interactions (if present), and the appearance of phase separated crystalline drug. Changes in these features can be studied using physical methods, such as DSC/MTDSC, DMA, dielectric spectroscopy, PXRD, IR, Raman, SS-NMR, and a range of imaging techniques such as SEM, AFM, and IR/Raman microscopies.

5.5.2 Glass Transition Temperature

Glass transition temperature (T_g) is a kinetic parameter associated with the molecular mobility (viscosity) in an amorphous state. T_g of an amorphous system can be detected using many thermal methods, such as DSC/MTDSC and DMA. Below T_g the amorphous materials are "kinetically frozen" (with great viscosity) into a thermodynamically unstable glassy state, and any further reduction in temperature has only a small effect on the decrease of the molecular mobility of amorphous solids. Above T_g, amorphous solids will enter a rubbery state with significantly increased molecular mobility and decreased viscosity [39]. Molecular mobility has been related to the occurrence of phase separation and recrystallization in amorphous solid dispersions [39, 40]. This is because the molecular mobility of drugs in amorphous solid dispersions can cause phase separation in amorphous solid dispersions and further recrystallization.

For amorphous solid dispersions, glass transition temperatures can be used as indicators to describe their physical state [15, 41, 42]. For a miscible drug–polymer dispersion, a single T_g should be observed. The presence of two glass transitions indicates that phase separation has occurred or the system is partially miscible [43]. The Gordon–Taylor (G-T) equation is a useful tool in predicting the glass transition temperatures of drug–polymer solid dispersions [41, 42]. For a completely miscible binary drug–polymer solid dispersion, the glass transition temperature of the system can be calculated by:

$$T_{gmix} = \left[\left(w_1 T_{g1}\right) + \left(K w_2 T_{g2}\right)\right] / \left[w_1 + \left(K w_2\right)\right] \tag{5.3}$$

where T_{g1}, T_{g2}, and T_{gmix} are the glass transition temperatures (Kelvin) of component 1, 2 and the mixture, and w_1 and w_2 are the weight fractions of each component and K is a constant which can be calculated by:

$$K \approx \left(\rho_1 T_{g1}\right) / \left(\rho_2 T_{g2}\right) \tag{5.4}$$

where $\rho 1$ and $\rho 2$ are the true densities of each component. The application of the equation in predicting the miscibility of solid dispersions lies in the comparison between the theoretical and experimental T_{gmix} values of the system. The Gordon–Taylor equation is based on the assumption that the mixing process is ideal and the molecules from the two components are blended completely [41]. Therefore, if the consistency of the comparison between the calculated and experimental T_{gmix} values is acquired, it may indicate that this system is miscible [44]. However, exceptions have been reported in previous studies [41, 42, 45]. The discrepancies of T_gs between the calculated values using the Gordon–Taylor equation and experimental data may be caused by two reasons. First, interactions between drugs and polymers in amorphous solid dispersions can result in a deviation [42]. A positive deviation, whereby the experimental value is higher than the G-T predicted one, could occur if the interaction between drugs and polymers is stronger than that between two drug molecules [45]; whereas a negative deviation could be observed if the drug–polymer interaction is weaker than that between two drug molecules [46]. Second, water sorption in amorphous solid dispersions can decrease the T_g value of the system, as water is a well-known plasticizer [44, 47].

5.5.3 Molecular Mobility and Structural Relaxation of Amorphous Drugs

The molecular mobility of amorphous drugs is commonly considered to be one of the key factors associated with the stability of amorphous solid dispersions, since a high molecular mobility of drugs in systems can lead to rapid phase separation and recrystallization in amorphous solid dispersions on aging [48]. This is because amorphous materials which remain in a non-equilibrium state at temperatures below T_g have extra enthalpy and configurational entropy (this results from the extra number of configurations of molecules in the amorphous state in comparison to the corresponding crystalline state). Therefore, on aging, non-equilibrium amorphous materials will approach the equilibrium state by releasing extra enthalpy and configurational entropy. This reducing extra energy process is termed structural relaxation and the time length where the structural relaxation occurs is termed relaxation time. Molecular mobility is in a reciprocal relationship to the relaxation time constant (τ). In the Adam–Gibbs model [49], it is calculated as:

$$\tau = \tau_0 \exp\left[C/(TS_c)\right] \quad (5.5)$$

where τ is a molecular relaxation time constant, τ_0 is a constant, T is the absolute temperature, S_c is the configurational entropy, and C is a material dependent constant. This equation was further modified as the Adam–Gibbs–Vogel equation:

$$\tau = \tau_0 \exp\left\{DT_0/\left[T(1 - T_0/T_f)\right]\right\} \quad (5.6)$$

where D is the strength parameter, T_0 is the temperature of zero molecular mobility, and T_f is the fictive temperature [50]. The fictive temperature is defined as the temperature of intersection between the equilibrium liquid line and the non-equilibrium glass line. In most cases, T_f values are very close to T_g values, and therefore in calculations the T_f value can be replaced by the T_g value [50–52].

Two types of relaxation of amorphous materials have been defined at temperatures below and above the glass transition temperatures [53]. For a molecule with a low molecular weight (such as many small molecular drugs), at a temperature below T_g, the dominant relaxation procedure is β-structural relaxation (termed local molecular mobility) which can take place by means of the spinning of atoms within the molecular structure. For amorphous polymers, at a temperature below T_g, β-structural relaxation refers to the vibrating of the side chains of the polymer [54]. At a temperature higher than T_g, the dominant relaxation is α-structural relaxation (termed global molecular mobility; β-structural relaxation still occurs at this temperature) which could occur for both amorphous drug and polymer whereby an intact molecule will be mobilized [54].

These two types of relaxation can be detected by techniques such as DSC, DMA, and dielectric spectroscopy [48, 55, 56]. For instance, stored under ambient condition for a certain time, the relaxation enthalpy of amorphous indomethacin at the glass transition region can be detected on heating in DSC, and the detected relaxation enthalpy is a contribution of both α and β relaxations on aging [48]. The detected relaxation enthalpy is attributed to the energy required to re-establish the liquid state on heating [48].

As discussed above, the molecular mobility of an amorphous drug changes with temperature. Increasing storage temperature can increase the molecular mobility and decrease the physical stability of amorphous solid dispersions. Therefore it has been proposed that amorphous solid dispersions should be stored at the temperature of $T_g - 50\,\mathrm{K}$ to reduce the molecular mobility of drugs in amorphous solid dispersions and thus to increase the physical stability [48, 57].

In addition to a high storage temperature, stressed humidity can also increase the molecular mobility of amorphous drugs in solid dispersions [58]. Two potential effects of moisture uptake on the molecular mobility have been suggested. First, absorbed water into the solid dispersions can act as a plasticizer (providing polymer chains with greater freedom, resulting in a reduction of glass transition temperature of the intact system) and thus, with decreased T_g, a faster molecular mobility and higher physical instability could be expected [59]. Second, for systems which drug molecules can hydrogen bond with polymer molecules, absorbed water molecules into the dispersions can disrupt the hydrogen bonding between drug and polymer molecules, as the water molecules are hydrogen bonding donor and accepter [58]. This can promote the phase separation of drug molecules from the amorphous solid dispersions, and further to recrystallizing out.

5.5.4 Interactions between Drug and Polymer in Solid Dispersions

Interactions between drugs and polymers in solid dispersions have been reported as another important factor that can influence the physical stability of amorphous solid dispersions [60, 61]. First, the interaction formed between drug and polymer molecules in amorphous solid dispersions (such as hydrogen bonding) can restrict the molecular motion of drugs and hence it can increase their physical stability. Second, the presence of interactions between drugs and polymers has been considered beneficial for increasing drug–polymer miscibility and the solubility of drugs in polymers. Therefore, using polymers which can potentially interact with the drug in a solid dispersion formulation can increase the stability of the formulation.

One of the common interactions between drugs and polymers is hydrogen bonding, which is formed by the presence of proton acceptors and donors. Hydrogen bonding is likely to occur among carbonyl groups (acceptors), amine groups (donors), and hydroxyl groups (donors and acceptors). Drugs and polymers with these groups have a high tendency to form hydrogen bonds in amorphous solid dispersions.

Besides hydrogen bonding, other interactions in solid dispersions such as acid–base interactions were also discovered to be favorable for enhancing the physical stability of solid dispersions [62]. These interactions share the same mechanism as hydrogen bonding which reduces molecular mobility, preventing phase separation and recrystallization. However, this type of interaction could be very limited and thus might not be widely utilized in solid dispersion formulation design.

Effective tools for confirming these interactions are spectroscopy-based methods such as FT-IR, Raman, and solid state NMR [41]. In FT-IR, the frequency of vibrations within a chemical structure is very sensitive to how the atoms and molecules interact with neighboring functional groups. Variation of peak positions and intensities in the IR spectrum can be indicators of the presence of hydrogen bonds [63]. In solid state NMR, comparing the chemical shift of amorphous drugs and drugs in solid dispersion could be useful in judging the formation of hydrogen bonds. Hydrogen bonding was found to have a significant effect on the physical stability of amorphous drug–polymer systems [64]. The mechanism as suggested by those studies was a restriction on the molecular mobility of drug molecules and the enhancement of drug–polymer miscibility.

Although the importance of hydrogen bonding in stabilizing solid dispersions has been widely reported in the literature, aging under high humidity may disrupt such drug–polymer interactions and lead to instability. Polymers that contain carbonyl groups can be hygroscopic, such as PVP and PVPVA. Solid dispersions prepared using these polymers often absorb moisture when aged upon exposure to high humidity [65–67]. The moisture uptake can disrupt the hydrogen bonds formed between drugs and polymers since a water molecule is strong hydrogen bond acceptor and donor [67]. Accordingly, the occurrence of phase separation followed by drug recrystallization can often be observed in these systems [68].

5.5.5 Characterization Phase Separation in Amorphous Solid Dispersion

The physical instability of an amorphous solid dispersion eventually leads to the appearance of phase separation in the formulation. During phase separation, the dimensions, quantity, and physical nature of the separated phases will change over time. In terms of characterizing and quantifying a phase separated drug in an amorphous solid dispersion, the selection of detection method depends on the stage and level of phase separation [69]. Initially the separate phases are often low in number and of submicron size. This makes the detection of the phase separation difficult as it may approach or be below the detection limits of many conventional characterization tools. At the latter stages of phase separation, when a considerable amount of drug has crystallized out, the presence of crystalline drug may be easily detected using conventional characterization methods, including thermal analysis (DSC, MTDSC, hyperDSC), spectroscopic methods (FTIR, Raman, Terahertz spectroscopy, solid state NMR), and PXRD.

Moreover, the distribution of the phase separation may be an important issue in some particular cases where there is significant drug concentration variation across the formulation. This may affect dissolution and be an important consideration for monitoring batch variation. To characterize early stage phase separation and distribution of separated drug phases, localized imaging-based characterization methods can provide important complementary information. Imaging techniques combined with chemical identification is a growing stream of new characterization techniques. These include IR microscopy and imaging, MRI imaging, confocal Raman microscopy, and AFM-based techniques [70–75]. However, these techniques all have different spatial resolutions. Among these, Raman imaging and AFM have the potential to reach submicron resolution [70–75]. These novel approaches can detect phase separations which are not identified by conventional analytical methods [76, 77]. However, conventional AFM-based techniques have a much lower capability for chemical identification compared with spectroscopic imaging techniques, such as IR and Raman imaging [73–75]. For example, phase imaging conducted using tapping mode AFM and PFM-AFM can only provide phase identification based on the measurement of the physical properties of the separated phases; this in turn requires further supplementary characterization for confirmation of the actual chemical composition of the phases (being either drug or polymer or mix). Some new developments of functionalized AFM strengthen the physical property-based phase identification and render chemical phase identification possible. For example, recently developed local thermal analysis (LTA), transition temperature microscopy, and photothermal FTIR (PT-FTIR) microspectroscopy can provide localized characterization of a sample at micron to submicron scale with no additional sample preparation [76–79]. Figure 5.5 demonstrates the potential of the combined use of conventional AFM and phase identification AFM, which is transition temperature microscopy in this case to achieve clear characterization of material distribution in solid dispersions with submicron resolution [80].

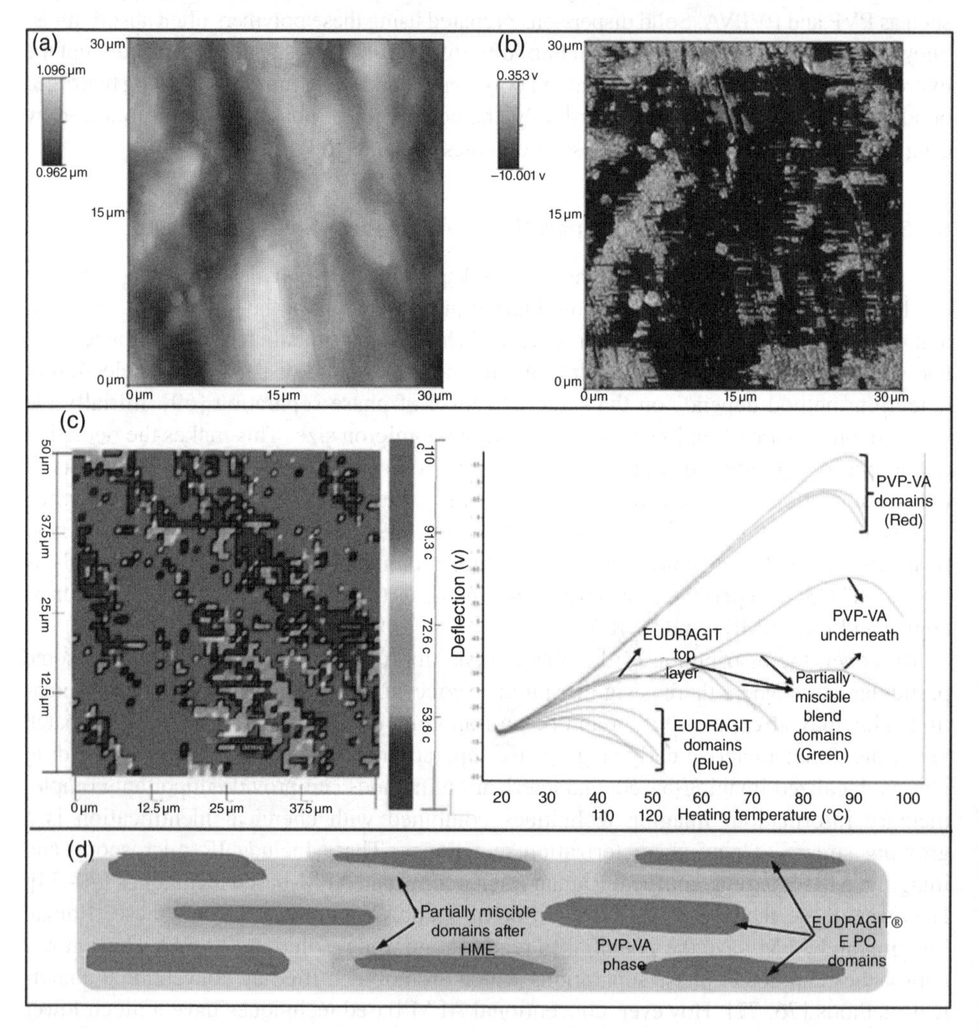

Figure 5.5 Using conventional AFM (a,b) and transition temperature microscopy (c–d) to understand the phase separation behavior of a complex solid dispersions. Reproduced with permission from Ref. [80]. Copyright 2013, American Chemical Society.

5.6 Future Prospects

Although half a century has passed, there only are few marketed products using solid dispersion strategy. A critical factor to maintaining the dissolution properties of a solid dispersion after storage is to prevent recrystallization of the amorphous drug. To develop a new theoretical model and methods of solid dispersions will be the key to the future success of solid dispersion formulations.

References

[1] Ke, P., Hasegawa, S., Al-Obaidi, H., Buckton G. Investigation of preparation methods on surface/bulk structural relaxation and glass fragility of amorphous solid dispersions. *International Journal of Pharmaceutics*, 2012. **422**(1/2): 170–178.

[2] Sekiguchi, K. and N. Obi, Studies on absorption of eutectic mixture. 1. A comparison of the behavior of eutectic mixture of sulfathiazole and that of ordinary sulfathiazole in man. *Chemical and Pharmaceutical Bulletin*, 1961. **9**(11): 866–872.

[3] Chiou, W.L. and Riegelman. S., Pharmaceutical applications of solid dispersion systems. *Journal of Pharmaceutical Sciences*, 1971. **60**(9): 1281–1302.

[4] Vippagunta, S.R., Z. Wang, S. Hornung, *et al.*, Factors affecting the formation of eutectic solid dispersions and their dissolution behavior. *Journal of Pharmaceutical Sciences*, 2007. **96**(2): 294–304.

[5] Janssens, S. and G. Van den Mooter, Review: physical chemistry of solid dispersions. *Journal of Pharmacy and Pharmacology*, 2009. **61**(12): 1571–1586.

[6] Vasconcelos, T., B. Sarmento, and P. Costa, Solid dispersions as strategy to improve oral bioavailability of poor water soluble drugs. *Drug Discovery Today*, 2007, **12**(23/24): 1068–1075.

[7] Kanig, J.L., Properties of fused mannitol in compressed tablets. *Journal of Pharmaceutical Sciences*, 1964. **53**: 188–192.

[8] Chiou, W.L. and S. Riegelman, Preparation and dissolution characteristics of several fast-release solid dispersions of griseofulvin. *Journal of Pharmaceutical Sciences*, 1969. **58**(12): 1505–1510.

[9] Dannenfelser, R.M., H. He, Y. Joshi, *et al.*, Development of clinical dosage forms for a poorly water soluble drug I: application of polyethylene glycol-polysorbate 80 solid dispersion carrier system. *Journal of Pharmaceutical Sciences*, 2004. **93**(5): 1165–1175.

[10] Vo, C.L.N., C. Park, and B.J. Lee, Current trends and future perspectives of solid dispersions containing poorly water-soluble drugs. *European Journal of Pharmaceutics and Biopharmaceutics*, 2013. **85**(3, Pt. B): 799–813.

[11] Serajuddin, A.T.M., Solid dispersion of poorly water-soluble drugs: early promises, subsequent problems, and recent breakthroughs. *Journal of Pharmaceutical Sciences*, 1999. **88**(10): 1058–1066.

[12] Patterson, J.E., M. B. James, A. H. Forster, *et al.*, Preparation of glass solutions of three poorly water soluble drugs by spray drying, melt extrusion and ball milling. *International Journal of Pharmaceutics*, 2007. **336**(1): 22–34.

[13] Karanth, H., V.S. Shenoy, and R.R. Murthy, Industrially feasible alternative approaches in the manufacture of solid dispersions: a technical report. *AAPS PharmSciTech*, 2006. **7**(4): 8.

[14] Claeys, B., Vervaeck, A., Hillewaere, X.K., Possemiers, S., De Beer, T., Remon, J.P., Vervaet, C, Thermoplastic polyurethanes for the manufacturing of highly dosed oral sustained release matrices via hot melt extrusion and injection molding. *European Journal of Pharmaceutics and Biopharmaceutics*. 2014. **90**: 44–52.

[15] Quinten, T., De Beer, T., Vervaet, C., Remon, J.P, Evaluation of injection moulding as a pharmaceutical technology to produce matrix tablets. *European Journal of Pharmaceutics and Biopharmaceutics*. 2009. **71**(1): 145–154.

[16] Maniruzzaman, M., Boateng, J.S., Snowden, M.J., Douroumis, D, A review of hot-melt extrusion: process technology to pharmaceutical products. *ISRN Pharmaceutics*. 2012. 436763.

[17] Won, D.H., Kim M.S., Lee S., *et al.*, Improved physicochemical characteristics of felodipine solid dispersion particles by supercritical anti-solvent precipitation process. *International Journal of Pharmaceutics*, 2005. **301**(1/2): 199–208.

[18] Willart, J.F. and M. Descamps, Solid state amorphization of pharmaceuticals. *Molecular Pharmaceutics*, 2008. **5**(6): 905–920.

[19] Gupta, M.K., A. Vanwert, and R.H. Bogner, Formation of physically stable amorphous drugs by milling with neusilin. *Journal of Pharmaceutical Sciences*, 2003. **92**(3): 536–551.

[20] Prudic, A., Y. Ji, and G. Sadowski, Thermodynamic phase behavior of API/polymer solid dispersions. *Molecular Pharmaceutics*, 2014. **11**(7): 2294–2304.

[21] Leuner, C. and J. Dressman, Improving drug solubility for oral delivery using solid dispersions. *European Journal of Pharmaceutics and Biopharmaceutics*, 2000. **50**(1): 47–60.

[22] Ouyang, D., Investigating the molecular structures of solid dispersions by the simulated annealing method. *Chemical Physics Letters*, 2012. **554**: 177–184.

[23] Leach, A.R., *Molecular Modelling: Principles and Applications*. 2nd edn. 2001: Prentice Hall.

[24] Langer, M., Höltje M., Urbanetz N.A., *et al.*, Investigations on the predictability of the formation of glassy solid solutions of drugs in sugar alcohols. *International Journal of Pharmaceutics*, 2003. **252**(1/2): 167–179.

[25] Yang, H., L. Ze-Sheng, H. Qian, *et al.*, Molecular dynamics simulation studies of binary blend miscibility of poly(3-hydroxybutyrate) and poly(ethylene oxide). *Polymer*, 2004. **45**(2): 453–457.

[26] Bogdanova, S., Pajeva, I. Nikolova, P., *et al.*, Interactions of poly(vinylpyrrolidone) with ibuprofen and naproxen: experimental and modeling studies. *Pharmaceutical Research*, 2005. **22**(5): 806–815.

[27] Gupta, P. and A.K. Bansal, Molecular interactions in celecoxib-PVP-meglumine amorphous system. *Journal of Pharmacy and Pharmacology*, 2005. **57**(3): 303–310.

[28] Gupta, P., Thilagavathi R., Chakraborti A.K., *et al.*, Role of molecular interaction in stability of celecoxib-PVP amorphous systems. *Molecular Pharmaceutics*, 2005. **2**(5): 384–391.

[29] Karavas, E., Georgarakis E., Sigalas M.P., *et al.*, Investigation of the release mechanism of a sparingly water-soluble drug from solid dispersions in hydrophilic carriers based on physical state of drug, particle size distribution and drug–polymer interactions. *European Journal of Pharmaceutics and Biopharmaceutics*, 2007. **66**(3): 334–347.

[30] Gashi, Z., Censi R., Malaj L., *et al.*, Differences in the interaction between aryl propionic acid derivatives and poly(vinylpyrrolidone) K30: a multi-methodological approach. *Journal of Pharmaceutical Sciences*, 2009. **98**(11): 4216–4228.

[31] Oliveira, B.G., Lima, M.C.A., Pitta, I.R., *et al.*, A theoretical study of red-shifting and blue-shifting hydrogen bonds occurring between imidazolidine derivatives and PEG/PVP polymers. *Journal of Molecular Modelling*, 2010. **16**(1): 119–127.

[32] Gupta, J., Nunes C., Vyas S., *et al.*, Prediction of solubility parameters and miscibility of pharmaceutical compounds by molecular dynamics simulations. *Journal of Physical Chemistry B*, 2011. **115**(9): 2014–2023.

[33] Khan, N. and D.Q.M. Craig, Role of blooming in determining the storage stability of lipid-based dosage forms. *Journal of Pharmaceutical Sciences*, 2004. **93**(12): 2962–2971.

[34] Prentis, R.A., Y. Lis, and S.R. Walker, Pharmaceutical innovation by the seven UK-owned pharmaceutical companies (1964–1985). *British Journal of Clinical Pharmacology*, 1988. **25**(3): 387–396.

[35] Crowley, M.M., S.K. Battu, S.B. Upadhye, *et al.*, Pharmaceutical applications of hot-melt extrusion: part I. *Drug Development and Industrial Pharmacy*, 2007. **33**(9): 909–926.

[36] Repka, M.A., Battu S.K., Upadhye S.B., *et al.*, Pharmaceutical applications of hot-melt extrusion: part II. *Drug Development and Industrial Pharmacy*, 2007. **33**(10): 1043–1057.

[37] Paudel, A., Worku Z.A., Meeus J., *et al.*, Manufacturing of solid dispersions of poorly water soluble drugs by spray drying: formulation and process considerations. *International Journal of Pharmaceutics*, 2012. **453**(1): 253–284.

[38] Bruce, C.D., Fegely K.A., Rajabi-Siahboomi A.R., *et al.*, The influence of heterogeneous nucleation on the surface crystallization of guaifenesin from melt extrudates containing EudragitÂ® L10055 or Acryl-EZEÂ®. *European Journal of Pharmaceutics and Biopharmaceutics*, 2010. **75**(1): 71–78.

[39] Duddu, S.P. and T.D. Sokoloski, Dielectric analysis in the characterization of amorphous pharmaceutical solids. 1. Molecular mobility in poly(vinylpyrrolidone)–Water systems in the glassy state. *Journal of Pharmaceutical Sciences*, 1995. **84**(6): 773–776.

[40] Zhou, D., Zhang G.G., Law D., *et al.*, Thermodynamics, molecular mobility and crystallization kinetics of amorphous griseofulvin. *Molecular Pharmaceutics*, 2008. **5**(6): 927–936.

[41] Tobyn, M., Brown J., Dennis A.B., *et al.*, Amorphous drug-PVP dispersions: application of theoretical, thermal and spectroscopic analytical techniques to the study of a molecule with intermolecular bonds in both the crystalline and pure amorphous state. *Journal of Pharmaceutical Sciences*, 2009. **98**(9): 3456–3468.
[42] Van den Mooter, G., Wuyts M., Blaton N., *et al.*, Physical stabilisation of amorphous ketoconazole in solid dispersions with polyvinylpyrrolidone K25. *European Journal of Pharmaceutical Sciences*, 2000. **12**(3): 261–269.
[43] Craig, D.Q.M., Barsnes M., Royall P.G., *et al.*, An evaluation of the use of modulated temperature DSC as a means of assessing the relaxation behaviour of amorphous lactose. *Pharmaceutical Research*, 2000. **17**(6): 696–700.
[44] Hancock, B.C. and G. Zografi, The relationship between the glass transition temperature and the water content of amorphous pharmaceutical solids. *Pharmaceutical Research*, 1994. **11**(4): 471–477.
[45] Khougaz, K. and S.D. Clas, Crystallization inhibiton in solid dispersions of MK-0591 and poly(vinylpyrrolidone) polymers. *Journal of Pharmaceutical Sciences*, 2000. **89**(10): 1325–1334.
[46] Shamblin, S.L., L.S. Taylor, and G. Zografi, Mixing behavior of colyophilized binary systems. *Journal of Pharmaceutical Sciences*, 1998. **87**(6): 694–701.
[47] Roos, Y.H., Frozen state transitions in relation to freeze drying. *Journal of Thermal Analysis*, 1997. **48**(3): 535–544.
[48] Hancock, B.C., S.L. Shamblin, and G. Zografi, Molecular mobility of amorphous pharmaceutical solids below their glass transition temperatures. *Pharmaceutical Research*, 1995. **12**(6): 799–806.
[49] Hodge, I.M., Effects of annealing and prior history on enthalpy relaxation in glassy polymers. 6. Adam–Gibbs formulation of nonlinearity. *Macromolecules*, 1987. **20**(11): 2897–2908.
[50] Aso, Y., S. Yoshioka, and S. Kojima, Molecular mobility-based estimation of the crystallization rates of amorphous nifedipine and phenobarbital in poly(vinylpyrrolidone) solid dispersions. *Journal of Pharmaceutical Sciences*, 2004. **93**(2): 384–391.
[51] Claudy, P., S. Jabrane, and J.M. Létoffé, Annealing of a glycerol glass: enthalpy, fictive temperature and glass transition temperature change with annealing parameters. *Thermochimica Acta*, 1997. **293**(1/2): 1–11.
[52] Badrinarayanan, P., W. Zheng, Q. Li, *et al.*, The glass transition temperature versus the fictive temperature. *Journal of Non-Crystalline Solids*, 2007. **353**(26): 2603–2612.
[53] Hancock, B.C. and G. Zografi, Characteristics and significance of the amorphous state in pharmaceutical systems. *Journal of Pharmaceutical Sciences*, 1997. **86**(1): 1–12.
[54] R. Bohmer, K.L. Ngai, C.A. Angell, D.J. Plazek, Nonexponential relaxations in strong and fragile glass formers. *Journal of Chemical Physics*, 1993. **99**(5): 4201–4209.
[55] Bhugra, C.,Shmeis R., Krill S.L., *et al.*, Predictions of onset of crystallization from experimental relaxation times I-correlation of molecular mobility from temperatures above the glass transition to temperatures below the glass transition. *Pharmaceutical Research*, 2006. **23**(10): 2277–2290.
[56] Hasegawa, S., P. Ke, and G. Buckton, Determination of the structural relaxation at the surface of amorphous solid dispersion using inverse gas chromatography. *Journal of Pharmaceutical Sciences*, 2009. **98**(6): 2133–2139.
[57] Qian, F., J. Huang, and M.A. Hussain, Drug–polymer solubility and miscibility: stability consideration and practical challenges in amorphous solid dispersion development. *Journal of Pharmaceutical Sciences*, 2010. **99**(7): p. 2941–2947.
[58] Konno, H. and L.S. Taylor, Ability of different polymers to inhibit the crystallization of amorphous felodipine in the presence of moisture. *Pharmaceutical Research*, 2008. **25**(4): 969–978.
[59] Ahlneck, C. and G. Zografi, The molecular basis of moisture effects on the physical and chemical stability of drugs in the solid state. *International Journal of Pharmaceutics*, 1990. **62**(2/3): 87–95.
[60] Wiranidchapong, C., Tucker I.G., Rades T., *et al.*, Miscibility and interactions between 17Î2-estradiol and EudragitÂ® RS in solid dispersion. *Journal of Pharmaceutical Sciences*, 2008. **97**(11): 4879–4888.
[61] Rumondor, A.C.F., Ivanisevic I., Bates S., *et al.*, Evaluation of drug–polymer miscibility in amorphous solid dispersion systems. *Pharmaceutical Research*, 2009. **26**(11): 2523–2534.
[62] Telang, C., S. Mujumdar, and M. Mathew, Improved physical stability of amorphous state through acid base interactions. *Journal of Pharmaceutical Sciences*, 2009. **98**(6): 2149–2159.

[63] Tang, X.C., M.J. Pikal, and L.S. Taylor, A spectroscopic investigation of hydrogen bond patterns in crystalline and amorphous phases in dihydropyridine calcium channel blockers. *Pharmaceutical Research*, 2002. **19**(4): 477–483.
[64] Huang, J., R.J. Wigent, and J.B. Schwartz, Drug–polymer interaction and its significance on the physical stability of nifedipine amorphous dispersion in microparticles of an ammonio methacrylate copolymer and ethylcellulose binary blend. *Journal of Pharmaceutical Sciences*, 2008. **97**(1): 251–262.
[65] Ng, Y.C., Yang Z., McAuley W.J., *et al.*, Stabilisation of amorphous drugs under high humidity using pharmaceutical thin films. *European Journal of Pharmaceutics and Biopharmaceutics*, 2013. **84**(3): 555–565.
[66] Konno, H. and L.S. Taylor, Influence of different polymers on the crystallization tendency of molecularly dispersed amorphous felodipine. *Journal of Pharmaceutical Sciences*, 2006. **95**(12): 2692–2705.
[67] Rumondor, A.C.F. and L.S. Taylor, Effect of polymer hygroscopicity on the phase behavior of amorphous solid dispersions in the presence of moisture. *Molecular Pharmaceutics*, 2010. **7**(2): 477–490.
[68] Qi, S., J.G. Moffat, and Z. Yang, Early stage phase separation in pharmaceutical solid dispersion thin films under high humidity: improved spatial understanding using probe-based thermal and spectroscopic nanocharacterization methods. *Molecular Pharmaceutics*, 2013. **10**(3): 918–930.
[69] Qi, S. and D.Q.M. Craig, Detection of phase separation in hot melt extruded solid dispersion formulations: global vs. localized characterization. *American Pharmaceutical Review*, 2010. **13**(6): 68–74.
[70] Zhang, L., M.J. Henson, and S.S. Sekulic, Multivariate data analysis for Raman imaging of a model pharmaceutical tablet. *Analytica Chimica Acta*, 2005. **545**(2): 262–278.
[71] Vajna, B., Farkas I., Szabó A., *et al.*, Raman microscopic evaluation of technology dependent structural differences in tablets containing imipramine model drug. *Journal of Pharmaceutical and Biomedical Analysis*, 2010. **51**(1): 30–38.
[72] Tajiri, T., Morita S., Sakamoto R., *et al.*, Release mechanisms of acetaminophen from polyethylene oxide/polyethylene glycol matrix tablets utilizing magnetic resonance imaging. *International Journal of Pharmaceutics*, 2010. **395**(1/2): 147–153.
[73] Weuts, I., Van Dycke F., Voorspoels J., *et al.*, Physicochemical properties of the amorphous drug, cast films, and spray dried powders to predict formulation probability of success for solid dispersions: etravirine. *Journal of Pharmaceutical Sciences*, 2011. **100**(1): 260–274.
[74] Zhang, J., Bunker M., Chen X., *et al.*, Nanoscale thermal analysis of pharmaceutical solid dispersions. *International Journal of Pharmaceutics*, 2009. **380**(1/2): 170–173.
[75] Qi, S., Gryczke A., Belton P., *et al.*, Characterisation of solid dispersions of paracetamol and EUDRAGITÂ® E prepared by hot-melt extrusion using thermal, microthermal and spectro] scopic analysis. *International Journal of Pharmaceutics*, 2008. **354**(1/2): 158–167.
[76] Qi, S., Belton P., Nollenberger K., *et al.*, Characterisation and prediction of phase separation in hot-melt extruded solid dispersions: a thermal, microscopic and NMR relaxometry study. *Pharmaceutical Research*, 2010. **27**(9): 1869–1883.
[77] Qi, S., Belton P., Nollenberger K., *et al.*, Compositional analysis of low quantities of phase separation in hot-melt-extruded solid dispersions: a combined atomic force microscopy, photothermal Fourier-transform infrared microspectroscopy, and localised thermal analysis approach. *Pharmaceutical Research*, 2011. **28**(9): 2311–2326.
[78] Harding, L., Qi S., Hill G., *et al.*, The development of microthermal analysis and photothermal microspectroscopy as novel approaches to drug-excipient compatibility studies. *International Journal of Pharmaceutics*, 2008. **354**(1/2): 149–157.
[79] Dai, X., Moffat J.G., Mayes A.G., *et al.*, Thermal probe based analytical microscopy: thermal analysis and photothermal Fourier-transform infrared microspectroscopy together with thermally assisted nanosampling coupled with capillary electrophoresis. *Analytical Chemistry*, 2009. **81**(16): 6612–6619.
[80] Yang, Z., Nollenberger, K., Albers, J., Craig, D., Qi, S., Microstructure of an immiscible polymer blend and its stabilization effect on amorphous solid dispersions. *Molecular Pharmaceutics*, 2013. **10**(7): 2767–2780.
[81] Yu, D.G., Zhu, L.-M., White, K., *et al.*, Electrospun nanofiber-based drug delivery systems. *Health*, 2009. **1**(2): 67–75.

6

Computer Simulations of Lipid Membranes and Liposomes for Drug Delivery

David William O'Neill[1], Sang Young Noh[1,2], and Rebecca Notman[1]

[1] Department of Chemistry and Centre for Scientific Computing, University of Warwick, UK
[2] Molecular Organization and Assembly in Cells Doctoral Training Centre, University of Warwick, UK

6.1 Introduction

In order for a drug to reach its target site of action it almost always has to cross one or more biological lipid membrane barriers, for example the lung, skin or gastrointestinal (GI) tract to reach the systemic circulation, and then generally the membrane of the target cell where it will execute its therapeutic mechanism of action. Both the partitioning of a drug into the membrane and its diffusion in the heterogeneous membrane environment will determine the rate at which it crosses the membrane and thus the bioavailability and distribution of the drug in the body. This in turn is dependent on the relative chemistry of the drug and the lipid and protein molecules that comprise the membrane. In addition, as the structure and function of membrane-bound proteins can be modulated by changes in the structural and mechanical properties of the host membrane, the membrane itself may be the therapeutic target. Additives may be included to make membranes more or less permeable to enhance or retard delivery and reduce toxicity. For drugs that are poorly soluble or that do not readily cross membranes, lipid-based nanocarriers (i.e. liposomes) can be used; it is desirable to be able to fine-tune both the encapsulation efficiency of a liposome and the rate and

Computational Pharmaceutics: Application of Molecular Modeling in Drug Delivery, First Edition.
Edited by Defang Ouyang and Sean C. Smith.

mechanism of release for the delivery of a particular drug. All things considered, it is clear that an understanding of lipid membrane biophysics and biochemistry plus the ability to rationalise, predict and control drug–membrane and additive–membrane interactions is integral to the design of new, optimised drug formulations.

Over the last 20 years, molecular simulation has played an increasingly important role in answering fundamental questions concerning lipid assemblies. There are two overarching approaches [1, 2]: (i) in *molecular dynamics* (MD) simulations, which are the focus of the work presented here, one uses the laws of classical mechanics to generate a trajectory of interacting particles in a system over time and (ii) in *Monte Carlo* (MC) simulations, an ensemble of microstates of the system is generated by making random perturbations to the system and the move is accepted or rejected depending on the Boltzmann-weighted probability of making the move. The framework of statistical thermodynamics is used to compute averages from the simulations that can be related to experimental measurements. Both methods yield a molecular level view of the system, akin to looking down a 'molecular microscope', that it is not possible to achieve by experiments. Thus molecular simulations offer a route to be able to rationalise and predict experimentally observed quantities and functions from molecular structure.

In this chapter we begin with a brief outline of some of the methodological considerations for lipid membrane simulations, including different representations of lipids and measurable properties. This is followed by reviews of simulations of different model membranes, small-molecule uptake and permeation across membranes and the relatively new discipline of NP-membrane interactions. We next consider simulations of chemical penetration enhancers, additive molecules that alter membrane permeability. Finally we identify some future challenges for simulations in this field.

6.2 Methodological Considerations

6.2.1 Representations of Model Lipids

There are three main levels at which a molecule may be represented in a molecular simulation: (i) fully atomistic (or all-atom), where each atom of the molecule is included in the simulation as an explicit interaction site, (ii) united atom, where all atoms except nonpolar hydrogen atoms are included explicitly in the simulation and nonpolar hydrogen atoms are included implicitly as part of so-called united atom carbon atoms and (iii) coarse-grained (CG), where groups of typically three to five atoms are treated as a single interaction site. The most commonly modelled lipids are phospholipids, in particular dipalmitoylphosphatidylcholine (DPPC). An example of DPPC in each of the three representations described above is presented in Figure 6.1.

The main packaged biomolecular force fields such as CHARMM [3–6], AMBER [7–9] and GROMOS [10–13] all contain a range of lipid parameters that are compatible with the corresponding parameters for other biomolecules (proteins, nucleic acids, etc.) and solvent molecules. The advantage of a fully atomistic or united atom representation is that full molecular detail is retained; however this comes at considerable computational expense and thus only time scales of the order of 100 ns are achievable.

In contrast, coarse-graining speeds up MD simulations, making longer time (up to microseconds) and length scales (10 nm) routinely accessible. This is due to the reduction in the total

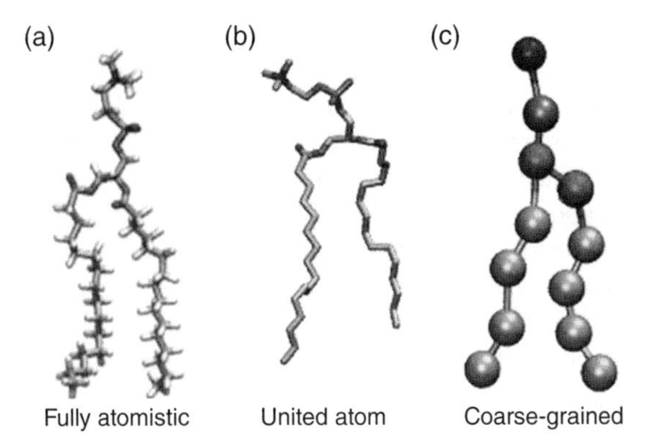

Figure 6.1 *Examples of the different representations of a DPPC lipid molecule in a molecular simulation. (a) Fully atomistic with 149 interaction sites. (b) United atom with 50 interaction sites. (c) CG with 12 interaction sites.*

number of particles, smoother potential energy landscape and, as the particles have a larger effective mass, one can use a larger MD time step. This of course comes with loss of molecular detail and thus coarse-graining is the most appropriate approach for problems where it is important to capture the essential physics of the system of interest rather than specific chemical details. There are a number of CG force fields for lipids (e.g. Refs. [14–19]), the most widely used of which is the MARTINI force field developed by Marrink and coworkers [18–20]. The MARTINI force field is distributed with parameters for a range of phospholipids, sphingolipids and cholesterol as well as water and other solvent molecules and ions.

An alternative approach is to take advantage of both worlds and adopt a dual resolution methodology that combines atomistic and CG representations of the system. The simplest way to do this involves running atomistic and CG simulations sequentially. For example, Brocos *et al.* used CG simulations to self assemble and equilibrate micelles from random mixtures of lysophospholipids of different chain length and then used a reverse-mapping procedure to switch back to atomistic resolution for the fine characterisation of structural and dynamic properties of the resulting aggregate [21]. Recently, there have been significant developments in hybrid multiscale methods where part of the system is represented at the atomistic level and the rest is considered at a lower resolution. For example, Orsi, Essex and coworkers have developed a hybrid scheme whereby a lipid membrane and the surrounding solvent are represented at a CG level and small permeants are represented atomistically [22]. This was achieved by developing mixing rules to handle the interactions between atomistic solutes and CG solvents. The authors successfully applied this approach to the permeation of small molecules [23] and the interactions of antimicrobials with lipid membranes [24].

6.2.2 Measurable Properties

6.2.2.1 Structure

The structure of a lipid bilayer may be characterised by a number of properties that can be computed from a molecular simulation trajectory. The area per lipid of a single component bilayer is most commonly determined by simply taking the average of the area of the simulation cell in the plane of the bilayer over time and dividing this number by the number of lipids per monolayer. Where a more detailed analysis of the distribution of different lipid components, additive molecules and free area is desired, approaches such as Voronoi tessellation or grid-based methods may be applied [25–27]. Bilayer density profiles (either mass or electron density) can be constructed by dividing the bilayer into slices along the normal direction and averaging the number of different system components (e.g. water, lipid headgroups, lipid tailgroups) in each slice. These density profiles can then be used to estimate quantities such as the bilayer thickness (e.g. from the distance between the two peaks in the headgroup distribution), the hydrophobic thickness or the water interfacial width. The alignment of the lipid tails can be determined by defining order parameters S_z per carbon atom C_n such as the following $S_z = \frac{3}{2}\langle \cos^2\theta_z \rangle - \frac{1}{2}$, where θ_z is the angle between the z-axis of the simulation box and the vector from C_{n-1} to C_{n+1} [28]. Here $S_z = 1$ represents perfect alignment of the tails with the bilayer normal, $S_z = 0$ indicates a random orientation and $S_z = -0.5$ demonstrates alignment of the tails in the plane of the bilayer.

It is also possible to probe the mechanical properties of lipid bilayers through simulation. The area compressibility modulus K_A, which is a measure of the energy needed to stretch or compress the bilayer per unit area can be computed from the mean square fluctuations in the membrane area over time or by performing a series of simulations at different fixed bilayer area and determining K_A from the slop of a plot of area versus surface tension [29]. The bending rigidity, which is the energy needed to bend the membrane away from its ideal curvature, can be determined by monitoring the height of the bilayer across its surface and analysing the undulatory modes [30–32]. The line tension, that is the free energy cost to form an edge of unit length, may be computed from simulations of a bilayer strip completely surrounded by solvent and analysis of the difference between the lateral and perpendicular pressures [33].

6.2.2.2 *Dynamics*

The mobility of lipids in the bilayer can be examined by computing the lateral diffusion coefficients D of different lipid components. This is typically achieved by determining the slope of the average mean squared displacement of lipid molecules (in the plane of the bilayer) at long times according to the Einstein relation. MD can also yield information about the dynamics of lipid molecules, such as rotational relaxation [34] and lipid flip flop between the leaflets of the bilayer [35].

6.2.2.3 *Molecule Permeation*

A key quantity used to characterise membrane transport is the membrane permeability P of the molecule of interest. Experimentally this is determined from membrane permeability assays that monitor the flux (amount of permeant passing through unit area per unit time), divided by the concentration difference of the permeant across the membrane. At the molecular level a lipid membrane is inhomogeneous as a function of depth along the bilayer normal direction (z) and the permeability of a molecule depends on the local partition coefficient $K(z)$ of the molecule in the membrane and the diffusion coefficient of the molecule at that depth $D(z)$. The permeability is expressed as an inverse of the resistance R:

$$R = \frac{1}{P} = \int_0^h \frac{dz}{K(z)D(z)} \tag{6.1}$$

where h is the barrier thickness. The partition coefficient, essentially a measure of the affinity of the molecule for the local environment, is related to the free energy profile of the molecules as a function of z, $G(z)$ or the potential of mean force (PMF), according to $K(z) = \exp(-\Delta G(z)/k_B T)$, where k_B is Boltzmann's constant and T is the temperature. In a molecular simulation, $G(z)$ and $D(z)$ can be obtained from a series of biased simulations where the molecule is inserted and simulated at discrete locations along the membrane normal (see Refs. [36–40] and references therein).

6.3 Model Membranes

6.3.1 Phospholipid Bilayers

Phospholipids are the dominant lipid species in cell membranes and are by far the most well simulated lipid membrane system. A wealth of simulation studies over the last 20 years have revealed the molecular basis for experimentally determined structural, dynamic and mechanical properties of phospholipid membranes and are reviewed extensively elsewhere [41–43]. Pertinent to our understanding of how molecules permeate membranes, and thus to drug delivery, molecular simulations have played an important role to elucidate the heterogeneous depth-dependent profile of phospholipid membranes [44–47] and how this is affected by factors such as phospholipid headgroup [48–51], chain length [48, 52, 53] and degree of unsaturation [6], the inclusion of cholesterol [54–58] and other additives (see below). Recent simulations have attempted to capture the behaviour of more complex membrane environments. For example, Javanainen *et al.* showed that lipid diffusion is hampered by several orders of magnitude in a membrane environment crowded by the inclusion of proteins [59].

6.3.2 Liposomes

Liposomal delivery systems may enhance drug solubility in formulations and allow for targeted delivery of drugs to their site of action. The development of a liposomal drug delivery system requires the optimisation of key factors such as encapsulation efficiency (how much drug can be loaded into the liposome) and controlled release of the drug (e.g. where is the drug released and how quickly). These factors are clearly dependent on the lipid composition of the liposome and the molecular level interactions between the drug molecule and the liposome.

Due to the large system size required to simulate a full liposome, a typical approach would be to infer results on liposomes from lipid bilayer simulations. For example, Magarkar *et al.* simulated a patch of PEGylated phospholipids in a bilayer surrounded by salt solutions as a representative model of a PEGylated liposome in the bloodstream [60].

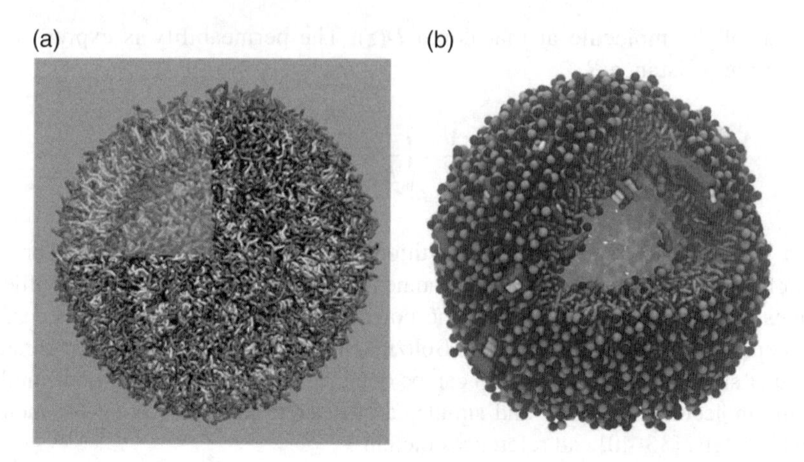

Figure 6.2 (a) Snapshot of an equilibrated DPPC–DLiPC liposome after 400 ns of simulation at 360 K. Reproduced from Ref. [61] with permission of The Royal Society of Chemistry. (b) Snapshot from a 10 μs MD simulations of a liposome loaded with hypericin molecules. Some lipids have been removed to illustrate the binding of hypericin; the hydrophilic parts of hypericin are shown in very dark grey, completely surrounding the hydrophobic parts (very pale grey). Reprinted with permission from Ref. [62]. Copyright 2013, American Chemical Society.

The drawback of this approach is that the effects of membrane curvature are not taken into account. Despite the computational expense, a limited number of full liposome simulations have been achieved. Risselada and Marrink used the MARTINI CG model (see Figure 6.2a) [63] and reduced the computational cost by applying special boundary conditions to simulate only the spherical shell of water molecules that surround the liposome explicitly (water molecules outside this region were treated using a mean field force representation [64]). Jämbeck *et al.* employed the same CG lipid model to investigate liposomal loading of the antitumour and antiviral drug hypericin [62]. It was shown that the hypericin molecules preferentially aggregate on the outer surface of the liposome (see Figure 6.2b), which may be due to the lower lipid density and curvature compared to the inner leaflet.

6.3.3 Skin–Lipid Membranes for Transdermal Drug Delivery

Delivery of drug molecules to or via the skin (transdermal delivery) can offer significant advantages over more conventional routes of delivery [65]. However, overcoming the natural barrier property of the stratum corneum (SC, the topmost layer of skin) remains a considerable challenge. The SC is often depicted as a 'bricks and mortar' arrangement with pancake-shaped skin cells (corneocytes) surrounded by a lipid matrix [66, 67]. The main barrier property of the skin is attributed to the SC lipids. These lipids form lamellar structures that are usually assumed to exist as bilayers or three-layer sandwich-type structures in a gel-like phase [68]. In actual fact, the molecular arrangement of the SC lipids remains a mystery. Therefore an understanding of the molecular structure and organisation of the SC skin barrier and the mechanisms by which molecules permeate the SC is essential for the development of new strategies for dermal and transdermal delivery.

The lipid mixture that forms lamellar layers surrounding the corneocytes is made up of three major components; ceramides, cholesterol and free fatty acids. Each of these components has specific properties that help give the SC its strong barrier properties but at the same time allow it to be soft and flexible. Out of the three components, it is known that the ceramide class of lipids plays a pivotal role in the barrier function of the skin [69]. One of the major properties that allows the SC to act as a strong barrier is the ceramides' apparent hydrogen bonding network between headgroups of neighbouring ceramides and other lamellar layers [70]. This network imparts structural integrity to the SC. The role of cholesterol is thought to be to provide a degree of fluidity to what would otherwise be a rigid structure [71]. Free fatty acids, in contrast, increase the density of the hydrocarbon chain packing [72], which suggests that they order the lipids.

Since we do not have a definitive model, experimental or computational, one has to find a way of developing a model acting as a skin lipid substitute that maintains the barrier properties characterised by the skin. To date there have been only a few attempts to model the skin lipid lamellae using molecular simulation. A selection of these model systems is presented in Figure 6.3. The simplest way of doing this is to use a ceramide bilayer, as ceramides are the dominant component of the lipid matrix. Pandit and Scott carried out the first MD simulation of a hydrated 16:0 ceramide lipid bilayer in the liquid crystalline phase [78]. They showed that there is increased hydrogen bonding between the ceramide carbonyl oxygen and the hydroxyl oxygen of a neighbouring ceramide compared to sphingomyelin, which highlights the role of the ceramide headgroup in forming a lateral hydrogen bonding network. Subsequent simulations by Notman *et al.*, using a united atom model, investigated ceramide 2 bilayers in the physiologically relevant gel phase [73]. In agreement with experiments, this study revealed that ceramide bilayers in the gel phase are characterised by close lipid packing, narrow interfacial width (almost no water penetrates the bilayer) and a lateral hydrogen bonding network between ceramide headgroups. These features are likely to play a major role in the barrier properties of the skin lipids and are thus putative targets for overcoming the skin barrier. The mismatch in the length of the hydrocarbon tails also gives rise to a region in the centre of the bilayer where the packing of the lipids is almost characteristic of the liquid–crystalline phase rather than the gel phase. More recently McCabe and coworkers parameterised an all-atom extended CHARMM force field for ceramides 2 and 3 [75]. Similar results to Notman *et al.* [73] are obtained for the structure of ceramide 2 bilayers at skin temperature, although the CHARMM-based model is able to better capture the phase transitions of the bilayers as a function of temperature [75].

Moving towards more realistic skin–lipid mixtures, Das *et al.* [74] showed the effects of both cholesterol and free fatty acids on ceramide bilayers by simulating various molar ratios of the standard three components. As part of a multicomponent system, the long asymmetric tails of the ceramides formed a dense bilayer phase, with major interdigitation of the tails. Addition of free fatty acids was found to increase the bilayer thickness and the ordering of the lipids, while cholesterol had the opposite effect due to the fact that it is smaller and more rigid than the ceramide molecules. The authors showed that the skin–lipid composition is optimal as it allows the lipids to withstand mechanical stresses while still keeping their barrier properties.

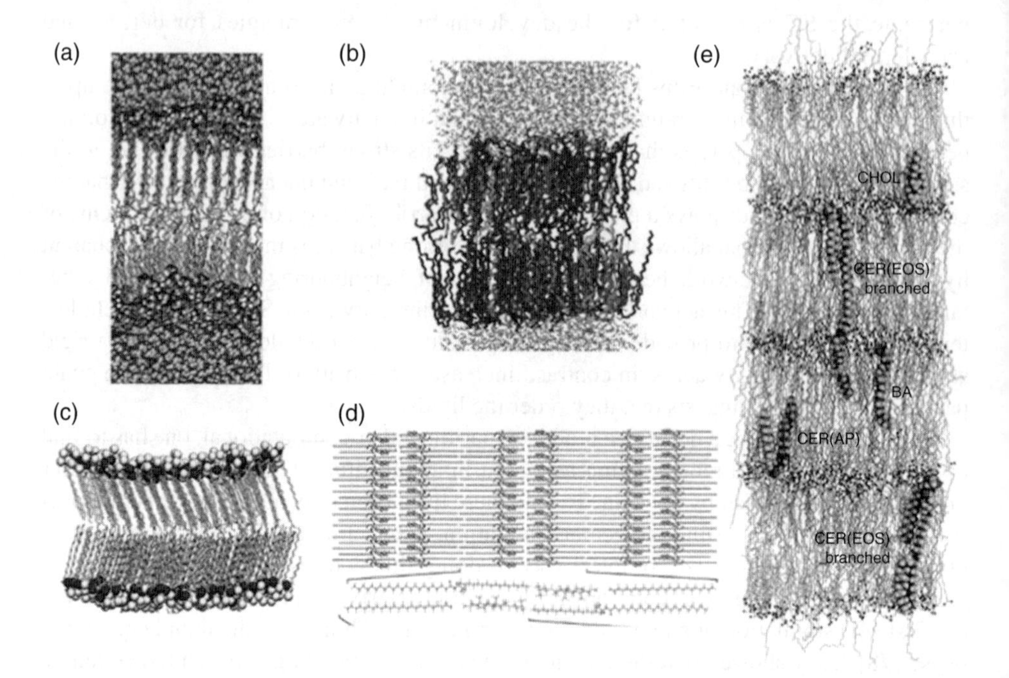

Figure 6.3 Representative snapshots of skin–lipid bilayer models from MD simulations. (a) United atom representation of a fully hydrated ceramide 2 bilayer with asymmetric lipid tails. Reprinted from Ref. [73], with permission from Elsevier. (b) Mixed ceramide 2: cholesterol: C24 fatty acid bilayer. From Ref. [74], with permission from Elsevier. (c) All-atom representation of a ceramide 2 bilayer with symmetric lipid tails. Reproduced with permission from Ref. [75]. Copyright 2013 American Chemical Society. (d) Anhydrous ceramide 2-cholesterol-C24 fatty acid multilayers with ceramides in the extended conformation. Reprinted by permission from Macmillan Publishers Ltd, Ref. [76], copyright 2012. (e) Mixed skin–lipid multilayers with low hydration including a long-chain ceramide that spans multiple layers. Reproduced with permission from Ref. [77], Royal Society of Chemistry.

6.4 Small Molecule Uptake and Permeation across Membranes

As discussed above, the measurable permeability of a molecule across a membrane represents the net effect of the affinity of the molecule to the lipid environment and the diffusion of the molecule over the entire heterogeneous membrane transport pathway. As well as direct permeation along the bilayer normal direction, the molecule may take a more tortuous route involving lateral diffusion pathways or transport through transient defects or pores [79, 80]. MD simulations of drugs or other small molecules as they traverse lipid bilayers, either spontaneously or via biased simulations are able to reveal the underlying mechanisms of drug partitioning and permeation and thus provide insights into how these mechanisms may be modified to achieve some desired outcome. The aim may be to maximise permeability, for example for drug delivery across a cell membrane, to encourage retention in the membrane, for example for liposomal delivery, or to shed light on toxicity, for example toxicity via direct topical interaction with the membranes that line the GI tract [81]. Clearly the molecular structure of the drug molecule plays the major role in membrane partitioning and permeability. A selection of the numerous MD simulations that have aimed to tackle this issue are discussed below.

In general small hydrophobic solutes are found to localise preferentially in the centre of phospholipid bilayers. For example the free energy profiles of O_2 [82], ethane [37], benzene [37] and hexane [83] show an *overall* decrease in free energy going from the bulk water phase to the centre of the bilayer (see Figure 6.4). Preferential partitioning into the hydrocarbon region of the bilayer is rationalised in terms of the classic hydrophobic effect and the free energy minimum corresponds to the centre of the bilayer where the tail particles are most disordered and less dense and the largest free volume pockets exist [84]. For ethane, benzene and hexane, a free energy barrier was observed in the headgroup-dense region of the bilayer. Interestingly, MacCallum and Tieleman computed the enthalpic and entropic contributions to the hexane free energy profile (Figure 6.4b) and showed that partitioning of hexane into the dense headgroup region was accompanied by a significant entropic cost and a counterbalancing enthalpic gain – the favourable enthalpy for partitioning was attributed to the tight packing in this region giving rise to more attractive van der Waals' interactions [83].

Direct interaction with the membrane may also be key to the mechanism of action of certain bioactive compounds. For example, Booker and Sum simulated the effect of the hydrophobic anaesthetic molecule xenon on the properties of dioleoylphosphatidylcholine (DOPC) bilayers [85]. At equilibrium, approximately 97% of the xenon atoms were located in the centre of the bilayer, which caused the bilayer to expand laterally and thicken. Interestingly xenon increased the order of the lipid tails in the centre of the bilayer, which was attributed to the increase in density in the region occupied by xenon. As a consequence of these structural changes, the lateral pressure profile of the bilayer was altered. It is well known that modulation in the lateral pressure profile can alter the structure and function of membrane proteins and thus modulation of the lateral pressure profile may be responsible for the anaesthetic mechanism of action of xenon.

In contrast, small hydrophilic solutes tend to show profiles with a free energy maximum in the hydrocarbon region. The profile of water is the most frequently studied (e.g. Refs. [23, 37, 86, 87] and Figure 6.4a). As water enters the bilayer there is an increase in the free energy associated with desolvation and entry into the dense membrane environment, this increases until the water is in the centre of the bilayer where there is a small decrease in the free energy as the lipid tails are less dense in this region. Other small hydrophilic solutes such as methanol, acetamide and methylamine [37] (Figure 6.4a) exhibit similar free energy profiles.

Amphiphilic solutes exhibit more complex position-dependent free energy profiles. For example, Tejwani *et al.* studied the effect of adding polar functional groups to aromatic drug-like molecules on their free energy profiles in DOPC bilayers [88]. The free energy minima correspond to the location of the lipid backbone, where the molecule orients such that hydrogen bonding between the solute and the lipid headgroups is maximised. However, it was also found that shielding the hydrophobic parts of the solutes from water (i.e. the hydrophobic effect) was an important driving force for binding to the interfacial region. Similar simulations of the antiviral drug amantadine and other adamantane derivatives also show a preferred binding domain in the headgroup region, with an orientational preference

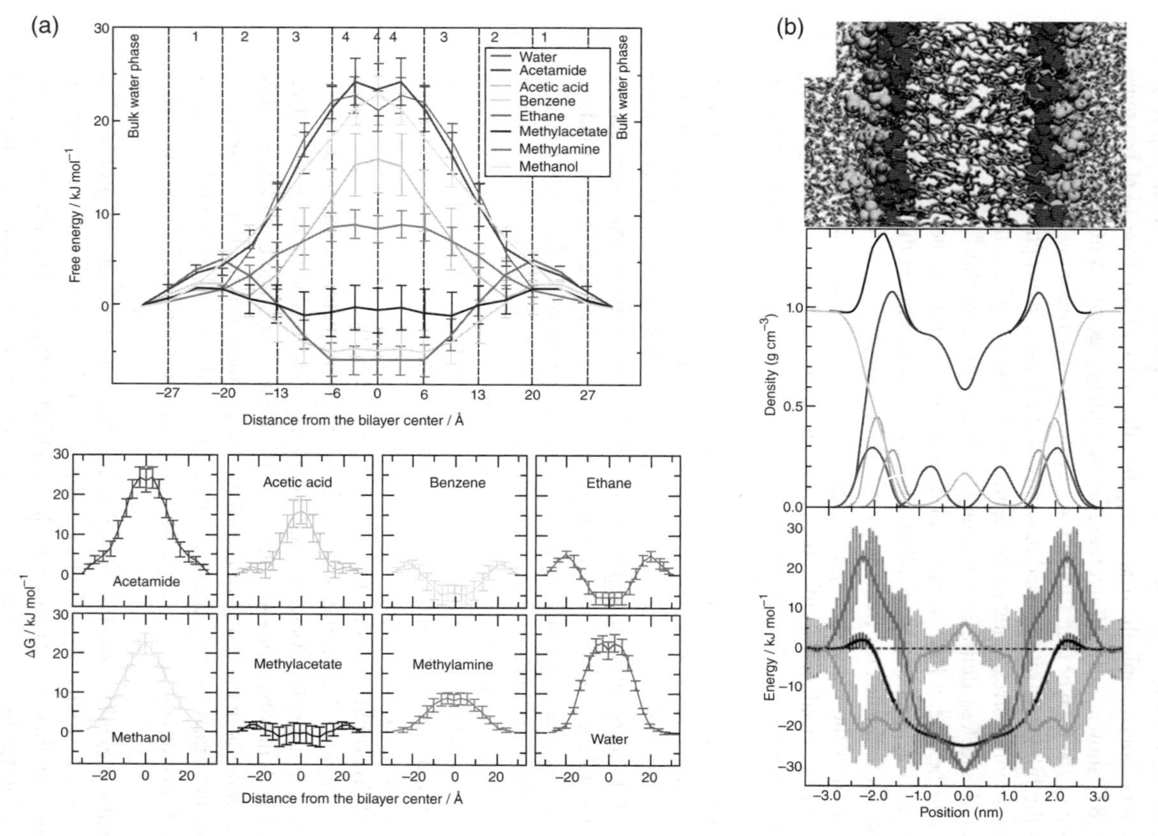

Figure 6.4 *(a) Free energy profiles of small molecules from constrained MD simulations. Top: the bilayer is divided into the four regions. Bottom: for clarity, each profile is plotted alone. Reprinted with permission from Ref. [37]. Copyright 2004 American Chemical Society. (b) Free energy of hexane in a DOPC bilayer. Top: snapshot of the lipid bilayer system used. Middle: average partial density profiles for various functional groups (black, total density; red, lipid; green, water; blue, choline; orange, phosphate; brown, glycerol; grey, carbonyl; purple, double bonds; cyan, methyl). Bottom: free energy of partitioning a hexane molecule from bulk water (black, free energy; red, entropic component of free energy, $-T\Delta S$; green, enthalpic component of free energy, ΔH). Reprinted with permission from Ref. [83]. Copyright 2005 American Chemical Society (see color plate section).*

that maximises favourable polar–polar and nonpolar–nonpolar interactions [89, 90]. MD simulations have been applied to a number of additional amphiphilic drug molecules interacting with membranes, including fluoroquinolones [91], benzocaine [92], chloropromazine [93] and others [94].

Boggara and Krishnamoorti studied the partitioning of the amphiphilic nonsteroidal anti-inflammatory drugs aspirin and ibuprofen into model DPPC membranes at different pH, that is considering both neutral and charged forms of the drugs [95]. It was found that both drugs preferentially partition into the bilayer however the neutral form is embedded deeper into the alkyl chains than the charged form. It was noteworthy that the charged forms of the drugs remain hydrated in the bilayer and thus transport water molecules into the membrane. This disrupts the adjacent lipid molecules and supports the idea of a transient pore mechanism for charged species, whereby the tilting of the lipids stabilises the entry of further water molecules and so on. The authors extended this work to consider the effect of multiple ibuprofen molecules on membrane transport [81]. The drug caused an overall thinning of the bilayer and, as reported previously, existed in a hydrated state – this local perturbation of the structure may be associated with NSAID-induced GI toxicity. Water intrusions were also observed in simulations of valproic acid, an anticonvulsant drug, traversing DPPC bilayers [96]. In particular, when the charged form of the drug was considered, the first hydration shell was retained all the way into the centre of the bilayer.

6.5 Nanoparticle–Membrane Interactions

Rapid advances in nanotechnology over the last 5–10 years have enabled the fabrication of multifunctional nanoparticles (NPs), which have the possibility to revolutionise the way medicines and other therapeutic drugs are administered to the body (see Refs. [97–99] for a selection of recent reviews). Nanotubes, nanoporous materials and hollow nanospheres have an innate ability to encapsulate and transport molecules. In addition the open ends of the pores serve as gates that can control the release of drugs [100–102]. An important feature is the ability to add multifunctionality, for example targeting moieties or imaging agents. Inorganic NPs also offer improved chemical stability, and therefore longer shelf life, greater choice of route of administration and lower susceptibility to biochemical attack inside the body. However to realise the potential of drug delivery using NPs, there needs to be an interdisciplinary understanding of the molecular level properties of these systems and the ways in which they interact with the components of biological systems.

As a NP approaches a cell membrane, a number of different transport processes may occur. For example, living cells can undergo a process known as endocytosis, where the membrane engulfs the NP and carries it across the membrane barrier [103]. This may be a specific (receptor-mediated) or nonspecific process. Alternatively, a NP might cross a membrane by passive diffusion across the bilayer or via membrane disruption [104]. Depending on the property of the NP, it can increase the toxicity in membranes by becoming entrapped inside the cell, eventually causing cell death. Alternatively, by modifying the cause of this encapsulation, it may be possible to design NPs that can bypass the membrane without causing permanent structural damage [105]. Hence, an understanding of how to optimise and control NP-membrane interactions (e.g. as a function of NP size, shape, surface chemistry and membrane composition) for specific applications is critical towards achieving the promises of NP-mediated drug delivery. To that end molecular simulation is starting to play an important role. Key challenges for simulations include the development of appropriate force field parameters that describe the interaction between the nanomaterial and the biological membrane and the large numbers of particles needed to model systems on the nanoscale.

A series of MD simulation studies have shown that hydrophobic NPs become embedded in the hydrophobic interior of the membrane, as one would expect due to the favourable interactions with the tailgroups of the lipids. The membrane partitioning of C_{60} fullerene, which has a diameter of ~0.7 nm, has been well studied by both atomistic and CG MD simulations [106–110]. The hydrophobic C_{60} molecule is seen to partition into the core of the bilayer where it causes minimal disruption to the membrane. This is consistent with a number of other simulations of small hydrophobic NPs with phospholipid bilayers, for example 3.0 nm diameter [111] and 1.3–2.8 nm diameter [112] generic hydrophobic NPs and 2.6–5.6 nm diameter hydrophobic polystyrene NPs [113]. Computed free energy profiles show that the most energetically stable position for the NP is in the centre of the membrane and that the free energy change of insertion becomes more negative (more favourable) with increasing particle size [106–110, 112, 113], which can largely be accounted for by simple hydrophobic burial [113]. Larger hydrophobic NPs, with diameter greater than the thickness of the bilayer, appear to be less readily accommodated by the bilayer, although the interaction with the bilayer core is still overall favourable. For example, Thake *et al.* showed that the membrane was required to bend around 7.4 nm diameter polystyrene NPs in order to maximise favourable contacts between the NP and the lipid hydrocarbon tails (see Figure 6.5) [113]. In addition Li *et al.* found slightly larger 10 nm hydrophobic NPs are not wrapped by the lipids, rather the NP plugs a hydrophobic pore in the bilayer, which causes a local thickening of the bilayer but does not compromise the membrane integrity [114]. The implication of this is that small hydrophobic NPs, with a diameter less than the bilayer thickness will become trapped in the interior of the membrane whereas hydrophobic NPs with a diameter greater than the thickness of the bilayer will have a higher probability of overcoming the barrier to exit and thus cross the membrane.

NPs functionalised with hydrophilic or charged moieties, in contrast, tend to exhibit a preference for the surface or interfacial regions of lipid membranes. D'Rozario *et al.* studied the interactions of hydroxylated C60 with a DPPC bilayer [106]. NPs with polar functional groups spread evenly over the surface of the NP intermittently adsorbed onto the surface of the membrane and the amount of time in the adsorbed state decreased with increasing polarity. NPs with an assymmetric distribution of polar groups, that is with amphiphilic character, partitioned beneath the lipid headgroups. Larger, 10 nm, semihydrophilic NPs were also shown to adsorb to the membrane surface and induced a local curvature of the surface [114]. Furthermore, simulations have suggested that both positive and negatively charged NPs absorb onto the surface of DPPC bilayers and distort the membrane surface [111, 115, 116]. The wrapping effect was most significant for NPs with a high negative surface charge density [115]. Ramalho *et al.* also suggested that negatively charged NPs decrease the fluid to gel phase transition temperature, however the underlying mechanism for this remains unclear [111]. Simulations have also been carried out for gold NPs functionalised with charged ligands interacting with neutral or negatively charged bilayers [117]. The cationic bilayer strongly adsorbed onto the surface of the negatively charged bilayer where it caused holes to form in the membrane.

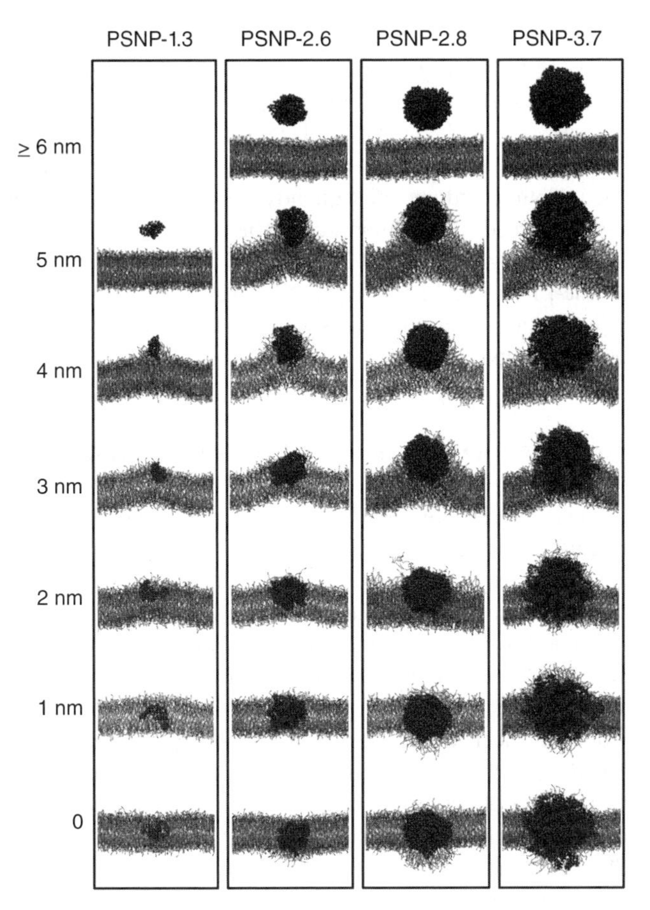

Figure 6.5 *Snapshots of different sized polystyrene NPs at a range of distances to the centre of a DPPC bilayer. For clarity, solvent particles are not shown. Reproduced from Ref. [113].*

The strength of NP binding to the lipid headgroups can control the particle-wrapping behaviour of the bilayer, which can lead to endocytosis-like uptake. For example, Vácha *et al.* included receptor molecules in the model membrane that have a strong affinity for the NP. Strong binding of the NP to the membrane was shown to induce particle uptake via an engulfing mechanism [118]. The size of the NP was also shown to play a role, with the uptake of larger (14.3 nm) NPs easier than for smaller (5.7 and 8.6 nm) – essentially there is a lower energetic cost to bend the membrane around a larger NP. While the most MD simulations to date have considered only a single NP interacting with a membrane, the pioneering work of Reynwar *et al.* demonstrated the effects of multiple 5 nm NPs that are attracted to the lipid headgroups – it was shown that the energetic penalty for bending the membrane provides a driving force for the NPs to aggregate and reduce the curvature free energy, which triggers a cooperative budding of NP-containing vesicles [119].

Finally we consider the effect of NP morphology on the mechanism of translocation through a bilayer. Carbon NPs, for example, may exist as spherical buckyballs, nanotubes or fragmented structures. Chang and Violi showed that flat fragment structures preferentially located in the dense hydrocarbon tail region, whereas spherical buckyballs resided in the centre of the bilayer [120]. Subsequent free energy calculations revealed the permeability coefficient *P* of curved and flat structures to be dependent on the curvature of the particle, with $P_{C60} > P_{open\text{-}C60} > P_{flat\ carbon\text{-}fragment}$ [121]. Wallace and Sansom investigated carbon nanotube translocation through a DPPC bilayer – it was observed that lipids bind to the interior and exterior of the nanotube and are subsequently extracted from the bilayer [122]. Lipid extraction was also observed in a similar study where the carbon nanotube contained an encapsulated paclitaxel drug molecule [123]. Nangia and Sureshkumar explored sphere-, pyramid-, rice-, cone-, rod- and cube-shaped NPs with a gold core and a charged shell [124]. NPs with flat surfaces were more easily internalised as they reorientated to maximise membrane-NP contacts.

6.6 Mechanisms of Action of Chemical Penetration Enhancers

Chemical penetration enhancers are chemicals that reversibly lower the barrier properties of the SC, allowing drug molecules to permeate through the skin with less resistance. Many different molecules have this property, such as water, alcohols, unsaturated fatty acids, sulfoxides, pyrrolidones, surfactants, polyols and terpenes [125, 126]. Despite this most of them are currently unsuitable for pharmaceutical use due to high toxicity levels [127]. There is also a lack of understanding at the molecular level of the mechanism of these chemicals; however molecular simulation offers an opportunity to address this and facilitate the rational design of penetration enhancer (or retarder) molecules for specific applications.

Dimethyl sulfoxide (DMSO) is an effective penetration enhancer of both hydrophobic and hydrophilic drugs. The effect of increasing concentration of DMSO on DPPC bilayers was investigated by Notman *et al.* using a CG model [128]. It was seen that the DMSO molecules partition into the bilayer just beneath the headgroups acting as spacer molecules between lipid headgroups. This causes many structural changes such as an expansion in lateral area thus leading to a reduction in the bilayer thickness. This spacer action also changes the mechanical nature of the membrane; the increase in headgroup area per lipid causes an increase in the volume that the lipid tails can expand into, thus reducing the tail density. These combined effects essentially make the membrane more affable to bending. At high concentrations, DMSO was seen to induce the formation of an hourglass-shaped water pore in the membrane. DMSO-induced water pores in the membrane would form transport routes for hydrophilic molecules in particular. These ideas were examined further by Gurtovenko and Anwar using atomistic scale MD simulations [129]. The authors investigated 14 different concentrations of DMSO to infer three distinct modes of action for the permeation enhancing effects of DMSO: (i) at low concentrations the DMSO molecules again act as spacer molecules, partitioning just below the headgroups of the lipids levering them apart, reducing the membrane thickness, allowing the tails to become more disordered and fluid, (ii) at medium concentrations water pores are formed, as was seen by Notman *et al.* [128] and (iii) a further increase in DMSO concentration leads to desorption of lipids from the membrane and thus leads to rupturing of the bilayer. From this information on the various modes of action of DMSO, one may now tune the DMSO concentration to a specific application, whether it be increasing permeability of hydrophilic molecules or acting as a cryoprotectant.

Moving beyond phospholipid bilayer models, Notman *et al.* investigated the concentration-dependent effects of DMSO on ceramide 2 bilayers [73]. Again it was seen that the DMSO molecules accumulate at the headgroups, where they weaken the lateral hydrogen bonding between the ceramides. At DMSO concentrations of 0.4 mol% (with respect to water) or higher the ceramide bilayers undergo a phase transition from the gel phase to the liquid crystalline phase. The liquid crystalline phase is characterised by highly disordered lipid tails and a decrease in bilayer thickness and is expected to be more permeable to solutes than the gel phase. The effect of DMSO on pore formation was investigated using constrained MD simulations to calculate the free energy of pore formation in both the gel phase and DMSO-induced fluidized state [130]. In the absence of DMSO, vapour-filled pores formed in the bilayer, where the pore was lined with the hydrophobic ceramide tails and no water entered the pore. This result further emphasises the barrier property of ceramides; even in the presence of small defects or pores, they still remain impermeable to solutes. In the presence of high concentrations of DMSO, the free energy barrier to pore formation was significantly reduced and hydrophilic pores were observed where the ceramide headgroups rearranged to shield the hydrocarbon tails from the water, in agreement with the predictions from CG simulations of DMSO with DPPC bilayers [128].

As mentioned above, small- to medium-chain alcohols such as ethanol are also known to be good penetration enhancers. An atomistic MD study on the effect of ethanol on 1-palmitoyl-2-oleoyl-phosphatidylethanolamine and 1-palmitoyl-2-oleoyl-phosphatidylcholine bilayers demonstrated that ethanol acts in a similar way to DMSO in that it partitions underneath the headgroups of the phospholipids at low concentrations [131]. But unlike DMSO these ethanol molecules do not just act as spacers but form hydrogen bonds with the headgroups. At higher concentrations, ethanol induced the formation of inverted micelles in the bilayer, containing small amounts of water trapped inside. It could be possible for these to act as delivery pockets, transporting polar molecules or ions from one leaflet to another within the membrane structure. No pores were induced, as was seen with high concentrations of DMSO.

A similar approach has been applied to the lipophilic penetration enhancer, oleic acid [132, 133]. One hypothesis is that oleic acid enhances permeability due to its kinked structure (due to the presence of an unsaturated bond), which disrupts the packing of the skin lipids. At physiological temperatures oleic acid is believed to exist in a separate phase within the SC [134, 135]. Therefore permeation through an oleic acid rich phase or boundary region may also be an important mechanism. CG simulations of oleic acid interacting with DPPC bilayers [136] suggest that oleic acid does not significantly affect the structure of phospholipid bilayers; however chemical potential calculations suggest that oleic acid does cause a small increase in the permeability of phospholipid bilayers to water. In terms of the skin lipids, the effect of oleic acid on a lipid bilayer containing a 1 : 1 : 1 mixture of ceramide 2, cholesterol and lignoceric acid in water was studied using atomistic MD at 300 and 340 K [137]. At the higher temperature, the diffusion of cholesterol was enhanced in the presence of oleic acid. It was suggested that small changes in the concentration of oleic acid could provide ways of altering the properties of the SC. For example, reductions in the density or thickness could allow hydrating water a greater freedom of movement between the multilamellar structure of the SC. Oleic acid does not seem to affect the hydrogen bonding within the bilayer structure, therefore any changes to the bilayer are believed to result from a change in the molecular configuration at the interface with water molecules.

6.7 Future Challenges

As illustrated above, the field of lipid membrane biophysics for drug delivery has benefitted from the recent considerable increase in computing power and methodological developments in MD simulations. We have reflected on a number of success stories where MD simulation has made useful predictions about the properties and behaviour of lipid systems and revealed previously undetermined mechanisms of action. Despite this, a number of key challenges remain for MD simulations in this discipline. To date the majority of simulations have been performed on single-component lipid bilayers or some binary or ternary systems. Clearly this is an oversimplification and future work should consider the development of more realistic model membrane environments, including different lipid compositions and taking into account the effect of crowding by proteins. When considering molecule or NP permeation through membranes, it is typical to assume that the route taken by a permeant is simply a direct pathway along the bilayer normal, however *lateral* diffusion and diffusion along defects or through transient pores are also likely to be important mechanisms. Features of the system that relax over long timescales, such as the orientation of the permeant, the curvature of the membrane and the formation of domains in the membrane may also need to be considered in a more rigorous way and will require further methodological developments. Finally, other important factors include the cooperative effects of molecules on membranes and the challenge of tackling multiscale problems, for example linking the molecular view of membrane permeation with pharmacokinetic models for predicting the absorption, distribution, metabolism and excretion of drugs in the body.

Acknowledgements

D.W.O. thanks the EPSRC for a PhD Studentship. R.N. acknowledges The Royal Society for a University Research Fellowship.

References

[1] Allen M.P., Tildesley D.J. (1987) *Computer Simulation of Liquids*. New York: Oxford University Press.

[2] Frenkel D., Smit B. (2002) *Understanding Molecular Simulation From Algorithms to Applications*. 2nd edn. San Diego, CA: Academic Press.

[3] MacKerell A.D., Bashford D., M. Bellott, Dunbrack R.L., Evanseck J.D., Field M.J., et al. (1998) All-atom empirical potential for molecular modeling and dynamics studies of proteins. Journal *of Physical Chemistry B* **102**(18):3586–3616.

[4] Mackerell A.D., Feig M., Brooks C.L. (2004) Extending the treatment of backbone energetics in protein force fields: limitations of gas-phase quantum mechanics in reproducing protein conformational distributions in molecular dynamics simulations. Journal of Computational Chemistry 25(11):1400–1415.

[5] Klauda J.B., Venable R.M., Freites J.A., O'Connor J.W., Tobias D.J., Mondragon-Ramirez C., et al. (2010) Update of the CHARMM all-atom additive force field for lipids: validation on six lipid types. Journal of Physical Chemistry B 114(23):7830–7843.

[6] Feller S.E., Yin D., Pastor R.W., MacKerell A.D. (1997) Molecular dynamics simulation of unsaturated lipid bilayers at low hydration: parameterization and comparison with diffraction studies. Biophys*ical Journal* **73**(5):2269–2279.

[7] Duan Y., Wu C., Chowdhury S., Lee M.C., Xiong G., Zhang W., et al. (2003) A point-charge force field for molecular mechanics simulations of proteins based on condensed-phase quantum mechanical calculations. Journal of Computational Chemistry 24(16):1999–2012.
[8] Yang L., Tan C.-H., Hsieh M.-J., Wang J., Duan Y., Cieplak P., et al. (2006) New-generation amber united-atom force field. *Journal of Physical Chemistry B* **110**(26):13166–13176.
[9] Hsieh M.-J., Luo R. (2010) Balancing simulation accuracy and efficiency with the amber united atom force field. Journal of Physical Chemistry B 114(8):2886–2893.
[10] Daura X., Mark A.E., van Gunsteren W.F. (1998) Parametrization of aliphatic CHn united atoms of GROMOS96 force field. Journal of Computational Chemistry 19(5):535–547.
[11] van Gunsteren W.F., Billeter S.R., Eising A.A., Hünenberger P.H., Krüger P., Mark A.E., et al. (1996) *Biomolecular Simulation: The GROMOS96 Manual and User Guide*. Zürich: ETH Zürich.
[12] Oostenbrink C., Villa A., Mark A.E., Van Gunsteren W.F. (2004) A biomolecular force field based on the free enthalpy of hydration and solvation: the GROMOS force-field parameter sets 53A5 and 53A6. *Journal of Computational Chemistry* **25**:1656–1676.
[13] Schmid N., Eichenberger A., Choutko A., Riniker S., Winger M., Mark A., *et al.* (2011) Definition and testing of the GROMOS force-field versions 54A7 and 54B7. *European Biophysics Journal* **40**(7):843–856.
[14] Hadley K.R., McCabe C. (2012) A simulation study of the self-assembly of coarse-grained skin lipids. Soft Matter 8(17):4802–4814.
[15] Izvekov S., Voth G. A. (2006) Multiscale coarse-graining of mixed phospholipid/cholesterol bilayers. Journal of Chemical Theory and Computation 2(3):637–648.
[16] Orsi M., Haubertin D.Y., Sanderson W.E., Essex J.W. (2008) A quantitative coarse-grain model for lipid bilayers. Journal of Physical Chemistry B 112(3):802–815.
[17] Shelley J.C., Shelley M.Y., Reeder R.C., Bandyopadhyay S., Klein M.L. (2001) A coarse grain model for phospholipid simulations. Journal of Physical Chemistry B 105(19):4464–4470.
[18] Marrink S.J., de Vries A.H., Mark A.E. (2004) Coarse grained model for semiquantitative lipid simulations. Journal of Physical Chemistry B 108(2):750–760.
[19] Marrink S.J., Risselada H.J., Yefimov S., Tieleman D.P., de Vries A.H. (2007) The MARTINI force field: coarse grained model for biomolecular simulations. Journal of Physical Chemistry B 111(27):7812–7824.
[20] Marrink S.J., Tieleman D.P. (2013) Perspective on the Martini model. Chemical Society Reviews 42(16):6801–6822.
[21] Brocos P., Mendoza-Espinosa P., Castillo R., Mas-Oliva J., Pineiro A. (2012) Multiscale molecular dynamics simulations of micelles: coarse-grain for self-assembly and atomic resolution for finer details. Soft Matter 8(34):9005–9014.
[22] Michel J., Orsi M., Essex J.W. (2008) Prediction of partition coefficients by multiscale hybrid atomic-level/coarse-grain simulations. Journal of Physical Chemistry B 112(3):657–660.
[23] Orsi M., Sanderson W.E., Essex J.W. (2009) Permeability of small molecules through a lipid bilayer: a multiscale simulation study. Journal of Physical Chemistry B 113(35):12019–12029.
[24] Orsi M., Noro M.G., Essex J.W. (2011) Dual-resolution molecular dynamics simulation of antimicrobials in biomembranes. Journal of the Royal Society Interface 8(59):826–841.
[25] Falck E., Patra M., Karttunen M., Hyvonen M.T., Vattulainen I. (2004) Lessons of slicing membranes: interplay of packing, free area, and lateral diffusion in phospholipid/cholesterol bilayers. Biophysical Journal 87(2):1076–1091.
[26] Lopez C.F., Nielsen S.O., Ensing B., Moore P.B., Klein M.L. (2005) Structure and dynamics of model pore insertion into a membrane. Biophysical Journal 88(5):3083–3094.
[27] Allen W.J., Lemkul J.A., Bevan D.R. (2009) GridMAT-MD: a grid-based membrane analysis tool for use with molecular dynamics. Journal of Computational Chemistry 30(12):1952–1958.
[28] van der Spoel, D., Lindahl, E., Hess, B. and the GROMACS Development Team (2013) GROMACS User Manual Version 4.6.4.
[29] Feller S.E., Pastor R.W. (1999) Constant surface tension simulations of lipid bilayers: the sensitivity of surface areas and compressibilities. Journal of Chemical Physics 111(3):1281–1287.
[30] Goetz R., Gompper G., Lipowsky R. (1999) Mobility and elasticity of self-assembled membranes. Physical Review Letters 82(1):221–224.

[31] Marrink S.J., Mark A.E. (2001) Effect of undulations on surface tension in simulated bilayers. *Journal of Physical Chemistry B* **105**(26):6122–6127.
[32] Lindahl E., Edholm O. (2000) Mesoscopic undulations and thickness fluctuations in lipid bilayers from molecular dynamics simulations. Biophysical Journal 79(1):426–433.
[33] Tolpekina T.V., den Otter W.K., Briels W.J. (2004) Simulations of stable pores in membranes: system size dependence and line tension. Journal of Chemical Physics 121(16):8014–8020.
[34] Moore P.B., Lopez C.F., Klein M.L. (2001) Dynamical properties of a hydrated lipid bilayer from a multinanosecond molecular dynamics simulation. Biophysical Journal 81(5):2484–2494.
[35] Gurtovenko A.A., Onike O.I., Anwar J. (2008) Chemically induced phospholipid translocation across biological membranes. Langmuir 24(17):9656–9660.
[36] Berendsen H.J.C., Marrink S.-J. (1993) Molecular dynamics of water transport through membranes: water from solvent to solute. Pure and Applied Chemistry 65(12):2513–2520.
[37] Bemporad D., Essex J.W., Luttmann C. (2004) Permeation of small molecules through a lipid bilayer: a computer simulation study. Journal of Physical Chemistry B 108(15):4875–4884.
[38] Notman R., Anwar J. (2013) Breaching the skin barrier — Insights from molecular simulation of model membranes. Advanced Drug Delivery Reviews 65(2):237–250.
[39] Trzesniak D., Kunz A.P.E., van Gunsteren W.F. (2007) A comparison of methods to compute the potential of mean force. Chemphyschem 8(1):162–169.
[40] Xiang T.X., Anderson B.D. (1994) The relationship between permeant size and permeability in lipid bilayer membranes. Journal of Membrane Biology 140(2):111–122.
[41] Scott H.L. (2002) Modeling the lipid component of membranes. Current Opinion in Structural Biology 12(4):495–502.
[42] Lyubartsev A.P., Rabinovich A.L. (2011) Recent development in computer simulations of lipid bilayers. Soft Matter 7(1):25–39.
[43] Marrink S.J., de Vries A.H., Tieleman D.P. (2009) Lipids on the move: simulations of membrane pores, domains, stalks and curves. Biochimica et Biophysica Acta, Biomembranes 1788(1):149–168.
[44] Venable R.M., Zhang Y., Hardy B.J., Pastor R.W. (1993) Molecular dynamics simulations of a lipid bilayer and of hexadecane: an investigation of membrane fluidity. Science 262:223–226.
[45] Feller S.E., Venable R.M., Pastor R.W. (1997) Computer simulation of a DPPC phospholipid bilayer: structural changes as a function of molecular surface area. Langmuir 13(24):6555–6561.
[46] Egberts E., Marrink S.J., Berendsen H.J.C. (1994) Molecular dynamics simulation of a phospholipid membrane. European Biophysics Journal 22(6):423–436.
[47] Berger O., Edholm O., Jahnig F. (1997) Molecular dynamics simulations of a fluid bilayer of dipalmitoylphosphatidylcholine at full hydration, constant pressure, and constant temperature. Biophysical Journal 72(5):2002–2013.
[48] Lopez C.F., Nielsen S.O., Klein M.L. (2004) Hydrogen bonding structure and dynamics of water at the dimyristoylphosphatidylcholine lipid bilayer surface from a molecular dyanmics simulation. Journal of Physical Chemistry B 108(21):6603–6610.
[49] Pitman M.C., Suits F. (2005) Molecular dynamics investigation of dynamical properties of phosphatidylethanolamine lipid bilayers. Journal of Chemical Physics 122(24):244715.
[50] Leekumjorn S., Sum A.K. (2007) Molecular studies of the gel to liquid-crystalline phase transition for fully hydrated DPPC and DPPE bilayers. Biochimica et Biophysica Acta (BBA): Biomembranes 1768(2):354–365.
[51] Pan J., Cheng X., Monticelli L., Heberle F.A., Kucerka N., Tieleman D.P., et al. (2014) The molecular structure of a phosphatidylserine bilayer determined by scattering and molecular dynamics simulations. Soft Matter 10(21):3716–3725.
[52] Chiu S.W., Jakobsson E., Subramaniam S., Scott H.L. (1999) Combined Monte Carlo and molecular dynamics simulation of fully hydrated dioleyl and palmitoyl-oleyl phosphatidylcholine lipid bilayers. Biophysical Journal 77(5):2462–2469.
[53] Sugii T., Takagi S., Matsumoto Y. (2005) A molecular-dynamics study of lipid bilayers: effects of the hydrocarbon chain length on permeability. Journal of Chemical Physics 123(18):184714.

[54] Hofsäß C., Lindahl E., Edholm O. (2003) Molecular dynamics simulations of phospholipid bilayers with cholesterol. Biophysical Journal 84(4):2192–2206.
[55] Tu K., Klein M.L., Tobias D.J. (1998) Constant-pressure molecular dynamics investigation of cholesterol effects in a dipalmitoylphosphatidylcholine bilayer. Biophysical Journal 75(5):2147–2156.
[56] Smondyrev A.M., Berkowitz M.L. (1999) Structure of dipalmitoylphosphatidylcholine/cholsterol bilayer at low and high cholesterol concentrations: molecular dynamics simulation. Biophysical Journal 77(4):2075–2089.
[57] Falck E., Patra M., Karttunen M., Hyvonen M.T., Vattulainen I. (2004) Impact of cholesterol on voids in phospholipid membranes. Journal of Chemical Physics 121(24):12676–12689.
[58] Bennett W.F.D., MacCallum J.L., Hinner M.J., Marrink S.J., Tieleman D.P. (2009) Molecular view of cholesterol flip-flop and chemical potential in different membrane environments. Journal of the American Chemical Society 131(35):12714–12720.
[59] Javanainen M., Hammaren H., Monticelli L., Jeon J.-H., Miettinen M.S., Martinez-Seara H., et al. (2013) Anomalous and normal diffusion of proteins and lipids in crowded lipid membranes. Faraday Discussions 161:397–417.
[60] Magarkar A., Karakas E., Stepniewski M., Róg T., Bunker A. (2012) Molecular dynamics simulation of PEGylated bilayer interacting with salt ions: a model of the liposome surface in the bloodstream. Journal of Physical Chemistry B 116(14):4212–4219.
[61] Risselada H.J., Marrink S.J. (2009) Curvature effects on lipid packing and dynamics in liposomes revealed by coarse grained molecular dynamics simulations. Physical Chemistry Chemical Physics 11(12):2056–2067.
[62] Jämbeck J.P.M., Eriksson E.S.E., Laaksonen A., Lyubartsev A.P., Eriksson L.A. (2013) Molecular dynamics studies of liposomes as carriers for photosensitizing drugs: development, validation, and simulations with a coarse-grained model. Journal of Chemical Theory and Computation 10(1):5–13.
[63] Risselada H.J., Marrink S.J. (2008) The molecular face of lipid rafts in model membranes. Proceedings of the National Academy of Sciences of the United States of America 105(45):17367–17372.
[64] Risselada H.J., Mark A.E., Marrink S.J. (2008) Application of mean field boundary potentials in simulations of lipid vesicles. Journal of Physical Chemistry B 112(25):7438–7447.
[65] Williams A.C. (2003) Transdermal and Topical Drug Delivery. Cornwall: Pharmaceutical Press.
[66] Elias P.M. (1983) Epidermal lipids, barrier function, and desquamation. Journal of Investigative Dermatology 80(Suppl. 6):44–49.
[67] Michaels A.S., Chandrasekaran S.K., Shaw J.E. (1975) Drug permeation through human skin: theory and in vitro experimental measurement. AIChE Journal 21(5):985–996.
[68] Bouwstra J.A., Dubbelaar F.E.R., Gooris G.S., Ponec M. (2000) The lipid organisation in the skin barrier. Acta Dermato-Venereologica 80(Suppl. 208):23–30.
[69] Bouwstra J.A., Dubbelaar F.E.R., Gooris G.S., Weerheim A.M., Ponec M. (1999) The role of ceramide composition in the lipid organisation of the skin barrier. Biochimica et Biophysica Acta, Biomembranes 1419(2):127–136.
[70] Moore D.J., Rerek M.E. (2000) Insights into the molecular organization of lipids in the skin barrier from infrared spectroscopy studies of stratum corneum lipid models. Acta Dermato-Venereologica 80(Suppl. 208):16–22.
[71] Wegener M., Neubert R., Rettig W., Wartewig S. (1997) Structure of stratum corneum lipids characterized by FT-Raman spectroscopy and DSC. III. Mixtures of ceramides and cholesterol. Chemistry and Physics of Lipids. 88(1):73–82.
[72] Bouwstra J.A., Gooris G.S., Dubbelaar F.E.R., Weerheim A., Ponec M. (1998) pH, cholesterol sulfate and fatty acids affect stratum corneum lipid organisation. Journal of Investigative Dermatology Symposium Proceedings 3(2):69–74.
[73] Notman R., den Otter W.K., Noro M.G., Briels W.J., Anwar J. (2007) The permeability enhancing mechanism of DMSO in ceramide bilayers simulated by molecular dynamics. Biophysical Journal 93:2056–2068.
[74] Das C., Noro M.G., Olmsted P.D. (2009) Simulation studies of stratum corneum lipid mixtures. Biophysical Journal 97(7):1941–1951.

[75] Guo S., Moore T.C., Iacovella C.R., Strickland L.A., McCabe C. (2013) Simulation study of the structure and phase behavior of ceramide bilayers and the role of lipid headgroup chemistry. *Journal of Chemical Theory and Computation* **9**(11):5116–5126.

[76] Iwai I., Han H., Hollander L.d., Svensson S., Ofverstedt L.-G., Anwar J., et al. (2012) The human skin barrier is organized as stacked bilayers of fully extended ceramides with cholesterol molecules associated with the ceramide sphingoid moiety. Journal of Investigative Dermatology 132(9):2215–2225.

[77] Engelbrecht T., Hau, Su K., Vogel A., Roark M., Feller S.E., et al. (2011) Characterisation of a new ceramide EOS species: synthesis and investigation of the thermotropic phase behaviour and influence on the bilayer architecture of stratum corneum lipid model membranes. Soft Matter 7(19):8998–9011.

[78] Pandit S.A., Scott H.L. (2006) Molecular-dynamics simulation of a ceramide bilayer. Journal of Chemical Physics 124(1):014708.

[79] Paula S., Volkov A.G., Deamer D.W. (1998) Permeation of halide anions through phospholipid bilayers occurs by the solubility-diffusion mechanism. Biophysical Journal 74(1):319–327.

[80] Bordi F., Cametti C., Naglieri A. (1998) Ionic transport in lipid bilayer membranes. Biophysical Journal 74(3):1358–1370.

[81] Boggara M.B., Mihailescu M., Krishnamoorti R. (2012) Structural association of nonsteroidal anti-inflammatory drugs with lipid membranes. Journal of the American Chemical Society 134(48):19669–19676.

[82] Marrink S.J., Berendsen H.J.C. (1996) Permeation process of small molecules across lipid membranes studied by molecular dynamics simulations. Journal of Physical Chemistry 100(41):16729–16738.

[83] MacCallum J.L., Tieleman D.P. (2005) Computer simulation of the distribution of hexane in a lipid bilayer: spatially resolved free energy, entropy, and enthalpy profiles. Journal of the American Chemical Society 128:125–130.

[84] Marrink S.J., Sok R.M., Berendsen H.J.C. (1996) Free volume properties of a simulated lipid membrane. Journal of Chemical Physics 104(22):9090–9099.

[85] Booker R.D., Sum A.K. (2013) Biophysical changes induced by xenon on phospholipid bilayers. Biochimica et Biophysica Acta (BBA): Biomembranes 1828(5):1347–1356.

[86] Marrink S.J., Berendsen H.J.C. (1994) Simulation of water transport through a lipid membrane. Journal of Physical Chemistry 98(15):4155–4168.

[87] Shinoda W., Mikami M., Baba T., Hato M. (2004) Molecular dynamics study on the effects of chain branching on the physical properties of lipid bilayers: 2. Permeability. Journal of Physical Chemistry B 108(26):9346–9356.

[88] Tejwani R.W., Davis M.E., Anderson B.D., Stouch T.R. (2011) An atomic and molecular view of the depth dependence of the free energies of solute transfer from water into lipid bilayers. Molecular Pharmaceutics 8(6):2204–2215.

[89] Li C., Yi M., Hu J., Zhou H.-X., Cross T.A. (2008) Solid-state NMR and MD simulations of the antiviral drug amantadine solubilized in DMPC bilayers. Biophysical Journal 94(4):1295–1302.

[90] Chew C.F., Guy A., Biggin P.C. (2008) Distribution and dynamics of adamantanes in a lipid bilayer. Biophysical Journal 95(12):5627–5636.

[91] Cramariuc O., Rog T., Javanainen M., Monticelli L., Polishchuk A.V., Vattulainen I. (2012) Mechanism for translocation of fluoroquinolones across lipid membranes. Biochimica et Biophysica Acta (BBA): Biomembranes 1818(11):2563–2571.

[92] López Cascales J.J., Oliveira Costa S.D. (2013) Effect of the interfacial tension and ionic strength on the thermodynamic barrier associated to the benzocaine insertion into a cell membrane. Biophysical Chemistry 172:1–7.

[93] Pickholz M., Oliveira Jr O.N., Skaf M.S. (2007) Interactions of chlorpromazine with phospholipid monolayers: effects of the ionization state of the drug. Biophysical Chemistry 125(2/3):425–434.

[94] Kopeć W., Telenius J., Khandelia H. (2013) Molecular dynamics simulations of the interactions of medicinal plant extracts and drugs with lipid bilayer membranes. FEBS Journal 280(12):2785–2805.

[95] Boggara M.B., Krishnamoorti R. (2010) Partitioning of nonsteroidal antiinflammatory drugs in lipid membranes: a molecular dynamics simulation study. Biophysical Journal 98(4):586–595.

[96] Ulander J., Haymet A.D.J. (2003) Permeation across hydrated DPPC lipid bilayers: simulation of the titrable amphiphilic drug valproic acid. Biophysical Journal 85(6):3475–3484.
[97] Bao G., Mitragotri S., Tong S. (2013) Multifunctional nanoparticles for drug delivery and molecular imaging. Annual Review of Biomedical Engineering 15(1):253–282.
[98] Rosenholm J.M., Sahlgren C., Linden M. (2010) Towards multifunctional, targeted drug delivery systems using mesoporous silica nanoparticles – opportunities and challenges. Nanoscale 2(10):1870–1883.
[99] Caruso F., Hyeon T., Rotello V.M. (2012) Nanomedicine. Chemical Society Reviews 41(7):2537–2538.
[100] Son S.J., Bai X., Lee S. (2007) Inorganic hollow nanoparticles and nanotubes in nanomedicine. Part 2: imaging, diagnostic, and therapeutic applications. Drug Discovery Today 12(15/16):657–663.
[101] Kam N.W.S., O'Connell M., Wisdom J.A., Dai H.J. (2005) Carbon nanotubes as multifunctional biological transporters and near-infrared agents for selective cancer cell destruction. Proceedings of the National Academy of Sciences of the United States of America 102(33):11600–11605.
[102] Lai C.Y., Trewyn B.G., Jeftinija D.M., Jeftinija K., Xu S., Jeftinija S., et al. (2003) A mesoporous silica nanosphere-based carrier system with chemically removable CdS nanoparticle caps for stimuli-responsive controlled release of neurotransmitters and drug molecules. Journal of the American Chemical Society 125(15):4451–4459.
[103] Sahay G., Alakhova D.Y., Kabanov A.V. (2010) Endocytosis of nanomedicines. Journal of Controlled Release 145(3):182–195.
[104] Smith P.J., Giroud M., Wiggins H.L., Gower F., Thorley J.A., Stolpe B., et al. (2012) Cellular entry of nanoparticles via serum sensitive clathrin-mediated endocytosis, and plasma membrane permeabilization. International Journal of Nanomedicine 7:2045–2055.
[105] Verma A., Uzun O., Hu Y.H., Hu Y., Han H.S., Watson N., et al. (2008) Surface-structure-regulated cell-membrane penetration by monolayer-protected nanoparticles. Nature Materials 7(7):588–595.
[106] D'Rozario R.S.G., Wee C.L., Wallace E.J., Sansom M.S.P. (2009) The interaction of C-60 and its derivatives with a lipid bilayer via molecular dynamics simulations. Nanotechnology 20(11):115102.
[107] Bedrov D., Smith G.D., Davande H., Li L.W. (2008) Passive transport of C-60 fullerenes through a lipid membrane: a molecular dynamics simulation study. Journal of Physical Chemistry B 112(7):2078–2084.
[108] Qiao R., Roberts A.P., Mount A.S., Klaine S.J., Ke P.C. (2007) Translocation of C-60 and its derivatives across a lipid bilayer. Nano Letters 7(3):614–619.
[109] Wong-Ekkabut J., Baoukina S., Triampo W., Tang I.M., Tieleman D.P., Monticelli L. (2008) Computer simulation study of fullerene translocation through lipid membranes. Nature Nanotechnology 3(6):363–368.
[110] Chiu C.-C., Shinoda W., DeVane R.H., Nielsen S.O. (2012) Effects of spherical fullerene nanoparticles on a dipalmitoyl phosphatidylcholine lipid monolayer: a coarse grain molecular dynamics approach. Soft Matter 8(37):9610–9616.
[111] Ramalho J.P.P., Gkeka P., Sarkisov L. (2011) Structure and phase transformations of DPPC lipid bilayers in the presence of nanoparticles: insights from coarse-grained molecular dynamics simulations. Langmuir 27(7):3723–3730.
[112] Lin X.B., Li Y., Gu N. (2010) Nanoparticle's size effect on its translocation across a lipid bilayer: a molecular dynamics simulation. Journal of Computational and Theoretical Nanoscience 7(1):269–276.
[113] Thake T.H.F., Webb J.R., Nash A., Rappoport J.Z., Notman R. (2013) Permeation of polystyrene nanoparticles across model lipid bilayer membranes. Soft Matter 9(43):10265–10274.
[114] Li Y., Chen X., Gu N. (2008) Computational investigation of interaction between nanoparticles and membranes: hydrophobic/hydrophilic effect. Journal of Physical Chemistry B 112(51):16647–16653.
[115] Li Y., Gu N. (2010) Thermodynamics of charged nanoparticle adsorption on charge-neutral membranes: a simulation study. Journal of Physical Chemistry B 114(8):2749–2754.

[116] da Rocha E.L., Caramori G.F., Rambo C.R. (2013) Nanoparticle translocation through a lipid bilayer tuned by surface chemistry. Physical Chemistry Chemical Physics 15(7):2282–2290.
[117] Lin J., Zhang H., Chen Z., Zheng Y. (2010) Penetration of lipid membranes by gold nanoparticles: insights into cellular uptake, cytotoxicity, and their relationship. ACS Nano 4(9):5421–5429.
[118] Vácha R., Martinez-Veracoechea F.J., Frenkel D. (2011) Receptor-mediated endocytosis of nanoparticles of various shapes. Nano Letters 11(12):5391–5395.
[119] Reynwar B.J., Illya G., Harmandaris V.A., Muller M.M., Kremer K., Deserno M. (2007) Aggregation and vesiculation of membrane proteins by curvature-mediated interactions. Nature 447(7143):461–464.
[120] Chang R., Violi A. (2006) Insights into the effect of combustion-generated carbon nanoparticles on biological membranes: a computer simulation study. Journal of Physical Chemistry B 110(10):5073–5083.
[121] Fiedler S.L., Violi A. (2010) Simulation of nanoparticle permeation through a lipid membrane. Biophysical Journal 99(1):144–152.
[122] Wallace E.J., Sansom M.S.P. (2008) Blocking of carbon nanotube based nanoinjectors by lipids: a simulation study. Nano Letters 8(9):2751–2756.
[123] Mousavi S., Amjad-Iranagh S., Nademi Y., Modarress H. (2013) Carbon nanotube-encapsulated drug penetration through the cell membrane: an investigation based on steered molecular dynamics simulation. Journal of Membrane Biology 246(9):697–704.
[124] Nangia S., Sureshkumar R. (2012) Effects of nanoparticle charge and shape anisotropy on translocation through cell membranes. Langmuir 28(51):17666–17671.
[125] Williams A.C., Barry B.W. (2004) Penetration enhancers. Advanced Drug Delivery Reviews 56(5):603–618.
[126] Barry B.W. (1987) Mode of action of penetration enhancers in human skin. Journal of Controlled Release 6(1):85–97.
[127] Karande P., Jain A., Mitragotri S. (2004) Discovery of transdermal penetration enhancers by high-throughput screening. Nature Biotechnology 22(2):192–197.
[128] Notman R., Noro M., O'Malley B., Anwar J. (2006) Molecular basis for dimethylsulfoxide (DMSO) action on lipid membranes. Journal of the American Chemical Society 128(43):13982–13983.
[129] Gurtovenko A.A., Anwar J. (2007) Ion transport through chemically induced pores in protein-free phospholipid membranes. Journal of Physical Chemistry B 111(47):13379–13382.
[130] Notman R., Anwar J., Briels W.J., Noro M.G., den Otter W.K. (2008) Simulations of skin barrier function: free energies of hydrophobic and hydrophilic transmembrane pores in ceramide bilayers. Biophysical Journal 95:4763–4771.
[131] Gurtovenko A.A., Anwar J. (2009) Interaction of ethanol with biological membranes: the formation of non-bilayer structures within the membrane interior and their significance. Journal of Physical Chemistry B 113(7):1983–1992.
[132] Cooper E.R. (1984) Increased skin permeability for lipophilic molecules. Journal of Pharmaceutical Sciences 73(8):1153–1156.
[133] Barry B.W., Bennett S.L. (1987) Effect of penetration enhancers on the permeation of mannitol, hyrocortisone and progesterone through human skin. Journal of Pharmacy and Pharmacology 39(7):535–546.
[134] Inoue T., Yanagihara S., Misono Y., Suzuki M. (2001) Effect of fatty acids on phase behaviour of hydrated dipalmitoylphosphatidylcholine bilayer: Saturated versus unsaturated fatty acids. Chemistry and Physics of Lipids 109(2):117–133.
[135] Busquets M.A., Mestres C., Alsina M.A., Anton J.M.G., Reig F. (1994) Miscibility of dipalmitoylphosphatidylcholine, oleic acid and cholesterol measured by DSC and compression isotherms of monolayers. Thermochimica Acta 232(2):261–269.
[136] Notman R., Noro M.G., Anwar J. (2007) Interaction of oleic acid with dipalmitoylphosphatidylcholine (DPPC) bilayers simulated by molecular dynamics. Journal of Physical Chemistry B 111:12748–12755.
[137] Hoopes M.I., Noro M., Longo M.L., Faller R. (2011) Skin lipids of the stratum corneum in the presence of oleic acid. Journal of Physical Chemistry B 115(12), 3164–3171.

7

Molecular Modeling for Protein Aggregation and Formulation

Dorota Roberts[1], Jim Warwicker[2], and Robin Curtis[1]

[1] *School of Chemical Engineering and Analytical Sciences, Manchester Institute of Biotechnology, University of Manchester, UK*

[2] *Faculty of Life Sciences, Manchester Institute of Biotechnology, University of Manchester, UK*

7.1 Introduction

Protein therapeutics, with their ability to deliver high affinity binding to defined targets, are of increasing importance in pharmacological intervention. The number of unmodified and modified proteins approved for clinical use by regulatory authorities of the United States and European Union runs into the hundreds, with many more in the pipeline, and monoclonal antibodies (mAbs) accounting for about one half of the sales revenue recorded in 2010 [1]. Muromonab-CD3 (trade name Orthoclone OKT3), an immune suppressant targeting the T cell receptor–CD3 complex, was the first monoclonal antibody approved for use as a therapeutic, in 1986 by the Food and Drug Administration (FDA) of the United States [2]. With their market share and potential for development as fragmented or multivalent molecules, it is convenient to focus on antibodies in this Introduction, but it should be remembered that the field of protein therapeutics is growing generally, not just with antibodies.

Despite the rapid growth of interest in protein therapeutics (also termed biologics), bringing the molecule to the market presents a challenge, as it does for small molecule therapeutics. Figure 7.1 outlines this process. First, there has to be a therapeutic requirement and a target identified, typically from a combination of academic research, industrial

Computational Pharmaceutics: Application of Molecular Modeling in Drug Delivery, First Edition.
Edited by Defang Ouyang and Sean C. Smith.

research, and clinical input. Increasingly these stages are informed by synthesis of genomic, transcriptomic, and proteomic datasets from human populations. Once a candidate protein therapeutic has been identified, relatively early in the process the amino acid sequence is locked in, so that there are no changes during the lengthy clinical trials stages (Figure 7.1). This provides a major challenge for handling solubility and aggregation problems that may become evident only later in development. Antibodies themselves are probably one of the better behaved classes of biologics in this respect, and generally problems can be circumvented with appropriate formulation. This may be because antibodies have evolved to circulate at relatively high concentrations *in vivo.* Other biologic platforms though may prove more troublesome, including antibody derivatives. For example, where naturally occurring proteins, such as antibodies, are engineered into individual domains (or domain combinations), previously unexposed surfaces will now need to be compatible with solvent and resistant to self-association or partial unfolding. In these cases there is a need to have an improved understanding of modeling for protein resistance to aggregation included earlier in the early stages of the design process. Where the emphasis for delivering a required

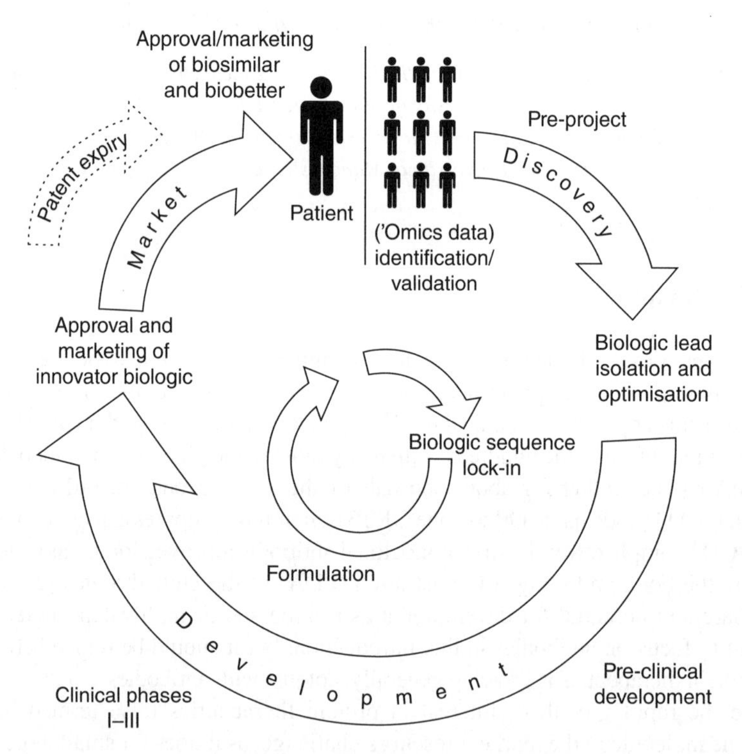

Figure 7.1 *The drug discovery, development, and marketing pipeline for protein therapeutics. Circular overall structure emphasizes that drug development starts at target identification that is increasingly being derived from population differences and 'omics data and that drug delivery occurs back at members within the population. A major issue, illustrated on the inner circle, is that biological (amino acid) sequence is fixed relatively early in the process. As a consequence, emphasis can be placed on formulation to stabilize the preparation of biologic for storage and distribution, and to render it sufficiently soluble for delivery at high concentration.*

solubility and aggregation resistance profile lies in formulation, there is also scope for a better understanding of how different formulation ingredients exert their influences. An additional factor is to appreciate what biologic–formulation combinations could lead to an undesired immunological response to treatment in a patient.

Returning specifically to mAbs, there have been many developments since the early days of the field [3], including humanization of nonhuman antibodies, and the use of display methods such as phage display to rapidly generate high affinity antibodies to a given antigen [4]. Modifications result in a lowering of immunogenicity [5]. In 2002, Adalimumab (trade name Humira), the first fully humanized mAb-based therapeutic, was approved for use in the treatment of rheumatoid arthritis, followed by approval for other disease treatments. Future developments will see sophisticated therapeutics developed from antibodies, including simultaneous targeting of two or more disease-linked molecules [6].

A Y-shaped antibody with modular architecture is shown in Figure 7.2, delineating the domains within the two heavy and two light chains. Not shown in this figure are the disulfide bonds that connect chains, or the glycosylated sites with covalently attached carbohydrate chains. Glycosylation can be important in the design of antibody [9] and other protein therapeutics, but is not considered further in this review. An antibody can be split into Fab and Fc fragments. Variable domains from light and heavy chains combine to form the antigen binding site in each Fab fragment. An example of the engineered constructs possible is a single chain variable fragment (scFv), formed by inserting a linker between the light and heavy chain variable domains, and removed from the remaining parts of the heavy and light chains.

During the development phase, the solubility and aggregation properties can only be optimized by controlling the solution environment of the protein, as the sequence will be locked in (see Figure 7.1). It is therefore crucial that we improve our understanding of how proteins behave in aqueous liquid formulations, which can contain a range of cosolvents including different types of buffers, salts, or other types of additives or excipients used to help stabilize the formulation. Some small molecule additives have specific abilities for the suppression of protein aggregation by means of weak interactions with the protein solvent accessible surface, such as arginine [10], although arginine is more an exception than a rule since most additives stabilize proteins against aggregation by stabilizing against unfolding through a preferential exclusion mechanism. Additionally, arginine can interact with the denatured state, which is probably linked to its ability to prevent aggregation. Other buffers and salt ions, used for controlling pH and ionic strength, may increase or decrease propensity of protein aggregation [11–15]. Experimentally, formulation cosolvents and their optimal concentration are screened with short-term stability trials in various storage conditions, packaging, and devices for drug administration. These short-term trials are often carried out at elevated temperatures, so as to accelerate the aggregation processes. However, many aggregation processes exhibit a nonmonotonic temperature dependence making it difficult to correlate with low temperature storage conditions [16]. The complexity of formulation development, correlation between the stability of bioformulation *in vitro* and *in vivo* and shelf life are major areas where predictability needs to be improved by integrating computational and experimental methods. Such improvement of protein aggregation prediction requires first determining how formulation cosolvents interact with proteins and their effects on weak protein–protein interactions.

This chapter proceeds to overview protein aggregation pathways in the next section, followed by discussion of protein interactions with cosolvents, and then protein–protein

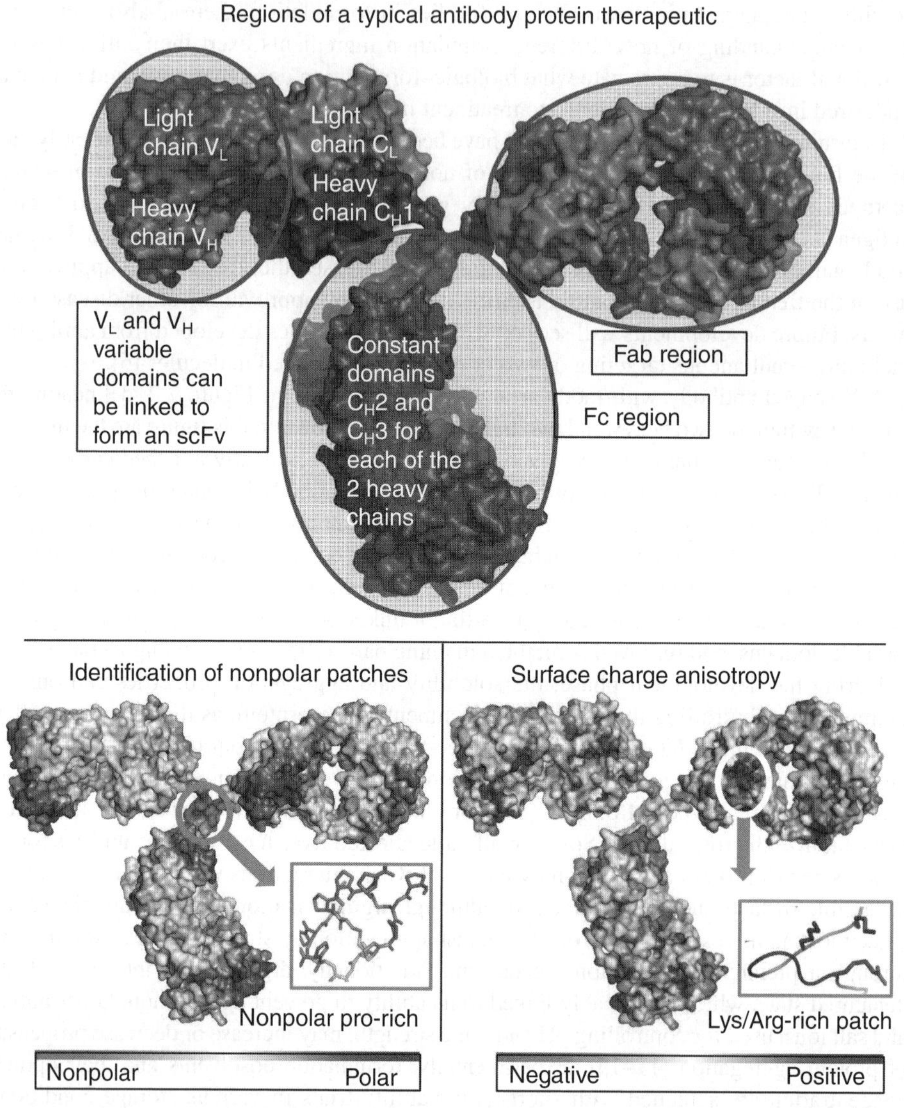

Figure 7.2 *Antibody and substituent structure, with graphical representation of non-polar and charged patches. IgG structure (code 1igt on the protein structural database) [7] is color-coded and labeled to show constituent domains, including, for example, the antigen-interacting V_L and V_H domains that can be linked to make a single chain variable fragment (scFv). Both of the lower panels use red–white–blue spectra to denote surface properties calculated with the same methods that give patch sizes [8]. The left-hand lower panel shows polarity, from most nonpolar (red) to most polar (blue), with a hydrophobic pro-rich sequence underlying the highlighted red patch. On the lower right-hand side is color-coding according to charge, from negative (red) to positive (blue), together with a lysine/arginine-rich sequence coincident with a particularly positive patch (see color plate section).*

interactions. After this, bioinformatics methods for analysis of protein solubility and aggregation are reviewed. The aim at all points is to discuss modeling, from detailed simulation methods to simple informatics, in the context of current experimental knowledge.

7.2 Protein Aggregation Pathways in Liquid Formulations

7.2.1 Multiple Pathways Can Lead to Protein Aggregation

At this point, it is important to describe what is meant by the phrase protein aggregation. In the context of bioprocessing, aggregation defines a protein association pathway, which is usually made irreversible due to conformational changes in the protein structure and formation of strong noncovalent inter- and intramolecular interactions. Several reviews cover this area [17–21]. Protein aggregation occurs by a complicated set of pathways with rates that are highly dependent on the protein and the solution conditions. The aggregation precursors are often partially folded proteins that associate to form a critically sized nucleus, which may or may not involve conformational rearrangements. It is important to note though that the solubility of aggregation-prone therapeutic molecules is not always controlled by conformational stability [22, 23], in which case protein surface properties become the key determinants of aggregation. Aggregate growth can occur by a variety of mechanisms, including monomer addition, by aggregate–aggregate condensation, further conformational changes, and the formation of insoluble precipitates. The aggregated states can only be broken up in these instances by adding a sufficient amount of denaturant to break both the intra- and intermolecular interactions holding the aggregate together. Association between native proteins is generally weaker and can lead to phase separation processes, such as the formation of a second liquid phase concentrated in the protein, amorphous precipitation, or crystallization. These processes are inherently reversible in cases where the protein is maintained in the native state. Then, redissolving the precipitated phase requires breaking the weak intermolecular interactions formed between protein surface groups, which can be brought about by subtle changes to the solution pH, ionic strength, or changing the salt or buffer type. It is clear though that consideration of native state and native state surface properties can be important factors, alongside conformational stability, in assessing solubility and aggregation properties of therapeutic candidates [22].

In the context of liquid protein formulations, protein aggregation is controlled by optimizing the solution conditions by judicious choice of pH, ionic strength, buffer and salt type, and the presence of excipients such as sugars, amino acids, and polyols. An especially challenging problem is to predict the shelf life of the formulation, which often requires using accelerated studies where aggregation is monitored under accelerated conditions, such as at higher temperatures or under agitation. For small globular proteins, often the rate-limiting step is the formation of the partially folded proteins, which occur at a very low relative population compared to the native state. Protein folding stability is thus a key solution property to quantify, which is usually done using melting temperature studies, either by temperature scanning calorimetry or fluorimetry [19, 24]. In addition, nonspecific protein–protein interactions (often referred to as colloidal stability) are commonly quantified in terms of the osmotic second virial coefficient obtained by static light scattering, self interaction chromatography, or by dynamic light scattering [25–28]. A strong correlation between colloidal stability and aggregation propensity is found at low ionic strength; for solutions removed from the pI of the protein, aggregation rates are reduced due to repulsive double layer forces [23, 29, 30], whereas near to the pI, aggregation is enhanced due to attractive electrostatic forces [31]. The effect of buffer and salt type on aggregation has also been rationalized in terms of how these alter the colloidal stability [13–15].

7.2.2 Overview of Cosolvent Effects on Protein–Protein Interactions

Practically speaking, predicting cosolvent effects on protein–protein interactions and unfolding requires using an additivity approximation, whereby the protein–cosolvent interactions can be decomposed into a sum of individual contributions arising from the different protein groups [32, 33]. This permits extrapolating studies of cosolvent effects on small model peptides or peptide similar compounds, for which interactions with polar, nonpolar, and charged protein groups can be delineated from each other [34–38]. Protein–solvent interactions are sufficiently weak that defining a dissociation constant is not possible as there is no well defined difference between an associated and dissociated state. Instead protein–solvent interactions are experimentally characterized in terms of a preferential interaction parameter, which characterizes the difference in cosolvent composition next to the protein surface versus in the bulk solution [39–42]. Using exact thermodynamic relationships, the preferential interaction parameter can be related to the perturbation of the protein thermodynamic activity by changing cosolvent concentration. Cosolvents that exhibit preferential adsorption to proteins decrease protein activity when added to the solution, whereas preferentially excluded cosolvents increase protein activity upon addition. Knowledge of the preferential interaction parameter can be used for predicting the free energy change for any process involving a change in protein solvent exposure. Cosolvents that increase protein activity favor processes that reduce the protein surface area, such as salting-out or protein collapse, whereas cosolvents that preferentially adsorb to proteins favor solvent exposure either by enhancing solubility (i.e., salting-in) or by protein unfolding. As a consequence, there is a strong correlation between the effect of cosolvent in salting-in/salting-out and destabilizing/stabilizing the protein fold.

Much progress has been made toward understanding these effects for small inorganic salts, which are a particular class of cosolvents used in formulations. Advancements in computational power and force field development have permitted accurate representations of salts in water solutions. The results have been benchmarked against experimental preferential interaction parameters that can be related at the molecular level to Kirkwood–Buff integrals, which correspond to integrals of the radial distribution functions between the different component pairs [43–45]. Experimental approaches based on X-ray absorption spectroscopy, NMR, and neutron scattering have also been used to test and refine the molecular simulations. The protein–cosolvent interactions section of this chapter covers these recent developments in the understanding of protein interactions with salts, and to a much lesser extent, with other excipients such as sugars, glycerol, or amino acids.

Understanding how protein–cosolvent interactions impact upon protein–protein interactions is only possible by using simplistic descriptions of proteins, such as Deryaguin, Landau, Verwey, and Overbeek (DLVO) theory. Thus, in the section on protein–protein interactions we first provide a review of DLVO theory and discuss modifications to incorporate cosolvent effects. Although DLVO theory has many shortcomings, it is still used as the starting point for determining the effect of long-ranged repulsive forces on aggregation kinetics providing further justification for covering the topic here [23, 29]. The protein–protein interactions section is concluded with a review of more realistic models and the insights gained into the molecular origin of short-ranged interactions between natively folded proteins.

7.3 Protein–Cosolvent Interactions

7.3.1 Lyotropic Series and Hofmeister Series Classifications of Ions

Interpreting the results of literature studies first requires knowing the nomenclature used for classifying salts and their ions [11, 46]. All salts lower the aqueous solubilities of nonpolar compounds with an effectiveness that, in general, follows the lyotropic series, which ranks ions according to their ability to interact with water. The series according to water binding affinity is given for cations $Ca^{2+} > Mg^{2+} > Li^{+} > Na^{+} > K^{+} > Cs^{+} > NH_4^{+}$ and for anions $SO_4^{2-} > HPO_4^{-} > OAc^{-} > Cl^{-} > Br^{-} > NO_3^{-} > I^{-} > SCN^{-}$. High lyotropic series ions are termed kosmotropes due to their water-structuring ability, which is linked to their high charge densities and low polarizabilities, whereas larger ions are more polarizable with lower charge densities, which is correlated with their water structure-breaking ability, so they are termed chaotropes. The lyotropic series should be distinguished from the Hofmeister series, which was originally based on a ranking for the salting-out effectiveness of globular proteins at high salt concentrations (i.e., greater than 1 M). There is some confusion between these classifications, as in some instances, the terms chaotropes and kosmotropes are used for describing ion positions in the Hofmeister series. For instance divalent cations, due to their salting-in effects for proteins are often termed chaotropes, even though the ions have high charge densities and are classified as water structure makers.

7.3.2 Modeling and Simulation of Ion–Interface Interactions

Much of what is known about ion interactions with nonpolar groups has been learned from studies aimed at elucidating salt effects on aqueous surface tensions, which also follows the lyotropic series. A positive surface tension increment indicates that there is a net exclusion of salt in the immediate domain of the air–water interface, which occurs due to repulsive image forces between ions and a low dielectric interface, but this mechanism alone cannot explain the dependence on salt type. Further insight was made accessible from explicit solvent simulations of an aqueous interface, which captured the dependence on the ion's position in the lyotropic series [47, 48]. Interestingly, preferential adsorption of large anions to the low dielectric interface was observed. Although counterintuitive, the adsorption was also found experimentally, providing further validation of the simulation [49]. Initial studies found the behavior was only observed by polarizable force fields, indicating the significance of dispersion interactions in controlling specific ion effects. More recently, similar effects have been captured using nonpolarizable force fields, which are carefully parameterized against the ion solvation properties in bulk, indicating that polarizability is not the determining factor [50, 51]. The ion-specific effects have now been rationalized in terms of general solvation forces, in which case a hydrophobic-like attraction drives the preferential adsorption of large (chaotropic) anions. Other more coarse-grained approaches have been developed using continuum models for water in the context of the Poisson–Boltzmann (PB) equation, where ion specificity is introduced by including an ion–surface dispersion potential self consistently [52–54].

Increasing salt concentration always reduces the aqueous solubility of nonpolar compounds indicating that the salt–solute preferential interaction parameter is positive, or there is a net exclusion of salt about the solute. This exclusion, as with the air–water interface, is driven by the ion preference to form ion–dipole interactions with water. However, analogous to the air–water interface, large chaotropic anions form preferential interactions with nonpolar solutes. These effects have been observed with nonpolarizable models, indicating dispersion forces are not a controlling factor, but instead the adsorption is driven by a hydrophobic-like attraction and ion interactions with polarized water molecules at the solute–water interface [55]. A simulation study on the pair potential of mean force for either methane or neopentane in salt solutions found that the ion specific effects correlated well with the effect of salt on the hydrogen bonding network in bulk water, further indicating direct interactions between the ion and solute are not significant [56]. This study also captured anomalous behavior observed for lithium ion, which exhibits a moderate salting-out effect that does not correlate with its position high in the lyotropic series. This was explained by the formation of linear clusters between anions and lithium, due to the cation high charge density. However, high charge density ions can polarize solvating water molecules leading to anomalous effects if not considered in simulations of nonpolarizable models of water [57].

7.3.3 Ion Interactions with Protein Charged Groups

The interactions of salt ions and protein charged groups are best rationalized in terms of the law of matching water affinities (LMWA) [58]. The strongest ion pair interactions formed by oppositely charged ions occur between ions with similar affinities for water; kosmotropic anions and cations form contact pairs due to direct electrostatic interactions that overcome the desolvation penalty, whereas chaotropic anions and cations form contact pairs due to a hydrophobic-like attraction [59]. Solvent-shared interactions between ions of different size are weaker due to the strong ion–water interactions of the small ion. The LMWA can be used to explain the preference of protein carboxylate groups for specific cations [60, 61], in particular, there is a preference for sodium over potassium. Carboxylates are weakly kosmotropic and have similar hydration enthalpies to sodium, also considered a weak kosmotrope, whereas potassium is a mild chaotrope. The small differences in water affinities of the cations lead to a twofold difference in the binding affinities to the carboxylates. Conversely, molecular simulations have been used to investigate specific anion effects on binding to positively charged protein groups. Ammonium cation has a higher binding affinity for smaller halide anions such as fluoride over larger chaotropic anions. However, the order is reversed when considering the interaction of anions with tetra-alkylated ammonium as larger chaotropic anions form preferential interactions with the methyl groups [62]. Similar ordering has been observed for interactions of anions with other charged protein groups; the preference of all charged groups for fluoride is greatest of the halide anions, and follows the order guanidinium > imidazolium > ammonium [63]. However proximal nonpolar groups to the positive charge preferentially interact with the larger anions such as iodide. The interactions of anions are sufficiently weak that the additivity approximation works well when decomposing the surface into different chemical groups. As such, the effects can be extrapolated to describe the behavior of proteins. As discussed later, protein–protein interactions are much more sensitive to larger halide anions, suggesting the preferential adsorption to nonpolar groups dominates over direct anion interactions with charged groups.

Most of what is known experimentally about interactions with charged protein groups has been elucidated from studies on model systems. Cation specific effects have been probed by examining the change in lower critical solution temperature (LCST) of an elastin like polypeptide containing 16 aspartic acid groups [64]. The LCST corresponds to the temperature above which a peptide or polymer undergoes a hydrophobic induced collapse as reflected by solution clouding. Peptide–salt binding interactions will stabilize the expanded peptide conformation and increase the LCST, whereas preferential exclusion of ions has the opposite effect. An increase in LCST for the aspartic acid ELP was used to calculate binding constants of monovalent cations to aspartic acid. The highest affinities correspond to the smallest cations, in agreement with predictions from the LMWA, except for ammonium and lithium ions. These ions formed stronger than expected ion pairing interactions, attributed to the hydrogen bonding capabilities of ammonium for the carboxylate, and to the high charge density of the lithium ion. The LMWA also fails to explain the dissociation constants (K_d) for a carboxylate with divalent cations, which range from 1 to 10 mM, reflecting much stronger affinities than that of monovalent cations, which exhibited K_d of 78–345 mM. X-ray absorption spectroscopy used to probe the contact pair formation by carboxylates of acetate or formate also found a preference for sodium over potassium, although anomalous behavior was observed with lithium [65, 66]. One explanation is that due to the large energetic penalty of dehydrating lithium, solvent-shared ion pairs are formed with carboxylates [67]. The contact pair formation follows the LMWA, but the net interaction is strongest for lithium. Further understanding highly charged ions such as lithium or divalent cations is progressing as force fields are being developed that account for water polarization. Force fields have been tested [57] for their ability to match experimentally obtained ion–ion distribution functions in solutions of either lithium chloride or lithium sulfate. Simulations using nonpolarizable force fields provided poor fits to the data and unphysical clustering of ions in the solution. A much better fit to the data was found by including polarizability effects using a so-called electronic continuum correction [68, 69]. A similar approach found ionic pairing to be overestimated when using nonpolarizable force fields to describe solutions of potassium dicarbonate, but was corrected for by using the electronic continuum correction [70].

Some salts preferentially adsorb to protein polar groups, in particular, the peptide bond, which is reflected by their solubilizing ability for uncharged peptides [35, 36], by increases to the LCST of peptide-mimic polymers and polypeptides [71–73], and via protein/polypeptide conformational destabilization [37]. Unraveling the determinants to the ion specificity remains a challenging problem. A solute partitioning model was used to extract ion–peptide interactions from a database of historical solubility data for a range of small molecules containing protein functional groups and found that cation binding to the amide group dominated over anion binding [32]. In contrast, LCST studies for salt solutions of poly-N-isopropylacrylamide (PNiPAM) or variants of the elastin-like polypeptide found anion binding to the peptide unit followed the reverse Hofmeister series and a minimal cation effect [71, 72]. Simulation studies have also yielded ambiguous findings. Molecular simulations of the peptide-mimic molecule N-methylacetamide (NMA) indicated the peptide bond only formed direct interactions to cations with a preference for sodium over potassium. The destabilizing effects of chaotropic anions were attributed to interactions with nonpolar side groups [74, 75], although a combined NMR and simulation study indicated chaotropic anions only interact at a hybrid binding site of a carbon atom attached to an electron withdrawing atom, such as the alpha carbon and amide nitrogen along the peptide backbone [71]. Binding at this location is also preferred because the amide hydrogen bonding ability to water is kept intact. The salting-in ability of chaotropic anions was also linked to their interactions with nonpolar moieties in a simulation study of pNiPAM [76]. Conversely, it was found [77] that cations have a much stronger interaction with the peptide group, attributing the reverse Hofmeister series dependence of the LCST for pNiPAM to an inverse relationship between the binding affinity of the cation for the peptide versus the cation–anion association. Association of the kosmotropic cation sodium is strongest with kosmotropic anions such as sulfate, and weakest with large chaotropic anions. In a follow-up study [78], the salting in ability of divalent and trivalent cations for the amide group is weaker than monovalent cations, although this is in the wrong direction expected from the solute partitioning model [32]. A combined experimental and theoretical study on butyramide did find that divalent cations have stronger interactions than monovalent cations, but the dissociation constants are on the order of M and much weaker than the binding of weakly hydrated anions [79]. Thus there is little general consensus so far gained from molecular simulation studies. The variation in determining the interaction of halide anions with the amide group is in part due to the sensitivity of the simulation findings to the force field [74, 76]. The interactions also appear to depend on the immediate chemical bonding environment of the functional group [71, 76], which means care must be taken when using the additivity approximation and extrapolating effects across different molecules. Part of the problem is that cation specific effects appear to be dominated by interactions with protein negatively charged groups [80]. Along these lines, it has been shown [81] that solubility studies of capped glycine peptides used for interpreting cation specific effects could have been flawed, in that the negatively charged C-terminus was not capped [81]. Thus, cation specific binding with the peptide group is difficult to delineate from effects arising from negatively charged protein groups. Some of these issues will be dealt with in the near future as ion force fields are developed to include polarizability, combined with further experimental studies and a re-examination of the historical specific ion effect datasets.

7.3.4 Protein Interactions with Other Excipients

Much less is known about other commonly used excipients, such as sugars, polyols, and single amino acids. Out of these, protein interactions with arginine have been studied the most, as arginine is a commonly used additive throughout processing, and is used extensively as an additive in refolding operations and in liquid formulations [82, 83]. The guanidinium group on arginine has a planar geometry with a delocalized charge, making it similar to an aromatic molecule [84, 85]. The stabilizing mechanism for arginine remains unknown, but is linked to its ability to form strong interactions with aromatic and negatively charged protein groups [86]. In addition, guanidinium groups can form stacking interactions with themselves leading to arginine cluster formation [85, 87], which could have both an indirect or direct effect on protein aggregation [88]. Force fields are currently being developed to include polarizability, as nonpolarizable force fields overpredict the ion pair energetics with carboxylates and the arginine clustering in solution [89]. Many studies of molecular simulations applied to protein–excipient interactions [88] need to be interpreted with care, as force fields used in the simulations have not been scrutinized with the same level of detail as in the specific ion effects literature.

7.4 Protein–Protein Interactions

7.4.1 The Osmotic Second Virial Coefficient and DLVO Theory

As with protein–salt binding, nonspecific protein–protein interactions cannot be represented by an on/off dissociation constant as there is no well defined self associated state. Instead protein–protein interactions are generally characterized in terms of an osmotic second virial coefficient, B_{22}, which is related through McMillan–Mayer theory to the solute–solute pair distribution function $g(r,\Omega)$ (taken at infinite dilution of protein):

$$B_{22} = -\frac{1}{2}\int_0^\infty \left[1-g\left(r,\Omega\right)\right]r^2 \mathrm{d}r\mathrm{d}\Omega \tag{7.1}$$

Because proteins have anisotropic shapes and surface properties, the integral is carried out over the center to center separation r and the set of Euler angles defining the relative orientations between a pair of proteins. Often the link to molecular interactions is made directly through the pair potential of mean force, W, given by

$$g(r,\Omega) = \exp[-\beta W(r,\Omega)]$$

W corresponds to the constrained free energy with two proteins held at a fixed orientation and separation integrated over all solvent degrees of freedom. Carrying out the integration given in Equation 7.1 requires making simplifications to the protein–protein interaction model.

DLVO theory provides a good starting point to understand the nature of protein–protein interactions. Within DLVO theory proteins are treated as uniformly charged polarizable spheres interacting through a dielectric continuum of water in which salt ions are included as point charges (e.g., [90]). The contributions to the protein–protein interaction include an excluded volume term, a repulsive electrical double-layer force, and a Hamaker dispersion attraction. The repulsive force is determined by solving the PB equation for two uniformly charged surfaces with the same dielectric as water. In the electrical double layer (EDL), there is a net accumulation of salt relative to the bulk due to the electrostatic forces between salt and the protein charges. When two double layers overlap, there is an osmotic repulsive force due to the increased concentration of ions located at the midplane relative to the bulk solution. Because the EDL ion concentration is proportional to protein net charge, the magnitude of the two-body force varies as the protein net charge squared. The range of the EDL force is determined by the Debye–Hückel screening length, which is proportional to the inverse square root of ionic strength. DLVO theory has been validated by its ability to capture the ionic strength and pH-dependence of protein–protein interactions for proteins ranging in size from lysozyme to mAbs [91–94]. Changing ionic strength attenuates the double layer repulsion due to ionic screening and changing pH away from pI increases the repulsion due to increased net charge. The experimental data fit well with independent measurements of net charge from potentiometric titrations or electrophoretic mobility determinations. The approach works only in solutions at low ionic strength (below 100 mM) and at pH values removed from the isoelectric pH. Under these conditions, the protein–protein interaction is sufficiently weak that there is no angular biasing to the integral of Equation 7.1, in which case the integration will be determined by averaged properties of the protein surface, consistent with DLVO theory.

7.4.2 Incorporating Specific Salt and Ion Effects

One of the shortcomings to DLVO theory is the inability to incorporate ion specific effects. The classical Hofmeister series effect corresponds to a salt-induced protein–protein attraction in concentrated salt solutions. The salting-out effect is due to image forces between ions and the protein surface, which leads to a layer of salt depletion with constant thickness given by the Bjerrum length, as shown using a variational approach that incorporated image forces self consistently into the PB equation [95]. An attractive force is generated between proteins when the layers of salt depletion overlap due to the lower ion density at the midplane between proteins relative to that in the bulk [96]. Although ion specificity has not been incorporated into the model directly, the depletion force will be greatest for salts with the largest exclusion, which follows the position of the ion in the Hofmeister series. This salting-out approach only works well for describing systems with strongly excluded salts. Different behavior is observed when salts preferentially adsorb to proteins. Preferential interactions of divalent cations with proteins have been linked to strong salting-in effects at salt concentrations above 1 M [97–99]. Conversely, the preferential interactions of chaotropic anions to positively charged proteins correlate with a reverse Hofmeister series dependence of protein solubility at low ionic strength [92, 100, 101]. The latter effect has been captured by incorporating ion-surface dispersion forces in the PB equation [52–54]. Within this approach, preferential adsorption to the protein surface is enhanced the greatest for chaotropic anions. Preferential adsorption screens the protein surface charge thereby lowering the EDL ion concentration and reducing the osmotic repulsion between proteins to give the reverse Hofmeister series. However the direct Hofmeister series dependence is recovered with further increasing salt concentration above 0.5 M. Increased anion preferential adsorption changes the surface charge from positive to negative. The negative surface charge draws cations into the double layer increasing the local salt concentration and creating an osmotic repulsion force thereby solubilizing the protein. The reversal in the Hofmeister series dependence of protein solubility has been observed experimentally from cloud point studies of lysozyme solutions [102]. Attributing ion specificity to dispersion forces has been criticized because previous studies have suggested that anion preferential adsorption is driven by solvation forces, or a hydrophobic-like attraction. This effect has been accounted for using explicit water simulations to extract the ion pair distribution function about a nonpolar surface [12, 103]. The surface–ion potential of mean force was then included in the PB equation to determine the surface–surface interaction, which reproduced the inversion of the Hofmeister series. It is likely that the salting-in effect of divalent cations can also be explained by cation preferential adsorption leading to an osmotic repulsion force.

7.4.3 Inclusion of Nonionic Excipients

Understanding deviations from DLVO theory also provides the starting point for predicting effects of excipients such as glycerols and sugars. Increasing glycerol concentration enhances protein–protein repulsion irrespective of the salt concentration [104–107]. This induced repulsion has been attributed to a decrease in dielectric constant and refractive index change for the glycerol solution, which, in turn, leads to an increase in the double layer force and a decrease in the Hamaker constant [104, 106]. Rationalizing the behavior in terms of the Hamaker constant can be misleading as short-ranged forces between proteins are not only due to dispersion forces, but can include hydration effects, hydrophobic interactions, and charge pairing forces. Understanding effects of excipients on short-range interactions requires first isolating their effects on protein polar, nonpolar, and charged groups. For instance, the effect of glycerol has been explained by its ability to preferentially hydrate polar and nonpolar surfaces thereby leading to repulsive hydration forces between proteins [105, 106]. Such a mechanism has also been used to describe the stabilization of protein–protein interactions by sugars [108], although similar cosolvent induced effects occur with monohydric alcohols, which do not preferentially hydrate proteins [105].

7.4.4 Models Accounting for Anisotropic Protein–Protein Electrostatic Interactions

Under solution conditions where the repulsive double layer force is sufficiently weak, the protein–protein interactions are short-ranged. In these instances, the interaction is likely anisotropic and sensitive to the distribution of charge and polarity on the protein surface. Anisotropic interactions have been inferred directly from measurements of attractive forces between proteins at low ionic strength. Under conditions where proteins carry a net charge, the net protein–protein interaction can only be attractive if there is an orientational bias for protein–protein configurations with charge complementarity. Attractive electrostatics were observed from experimental measurements of chymotrypsinogen over a pH range from 5.2 to 8.0, with a large decrease in protein–protein attraction occurring over an ionic strength range of 5–100 mM, characteristic of electrostatic screening [109]. These measurements were fit to an all-atomistic interaction model for the protein, which included electrostatic interactions via an approximate solution to the PB equation. Dispersion forces were included using a hybrid model, with an atomic Lennard–Jones force-field for surfaces separated by less than a solvent layer and a continuum expression used for larger separations [109, 110]. The results of fitting the model indicated that the protein–protein interaction is dominated by a few highly attractive configurations with well depths on the order of 10–20 k_BT. More interestingly, changing pH and ionic strength did not alter the low energy configurations, indicating that shape complementarity and dispersion forces are the controlling factors. The attractive electrostatics only result from the charge asymmetry in the interacting configurations. Similar conclusions have been drawn using a more detailed force field with a grid-based electrostatic desolvation energy to account for dehydrating polar surface groups, and a solvent accessible surface area potential to describe nonpolar interactions and other contributions to short-ranged forces [111]. These approaches still rely on using a semi-implicit model for the solvent such that it is computationally feasible to sample enough of the orientation space to provide an accurate estimate for the B_{22} value. Alternatively, explicit solvent simulations can be used to determine the energetics for surfaces buried in protein crystal contacts, which correspond to the low energy configurations. Using explicit water molecular dynamics, it has been shown [112] that the effective lysozyme–lysozyme interaction corresponding to a hydrophobic crystal contact does not exceed 3–4 $k_B T$. Interactions of this strength are not strong enough to constrain protein molecule orientations in solution. There is an order of magnitude decrease in the estimate of the low energy configurations when using explicit solvent force fields, which highlights the difficulties in identifying the short-ranged nature of protein–protein interactions using molecular simulation approaches.

7.5 Informatics Studies of Protein Aggregation

7.5.1 Comparison with Modeling Used for Small Molecule Pharmaceutics

In the area of small molecule drug design, development of therapeutic leads depends on affinity for the biological target and assessment/testing of the ADME, (absorption, distribution, metabolism, and excretion), and toxicity properties for drug candidates. Although there is still much that is unknown about *in vivo* drug delivery and toxicity, with, for example, systems pharmacology making the link between a candidate molecule and cellular metabolic networks [113], there is a commonality in some of the considerations for drug design and *in silico* ADME analysis [114]. Lipinski's rules express physicochemical properties that have a direct bearing on the design and development process [115]. The situation is somewhat different for protein therapeutics. A major difference is the current restriction to largely extracellular targets, and therefore the lack of a transmembrane transport step and interactions with intracellular components. Whilst this emphasis is likely to change with the development of import pathways for protein therapeutics [116], the major focus, after development of affinity for the biological target, is aqueous phase solubility. Higher solubility allows delivery of greater molar amount of protein from a given volume, typically 1.5 ml for convenient administration of a subcutaneous injection. Although some biologics can achieve this solubility (e.g., many mAbs), others cannot [117]. In contrast to the development of small molecule therapeutics, design for administration of biologics (in this case solubility) is at the current time largely uncoupled from development of the required target-binding affinity. Thus, the field of formulation assumes the important task of developing optimal solution conditions for solubilizing biologics and preventing aggregation. A part of the input to modeling and informatics studies relies on the underlying physicochemical properties of proteins, which will influence both the innate solubility of a particular biologic, or scaffold behind a series of biologics, and the response of a biologic to the excipients used in formulation. Accordingly there is scope for improved modeling ultimately leading to an increase in understanding that will bridge the design and formulation areas for biologics, which are largely separated at present (Figure 7.1). This section will discuss modeling and informatics for the solubility of protein therapeutics. It should be noted that other factors, not covered here, are also suitable for integration of modeling and experimental analysis. For example, as with small molecules, there is the ever-present issue of design for improved affinity and specificity [118]. Additionally, little is known at present about the processes involved in uptake of biologics from subcutaneous injection, although electrostatic interactions appear to play a role [119].

7.5.2 Prediction Schemes Deriving from Amyloid Deposition

A major influence on studies of protein aggregation has been the observation that proteins under fully or partially denaturing conditions use an exposed sequence with a propensity to form β-structures, to associate [120]. This observation followed X-ray fiber diffraction studies reporting what appears to be a general β-structure for proteins associated with protein deposition diseases [121], including neurodegenerative disorders such as Alzheimer's disease and spongiform encephalopathies. The confluence of these results gave rise to an interest in whether a relatively simple property, such as the propensity of an amino acid sequence (or subsequence window) to form β-strands, could form the basis for a predictor of amyloid-like aggregation [122]. Such properties can be easily calculated from protein structural databases, and have been used extensively in predictions of protein secondary structure [123]. More generally, the properties of β-strands in 3D protein structure have been studied in the context of amino residues (e.g., charged amino acids), that mitigate against β-mediated association [124]. As a result of the relative simplicity of applying sequence-based analysis of amyloid propensity, several online tools have been made available for prediction. These include TANGO [125], PASTA [126], and Zyggregator [127]. Sequence-based prediction can be applied to whole proteomes and transcriptomes. An example is the observation that mRNA levels in *Escherichia coli* are lower, on average, for proteins with sequences predicted to be more prone to aggregation [128], consistent with solubility being an evolutionary constraint in the crowded environment of the cytoplasm [129].

The simplicity of the amyloid aggregation hypothesis, and its convenient application to sequence-based prediction, has lead to an interest in applying these methods to the field of protein therapeutics [130, 131]. However, at this point no consensus has been reached in the literature concerning the ubiquity of amyloid and sequence-based prediction schemes in the design and formulation of protein therapeutics. It is known that such schemes are not uniformly applicable, for example, in a high-throughput study of the solubility of *E. coli* proteins in cell-free expression, amyloid-based prediction did not correlate with measured solubilities [132]. Reasons behind the domain-specific success of prediction methods presumably lie in the varied molecular mechanisms. Thus, a process such as self-association without a denaturing step, may not be amenable to the amyloid prediction schemes. Equally, partial denaturation, where exposure of potential β-strand interacting regions is limited, might also be less in scope for these prediction methods. Experimental data already point to the role of surface regions in association properties, for example, for mAbs [22]. The remainder of this section discusses methods that do not focus on amyloid-based prediction.

7.5.3 Solubility Prediction Based on Sequence, Structural, and Surface Properties

The best starting points for prediction schemes are datasets of measured solubilities. Since the early work of Wilkinson and Harrison [133], benchmark sets have been developed for protein solubility. It has for the most part though proven difficult to obtain data under a consistent set of conditions, a problem amplified when protein expression is considered alongside solubility. Wilkinson and Harrison distinguished between proteins known to form inclusion bodies (IBs) and those that do not, a broad distinction that covers a variety of conditions for expression. Subsequent study has also used the IB/nonIB separation [134], or even defined soluble proteins as all those for which a structure has been solved and released [135, 136], which fits into the framework of structural genomics programs of the early 2000s [137]. Encouragingly, the 'omics era of high-throughput datasets is leading to standardized measurements of solubility, for example, in the study of *E. coli* proteins in cell-free expression [132]. Such information provides data, not just to test hypotheses against, but also to generate hypotheses. Well constituted experimental datasets contain positive and negative subsets (most and least soluble proteins), so that features which distinguish between the subsets become candidates for prediction. In terms of biologics, a key aim for the future will be to generate these data. Underpinning much work for predictive algorithms applied to association of protein therapeutics, is an understanding that the physicochemical properties responsible for solubility of nonbiologic globular proteins, will be largely transferrable to biologics. Some of the features suggested as separating proteins, based on solubility, are net charge and turn-forming residue fraction [133], thermostability and lack of β-sheet [134]. Several studies have used machine-learning methods to derive the best distinction between soluble and insoluble subsets (e.g., [136]). With these techniques it can be difficult to extract the physicochemical interpretation.

An example usage of experimental datasets to determine features that separate soluble and insoluble proteins is demonstrated in analysis of *E. coli* protein cell-free expression data [132]. Proteins with known 3D structures were selected, and structural properties calculated. The feature that best separated datasets was size of the largest patch of calculated positive electrostatic potential [8]. There was no clear rationalization of why this may be the case, but with a similarity to the DNA-binding surfaces of proteins (also positively charged), it was suggested that positive surfaces may reflect an intermediate protein-nucleic acid binding step that somehow destabilizes protein and facilitates nonspecific protein–protein interactions [8]. Such an interpretation was supported by the observation that the equivalent parameter for negatively charged patches was unable to separate soluble and insoluble subsets, and by other reports of a role for positive charge in expression [138, 139]. This example illustrates the problem in making general conclusions about protein solubility and propensity to associate. Thus far, there is no evidence that the size of positively charged patches have a specific relationship with solubility for purified proteins, that goes beyond a general effect that would also be exhibited by negatively charged patches. Indeed, the speculated mechanism [8] invokes the rich nucleic acid content of an expression system, and thus is not directly applicable to protein solutions downstream from expression. It is therefore possible that parameters implicated for prediction algorithms in protein expression (e.g., from structural genomics pipelines) are not uniformly applicable to storage and formulation of protein therapeutics. Nevertheless, any general understanding gained in the role of physicochemical properties should be applicable between the systems under study.

An example of a feature, derived from informatics studies, of potentially uniform importance, is the ratio of lysine to arginine residue content in a protein. It has been found [117] that a higher lysine content tends to segregate with increased solubility in the *E. coli* protein set [132]. In this case the proposed molecular underpinning lies in the different chemical properties of amino (lysine) and guanidinium (arginine) groups. It is thought that arginine is more likely to be involved in weak nonspecific protein–protein interactions, consistent with what is known for specific interfaces [140]. Interestingly, it was also found that proteins known to be at high concentrations *in vivo,* (e.g., myoglobin, serum albumin, circulating antibodies), possess relatively high ratios of lysine to arginine content [117]. If the basis for this observation lies in the protein–protein interaction propensities of amino acid side chains, then such a property may be uniform across systems, and not specific to expression. Whilst further experimental work will establish the molecular details underlying the observed propensities, there is clearly scope for more informatics analysis, in part to see whether other features become apparent as distinguishing more and less soluble proteins.

Surface polarity is a feature well known in defining specific protein–protein interactions [140]. It is presumed that the high affinity of specific protein–protein interactions is often the result of surface complementarity of nonpolar regions, tailored by evolution. The analogous presumption for lower affinity nonspecific interactions is that such relatively nonpolar patches can also interact with other copies of the same patch, but with decreased shape complementarity, solvent exclusion, and entropic free energy gain upon interface formation. Nonpolar patches, as charged patches, are conveniently viewed using molecular graphics software, but assessing precisely the size of nonpolar patches that could mediate deleterious associations requires further analysis, including a ranking of patches. Such methods have been routinely developed in the field of protein structural bioinformatics, and in recent years applied to protein aggregation [8, 141, 142]. Figure 7.2 illustrates how nonpolar and positively or negatively charged patches can be conveniently displayed (using an antibody in this example) and shows that the display relates back to the underlying amino acid content of a surface region. Thus, even without an algorithm to predict patch sizes and sites for modification (e.g., the introduction of charge into a nonpolar region), molecular graphics can be used to qualitatively assess potential problem locations.

One of the key aims in future work will be to incorporate estimates of structural stability and local protein unfolding into predictive methods for aggregation that is associated with unfolding events and exposure of otherwise buried protein. This has been an important area in protein structural bioinformatics for many years, so there is a wealth of methodology for transfer. There are also objectives associated with surface analysis alone, without considering unfolding. These largely revolve around combining the properties of net charge, charge asymmetry (positively and negatively charged patches), nonpolar patches, and specific amino acid properties. With regard to specific amino acid properties, the example of lysine and arginine has been used. They contribute more or less equally to electrostatic patch properties, and yet are clearly used quite differently in evolution of proteins at different naturally occurring abundances [117]. There may well be other differences between amino acid sidechain chemistries, in the context of protein association and aggregation, that remain to be discovered. Equally, it is apparent now in several studies that positive and negative charges are not equivalent in their contributions to solubility. This is not just the case in expression systems, where specific charge–charge interactions may play a role [8], but also for purified proteins, where negative surface charge correlates with solubility [143]. Other studies support this observation: a strong preference for aspartic and glutamic acids over lysine and arginine was seen in a phage display screen for substitutions that enhance aggregation resistance in the variable domains of antibodies [144], the addition of an acidic tag enhanced the expression of a positively charged intrabody [145], and many chaperones possess negatively charged regions, with acidic segments modulating the anti-aggregation activity of Hsp90 [146].

Protein solubility often decreases near to the isoelectric point as net charge and electrostatic repulsion decreases, and near the pI there is more scope for proteins with anisotropic charge distributions to sample attractive interactions between patches of opposite charge, interactions that are screened with increasing ionic strength [147]. The challenge is for predictive algorithms to combine as many features as possible with appropriate weightings, into a

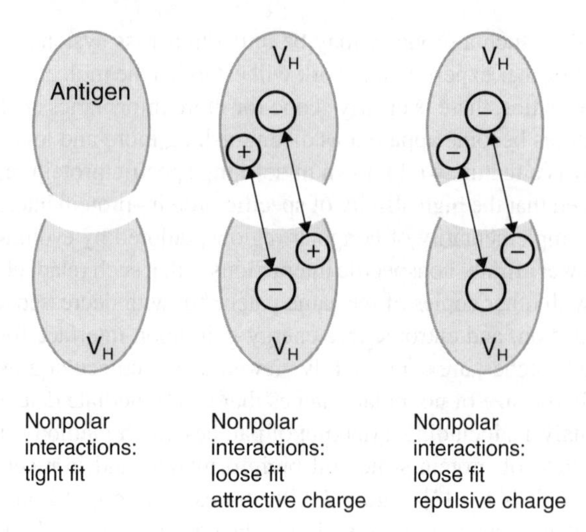

Figure 7.3 *Combination of charge–charge and nonpolar patch interactions. A schematic model for interactions between V_H domains that is consistent with experimental data [22]. This V_H domain can act as a single domain antibody, illustrated with antigen binding via complementary nonpolar surfaces, in the left-hand panel. It is hypothesized that the same V_H nonpolar patch mediate self-association, albeit with a loose fit, less water exclusion, and reduced binding affinity. Charge–charge interactions between residues around the antigen-binding region and the net charge of the V_H domain are included in the central and right-hand panels. Attractive charge interactions (central panel) will lead to higher association than will repulsive charge interactions (right panel).*

single model. Experimental data will play a crucial role in guiding this process. For example, aggregation is reduced more by charge mutations at the edges of CDRs that are of the same sign as net charge on the V_H experimental scaffold [22], than when the charges are of opposite sign. One interpretation is that association is driven by a combination of nonpolar and charge–charge interactions (Figure 7.3). This type of experiment gives a framework for calibrating non-polar and charge–charge interactions, which will be invaluable for model development.

7.6 Future Prospects

This chapter has gathered together diverse modeling areas which all bear on our understanding of the solution behavior of protein therapeutics. There are some detailed questions of force field parameterization that can yield insights into protein–small molecule interactions. At the same time, other work seeks associations of protein features with solubility, in a less atomistic, more informatics-based approach. Common factors are comparison with existing experimental data, and the understanding that improved models will lead to improved predictions for bioprocessing and formulation. In future it will be important to combine models for protein–protein and protein–small molecule interactions, with folded state stability. The relative roles of unfolding, or partial unfolding, and native state association are case-dependent, and this provides an appropriate challenge for the modeling community, that is, predicting that case-dependence. Whilst some biologic frameworks may be well developed in terms of formulation for delivery at sufficient concentrations, with suitable storage stability, others are less well characterized. Another challenge is for modeling to make an impact early in the design process for these more problematic biologic frameworks, before the amino acid sequence is finalized. Equally, in the formulation process, there may be excipients deriving from a more mature modeling field that can profitably extend the current repertoire. In all cases, a close coupling of modeling and experiment is crucial, from high-throughput development phases to targeted tests of hypotheses generated by modeling for improved aggregation-resistance in protein therapeutics. An example of a requirement for improved models lies in the area of storage, since extrapolation from experimental accelerated stability studies can be problematic. Ultimately, it should be possible to extend modeling to processes beyond aqueous phase formulations, and additionally to incorporate the behavior of protein therapeutics upon administration. Currently though there is still significant challenge in adapting decades of research into protein chemistry to produce reliable models for protein therapeutic behavior in liquid formulations.

References

[1] Dimitrov, D.S. (2012) Therapeutic proteins. *Methods in Molecular Biology (Clifton)*. **899**: 1.
[2] Goldstein, G. (1987) Overview of the development of Orthoclone OKT3: monoclonal antibody for therapeutic use in transplantation. *Transplantation Proceedings* **19**(2, Suppl. 1): 1.
[3] Kohler, G. and C. Milstein (1975) Continuous cultures of fused cells secreting antibody of predefined specificity. *Nature*. **256** (5517): 495.
[4] Vaughan, T.J., Williams AJ, Pritchard K, *et al.* (1996) Human antibodies with sub-nanomolar affinities isolated from a large non-immunized phage display library. *Nature Biotechnology* **14**(3): 309.
[5] Harding, F.A., Stickler, M.M., Razo, J., DuBridge, R.B. (2010) The immunogenicity of humanized and fully human antibodies: residual immunogenicity resides in the CDR regions. *MAbs* **2**(3): 256.
[6] Dimasi, N., Gao C, Fleming R, *et al.* (2009) The design and characterization of oligospecific antibodies for simultaneous targeting of multiple disease mediators. *Journal of Molecular Biology* **393**(3): 672.
[7] Harris, L.J., Larson SB, Hasel KW, McPherson A. (1997) Refined structure of an intact IgG2a monoclonal antibody. *Biochemistry* **36**(7): 1581.
[8] Chan, P., R.A. Curtis, and J. Warwicker (2013) Soluble expression of proteins correlates with a lack of positively-charged surface. *Scientific Reports* **3**: 3333.
[9] Wright, A. and S.L. Morrison (1997) Effect of glycosylation on antibody function: implications for genetic engineering. *Trends in Biotechnology* **15**(1): 26.
[10] Tsumoto, K., Ejima, D.; Kita, Y.; Arakawa, T. (2005) Review: why is arginine effective in suppressing aggregation? *Protein and Peptide Letters* **12**(7): 613.
[11] Baldwin, R.L. (1996) How Hofmeister ion interactions affect protein stability. *Biophysical Journal* **71**(4): 2056.
[12] Lund, M. and P. Jungwirth (2008) Patchy proteins, anions and the Hofmeister series. *Journal of Physics: Condensed Matter* **20**(49): 494218.
[13] Saluja, A., Matthew Fesinmeyer, R., Hogan, S., *et al.* (2010) Diffusion and sedimentation interaction parameters for measuring the second virial coefficient and their utility as predictors of protein aggregation. *Biophysical Journal* **99**(8): 2657.
[14] Le Brun, V., Friess W., Bassarab S., *et al.* (2010) A critical evaluation of self-interaction chromatography as a predictive tool for the assessment of protein–protein interactions in protein formulation development: a case study of a therapeutic monoclonal antibody. *European Journal of Pharmaceutics and Biopharmaceutics* **75**(1): 16.

[15] Kameoka, D., Masuzaki E., Ueda T., Imoto T. (2007) Effect of buffer species on the unfolding and the aggregation of humanized IgG *Journal of Biochemistry* **142**(3): 383.
[16] Wang, W. and C.J. Roberts (2013) Non-Arrhenius protein aggregation. *AAPS Journal* **15**(3): 840.
[17] Roberts, C.J. (2007) Non-native protein aggregation kinetics. *Biotechnology and Bioengineering* **98**: 927.
[18] Roberts, C.J., T.K. Das, and E. Sahin (2011) Predicting solution aggregation rates for therapeutic proteins: approaches and challenges. *International Journal of Pharmaceutics* **418**(2): 318.
[19] Chi, E.Y., Krishnan S., Carpenter J.F. (2003) Physical stability of proteins in aqueous solution: Mechanism and driving forces in nonnative protein aggregation. *Pharmaceutical Research* **20**(9): 1325.
[20] Jahn, T.R. and S.E. Radford (2008) Folding versus aggregation: polypeptide conformations on competing pathways. *Archives of Biochemistry and Biophysics* **469**(1): 100.
[21] Wang, W., S. Nema, and D. Teagarden (2010) Protein aggregation-pathways and influencing factors. *International Journal of Pharmaceutics* **390**(2): 89.
[22] Perchiacca, J.M., C.C. Lee, and P.M. Tessier (2014) Optimal charged mutations in the complementarity-determining regions that prevent domain antibody aggregation are dependent on the antibody scaffold. *Protein Engineering Design and Selection* **27**(2): 29.
[23] Chi, E.Y., Krishna, S., Kendrich, B.S., *et al.* (2003) Roles of conformational stability and colloidal stability in the aggregation of recombinant human granulocyte colony-stimulating factor. *Protein Science* **12**(5): 903.
[24] He, F., Hogan S, Latypov RF, *et al.* (2010) High throughput thermostability screening of monoclonal antibody formulations. *Journal of Pharmaceutical Sciences* **99**(4): 1707.
[25] Goldberg, D.S., Bishop SM, Shah AU, Sathish HA. (2011) Formulation development of therapeutic monoclonal antibodies using high-throughput fluorescence and static light scattering techniques: role of conformational and colloidal stability. *Journal of Pharmaceutical Sciences* **100**(4): 1306.
[26] He, F., Woods CE, Becker GW, *et al.* (2011) High-throughput assessment of thermal and colloidal stability parameters for monoclonal antibody formulations. *Journal of Pharmaceutical Sciences* **100**(12): 5126.
[27] Lehermayr, C., Mahler HC, Mäder K, Fischer S. (2011) Assessment of net charge and protein-protein interactions of different monoclonal antibodies. *Journal of Pharmaceutical Sciences* **100**(7): 2551.
[28] Connolly, B.D., C. Petry, S. Yadav, B. Demeule, *et al.* (2012) Weak interactions govern the viscosity of concentrated antibody solutions: high-throughput analysis using the diffusion interaction parameter. *Biophysical Journal* **103**(1): 69.
[29] Olsen, S.N., Andersen KB, Randolph TW, *et al.* (2009) Role of electrostatic repulsion on colloidal stability of Bacillus halmapalus alpha-amylase. *Biochimica et Biophysica Acta* **1794**(7): 1058.
[30] Saito, S., Hasegawa J, Kobayashi N, *et al.* (2012) Behavior of monoclonal antibodies: relation between the second virial coefficient (B-2) at low concentrations and aggregation propensity and viscosity at high concentrationsbrummi. *Pharmaceutical Research* **29**(2): 397.
[31] Saluja, A., Badkar, AV, Zeng, DL, *et al.* (2007) Ultrasonic rheology of a monoclonal antibody (IgG2) solution: implications for physical stability of proteins in high concentration formulations. *Journal of Pharmaceutical Sciences* **96**(12): 3181.
[32] Pegram, L.M. and M.T. Record (2008) Thermodynamic origin of Hofmeister ion effects. *Journal of Physical Chemistry B* **112**(31): 9428.
[33] Auton, M. and D.W. Bolen (2004) Additive transfer free energies of the peptide backbone unit that are independent of the model compound and the choice of concentration scale. *Biochemistry* **43**(5): 1329.
[34] Nandi, P.K. and D.R. Robinson (1972) Effects of salts on free-energies of nonpolar groups in model peptides. *Journal of the American Chemical Society* **94**(4): 1308.
[35] Nandi, P.K. and D.R. Robinson (1972) Effects of salts on free-energy of peptide group. *Journal of the American Chemical Society* **94**(4): 1299.

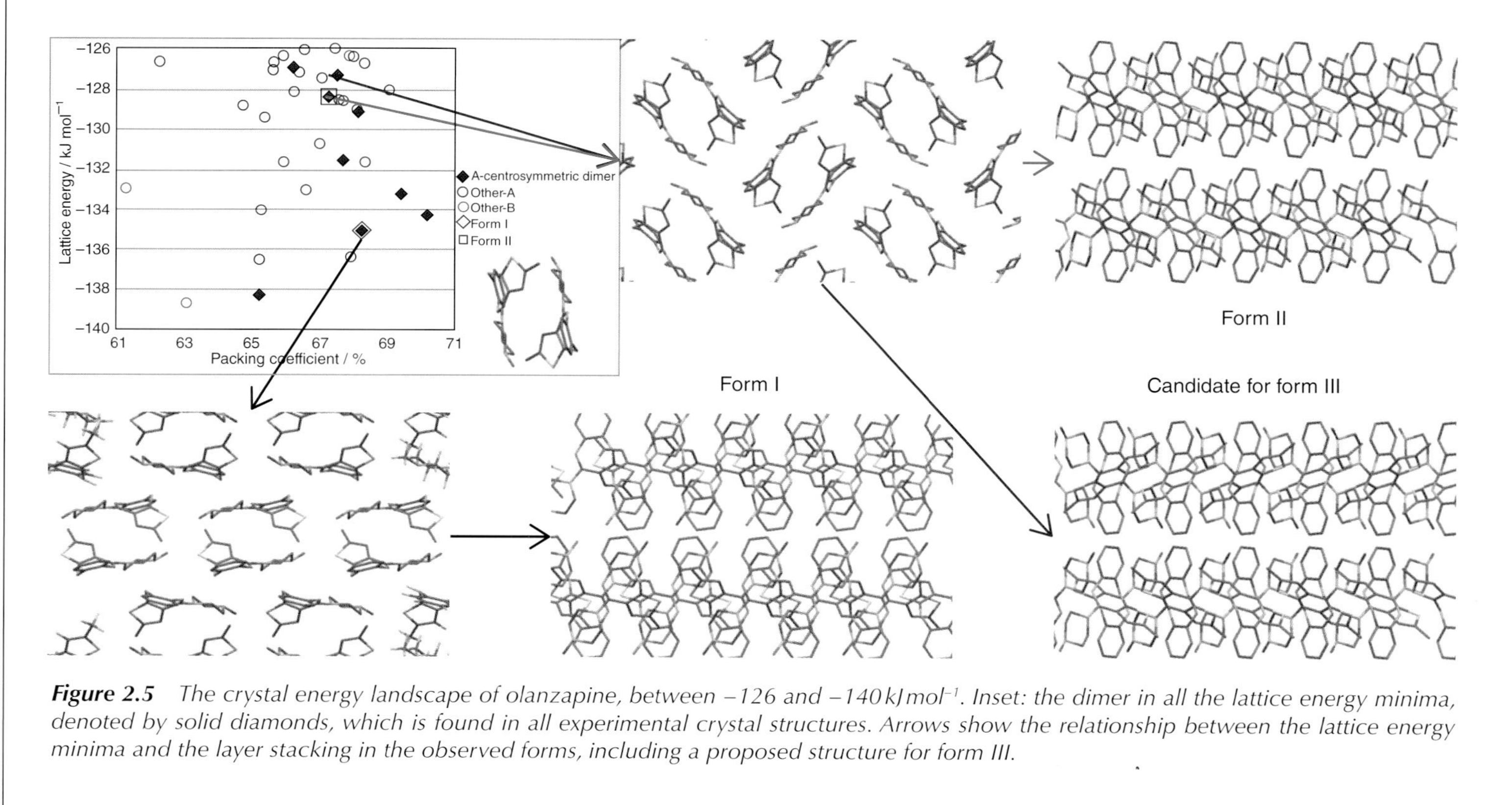

Figure 2.5 *The crystal energy landscape of olanzapine, between −126 and −140 kJ mol^{-1}. Inset: the dimer in all the lattice energy minima, denoted by solid diamonds, which is found in all experimental crystal structures. Arrows show the relationship between the lattice energy minima and the layer stacking in the observed forms, including a proposed structure for form III.*

Computational Pharmaceutics: Application of Molecular Modeling in Drug Delivery, First Edition.
Edited by Defang Ouyang and Sean C. Smith.

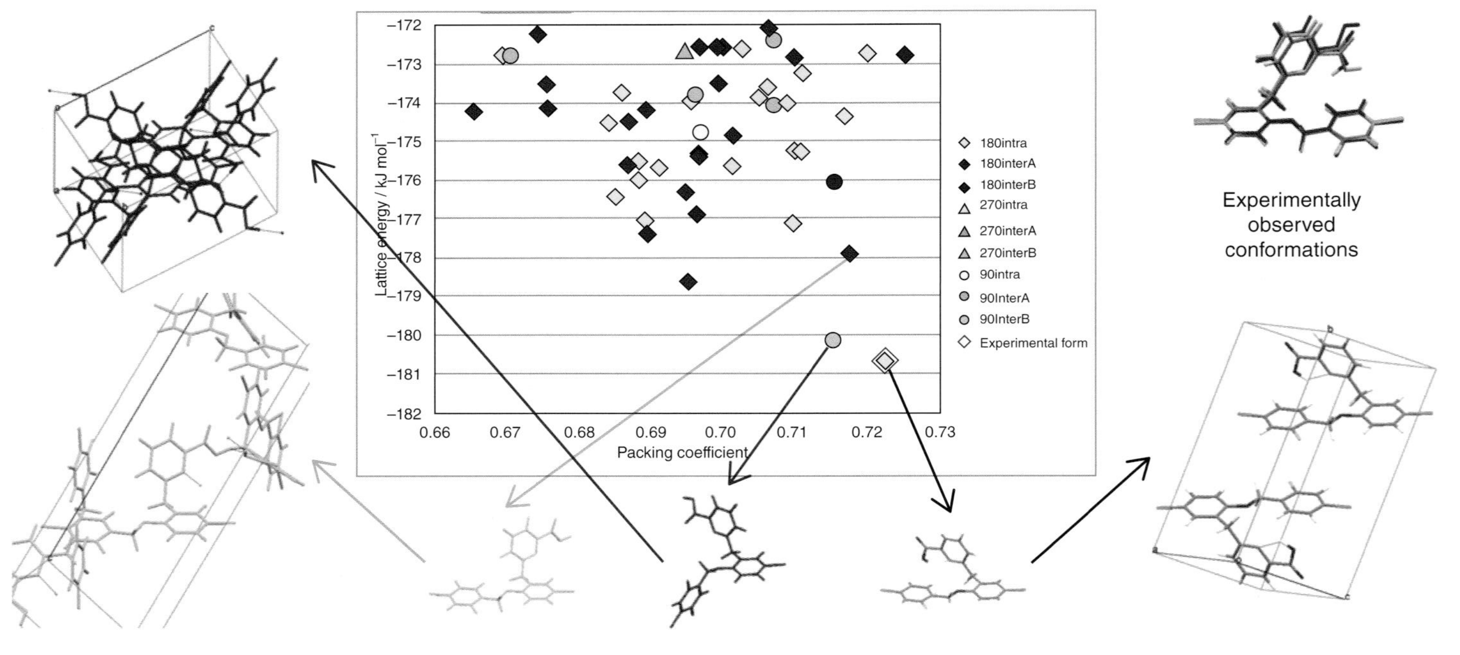

Figure 2.6 *Summary plot of the CSP study of GSK269984B. Each point represents a crystal structure classified by the approximate* θ_{CO} *(defined in Figure 2.2) angle and acid conformation (Intra/InterA/InterB). The three lowest energy structures and their conformations are linked by arrows, and should be contrasted with the overlay of the conformations in the experimental forms: form I and* ab initio *minimum (Intra carboxylic acid conformation, in element colours), the DMSO solvate (InterA, in orange) and the NMP solvate (InterB, in blue).*

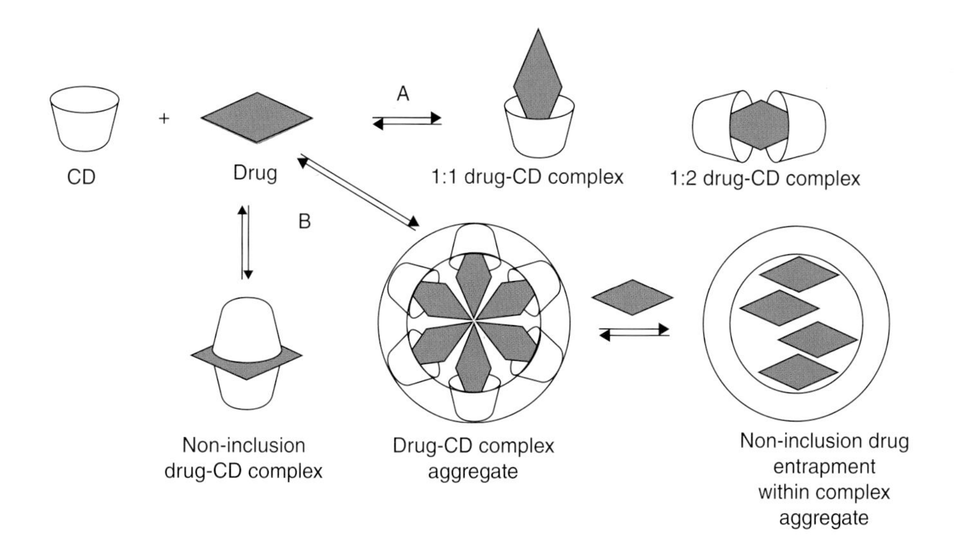

Figure 3.1 *Illustration of drug-CD complexes showing: path A – the ability to form 1 : 1 and 1 : 2 inclusion drug-CD complexes [83, 84]; B paths – possible mechanisms of non-inclusion drug-CD complex formation [85, 86]. Path A reprinted by permission from Macmillan Publishers Ltd: Nature Drug Reviews [83], copyright 2004 and reprinted from [84] with permission from Elsevier. B paths reprinted with permission from [85], with kind permission from Springer Science and Business Media and [86] with permission from Elsevier.*

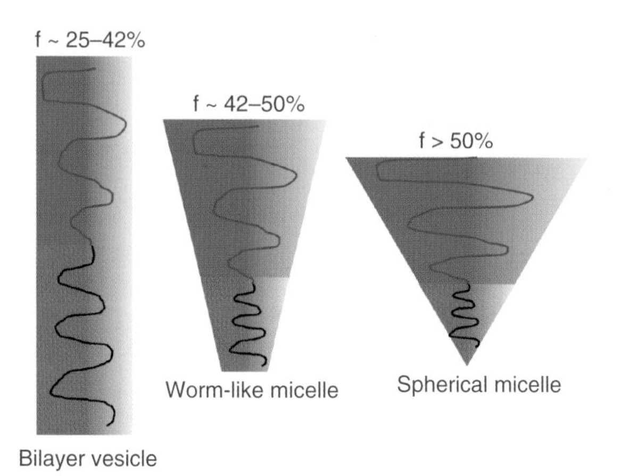

Figure 4.1 *As the hydrophilic mass fraction, f, increases the preferred morphology changes from a bilayer vesicle to worm-like micelle to spherical micelle morphology.*

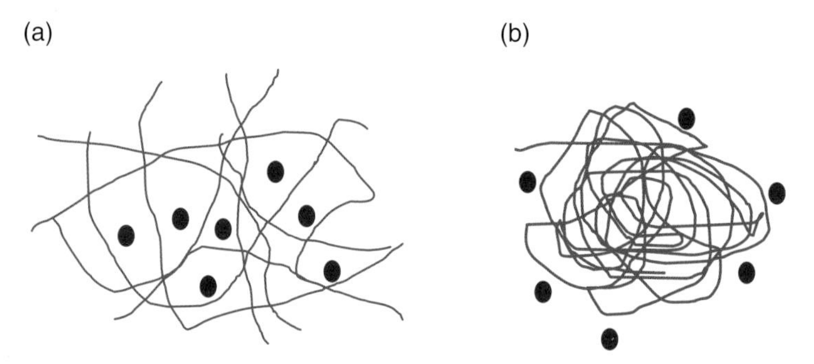

Figure 5.4 *Schematic representation of amorphous solid dispersions with polymer carriers: (a) the conventional model of amorphous solid solution [21] and (b) a new model of amorphous solid dispersions [22].*

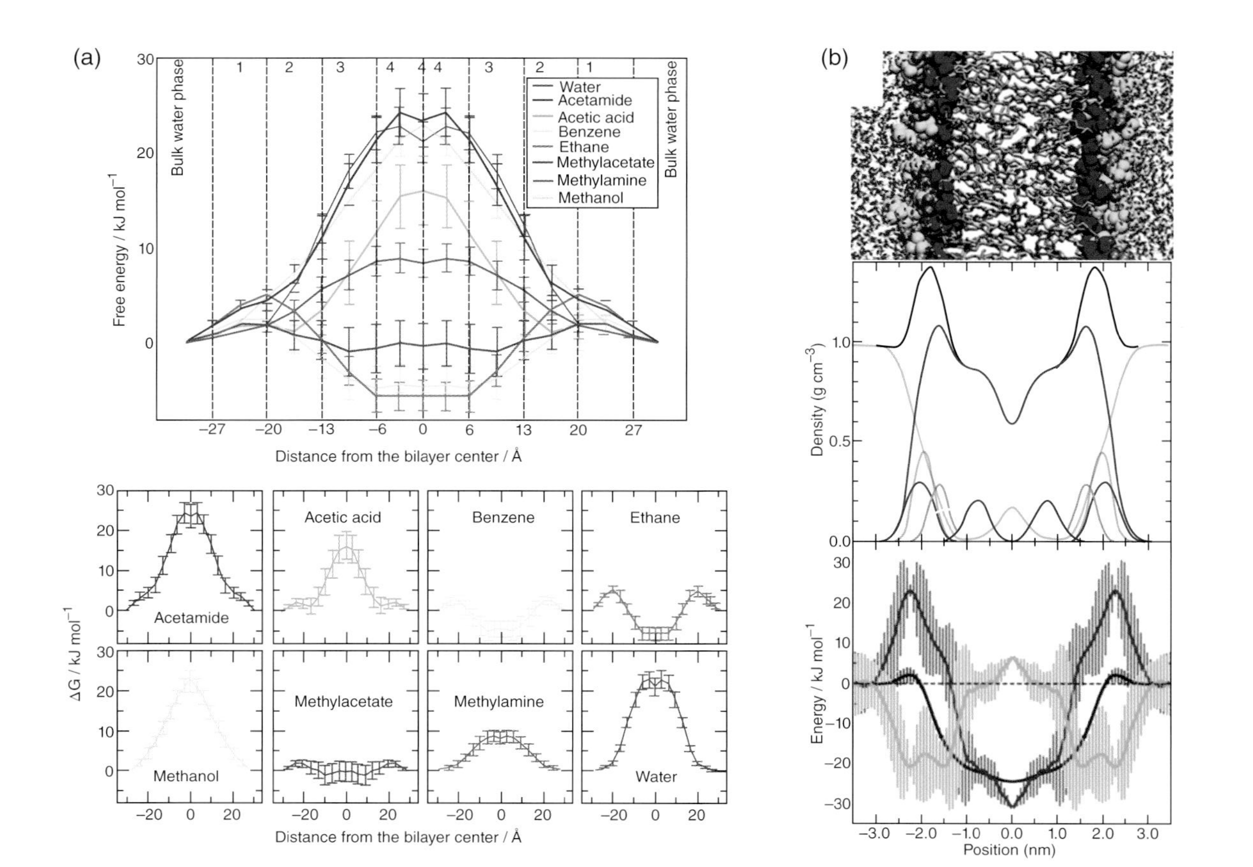

Figure 6.4 *(a) Free energy profiles of small molecules from constrained MD simulations. Top: the bilayer is divided into the four regions. Bottom: for clarity, each profile is plotted alone. Reprinted with permission from Ref. [37]. Copyright 2004 American Chemical Society. (b) Free energy of hexane in a DOPC bilayer. Top: snapshot of the lipid bilayer system used. Middle: average partial density profiles for various functional groups (black, total density; red, lipid; green, water; blue, choline; orange, phosphate; brown, glycerol; grey, carbonyl; purple, double bonds; cyan, methyl). Bottom: free energy of partitioning a hexane molecule from bulk water (black, free energy; red, entropic component of free energy, $-T\Delta S$; green, enthalpic component of free energy, ΔH). Reprinted with permission from Ref. [83]. Copyright 2005 American Chemical Society.*

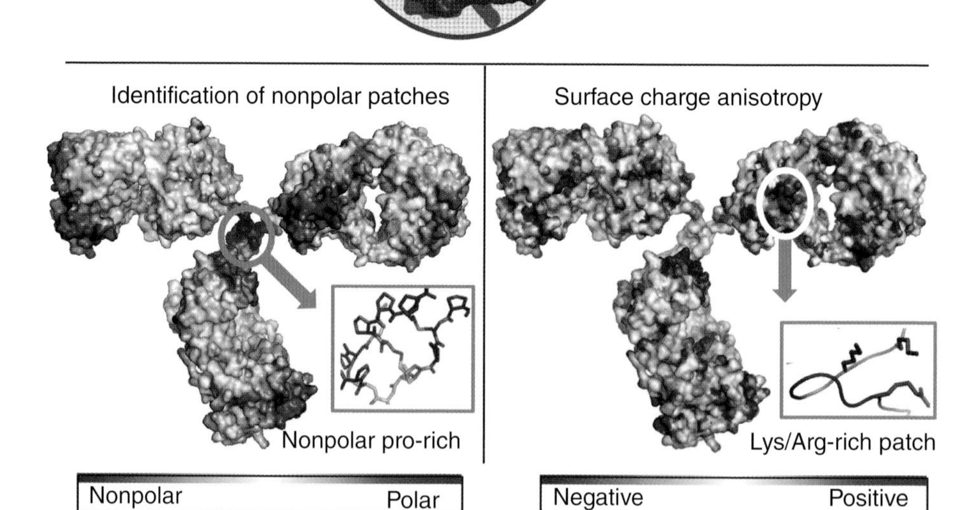

Figure 7.2 *Antibody and substituent structure, with graphical representation of non-polar and charged patches. IgG structure (code 1igt on the protein structural database) [7] is color-coded and labeled to show constituent domains, including, for example, the antigen-interacting* V_L *and* V_H *domains that can be linked to make a single chain variable fragment (scFv). Both of the lower panels use red–white–blue spectra to denote surface properties calculated with the same methods that give patch sizes [8]. The left-hand lower panel shows polarity, from most nonpolar (red) to most polar (blue), with a hydrophobic pro-rich sequence underlying the highlighted red patch. On the lower right-hand side is color-coding according to charge, from negative (red) to positive (blue), together with a lysine/arginine-rich sequence coincident with a particularly positive patch.*

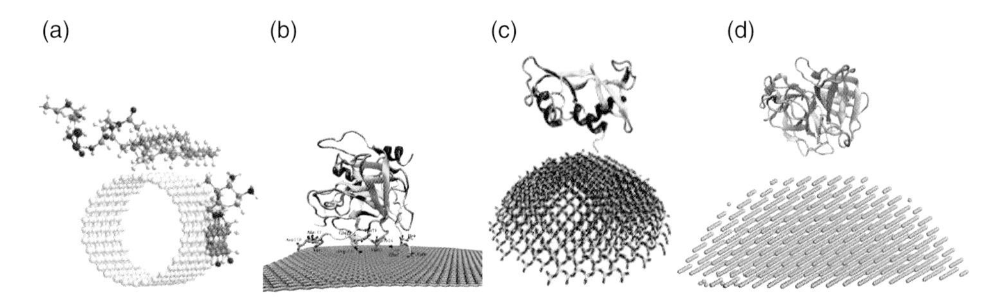

Figure 8.1 *Drug delivery strategies with: (a) carbon nanotubes, (b) graphene/graphene oxide, (c) silica nanoparticles, and (d) Au nanoparticles.*

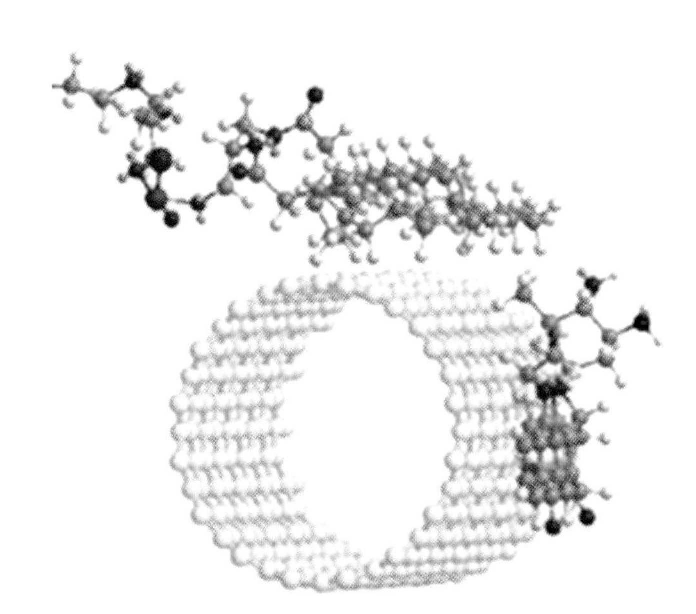

Figure 8.2 *Drug delivery strategies with carbon nanotubes. The hydrophobic drug molecules can be attached to the surface or the inner part of carbon nanotubes directly by van der Waals interaction. The drug molecules can also be attached to carbon nanotubes by its hydrophobic tail.*

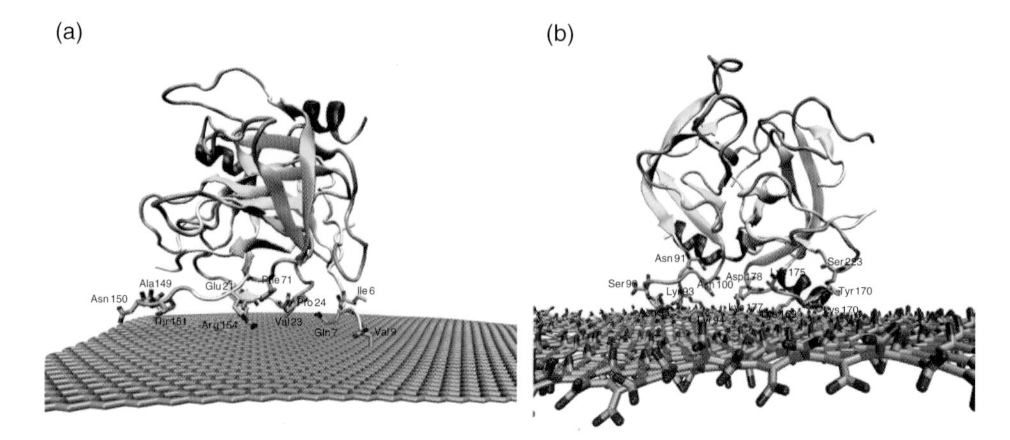

Figure 8.3 *Drug delivery strategies with: (a) graphene and (b) graphene oxide (GO). The hydrophobic drug molecules can be attached to the surface of graphene/GO directly.*

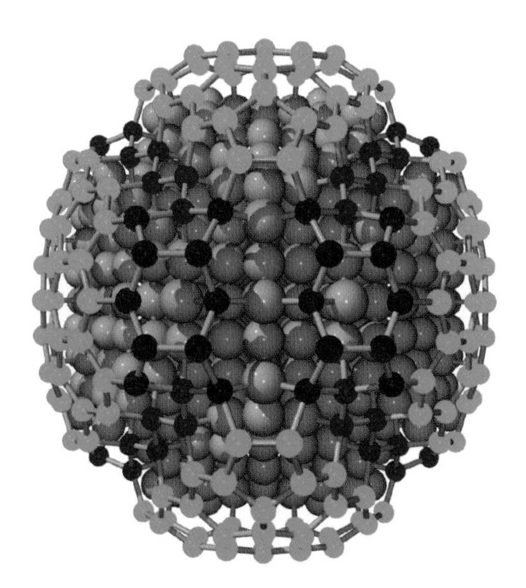

Figure 9.1 *Fully relaxed structure of a bare C_{705} ND. The outmost C atoms on the {100}, {110}, and {111} facets are highlighted in red, blue, and green, respectively [21]. Adapted from [21] with permission from The Royal Society of Chemistry.*

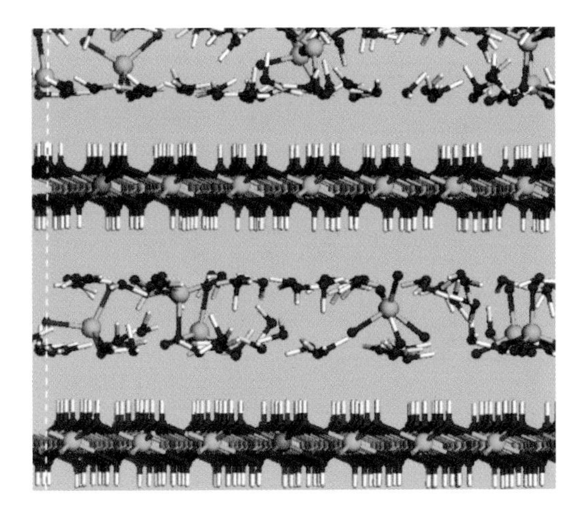

Figure 10.1 *Hydrotalcite 9×10×2 supercell intercalated with CrO_4^{2-} anions and water molecules. Green balls represent Mg atoms, pink=Al, red=oxygen, gray=hydrogen, and blue=chromium.*

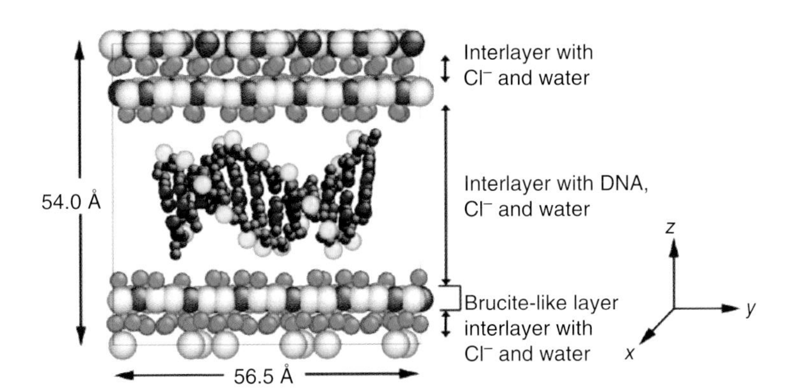

Figure 10.10 *Initial structure of LDH intercalated with DNA molecule with 12 bp [19]. Magnesium, aluminum, chlorine, phosphorus, carbon, and nitrogen atoms are represented as light-gray, pink, green, yellow, dark-gray, and blue spheres, respectively. Oxygen and hydrogen atoms have been removed in order to aid viewing. Reprinted with permission from Ref. [19]. Copyright 2008 American Chemical Society.*

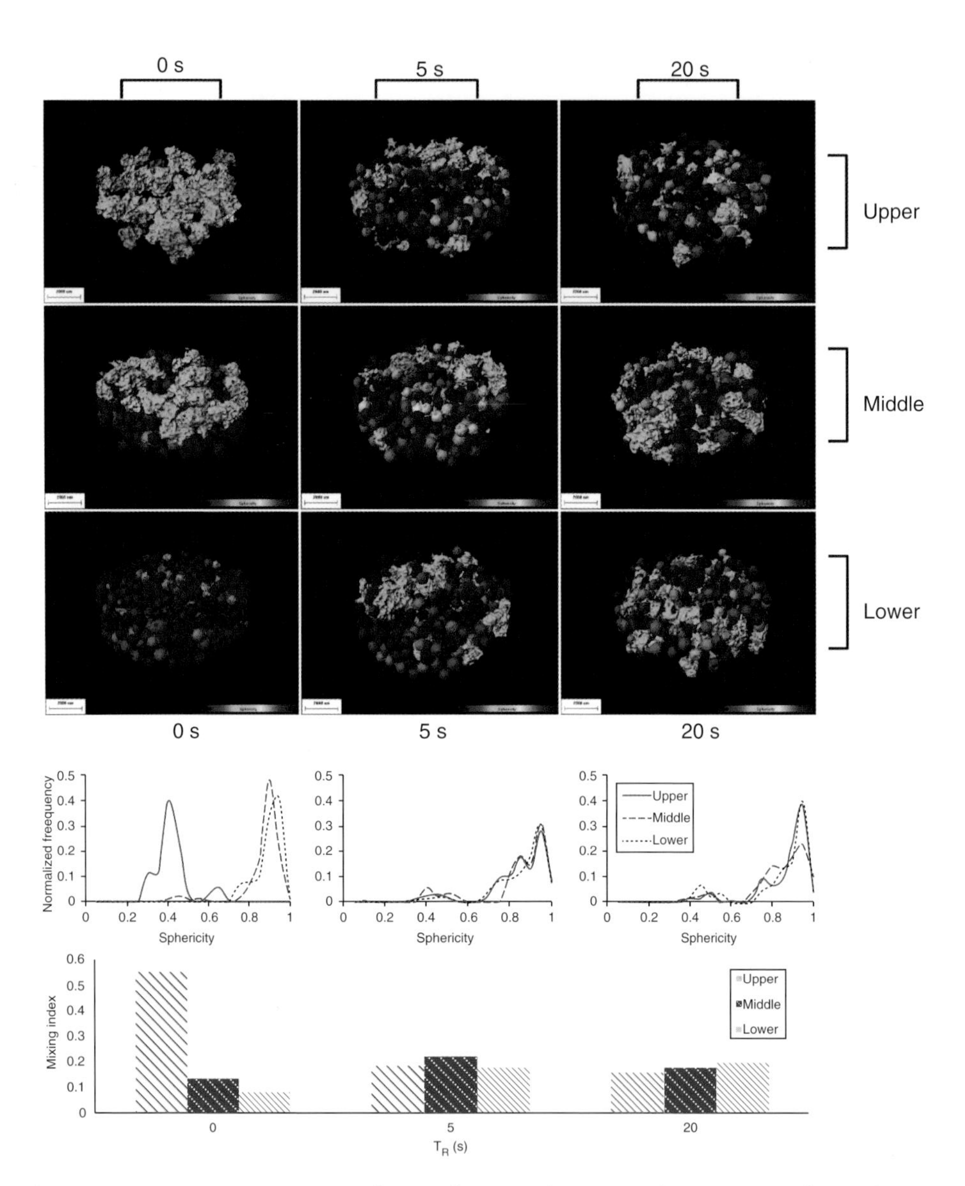

Figure 12.3 *3D reconstruction and quantification of mixing and segregation of granules [20]. Reproduced from Ref. [19], with permission from Elsevier.*

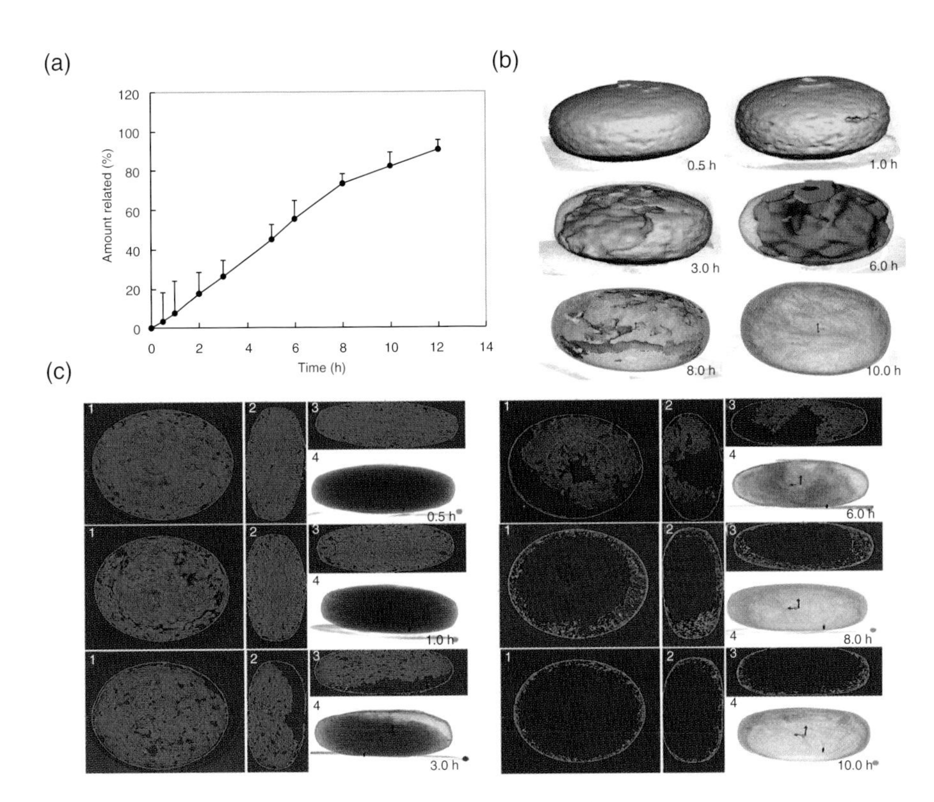

Figure 12.4 *(a) Amount of felodipine MOTS released. (b) Reconstructed 3D images of felodipine MOTS at different sampling times (yellow represents the solid moiety of the tablet core, air appears gray). (c) 2D monochrome X-ray CT images of felodipine MOTS viewing from four different aspects: 1. top, 2. front, 3. back, and 4. reconstructed image. Air appears dark, gray represents the solid moiety of the tablet core, and gray edge represents the semipermeable membrane [31]. Reproduced from [31], with permission from Elsevier.*

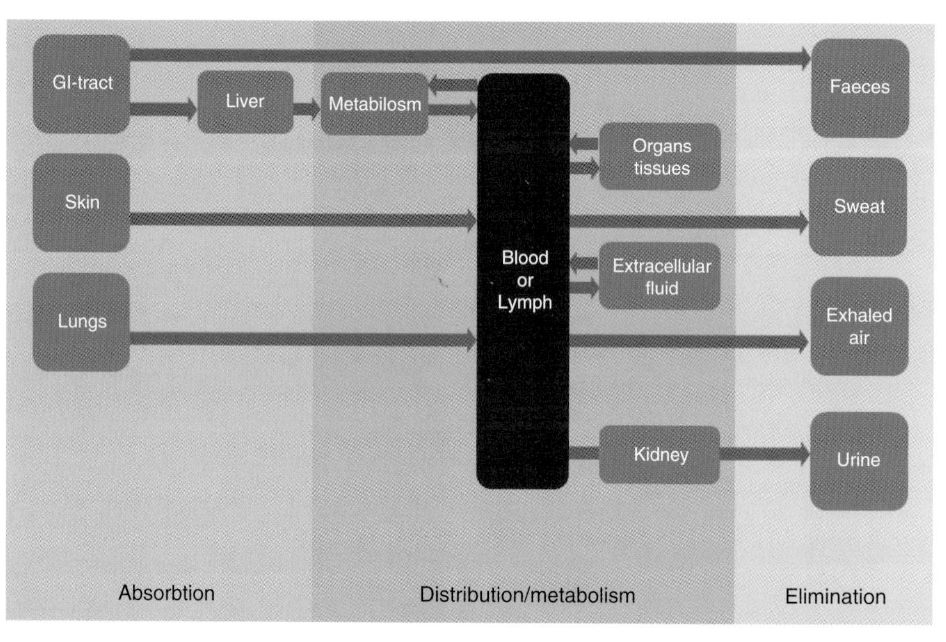

Figure 13.1 *Processes involved in absorption, distribution, metabolism and elimination.*

[36] Robinson, D.R. and W.P. Jencks (1965) Effect of concentrated salt solutions on activity coefficient of acetyltetraglycine ethyl ester. *Journal of the American Chemical Society* **87**(11): 2470.
[37] Schleich, T. and P.H. von Hippel (1969) Specific ion effects on solution conformation of Poly-l-proline. *Biopolymers* **7**(6): 861.
[38] Schrier, E.E. and E.B. Schrier (1967) Salting-out behavior of amides and its relation to denaturation of proteins by salts. *Journal of Physical Chemistry* **71**(6): 1851.
[39] Timasheff, S.N. (1993) The control of protein stability and association by weak-interactions with water – how do solvents affect these processes. *Annual Review of Biophysics and Biomolecular Structure* **22**: 67.
[40] Timasheff, S.N. (1998) Control of protein stability and reactions by weakly interacting cosolvents: the simplicity of the complicated. *Advances in Protein Chemistry*, **51**: 355.
[41] Record, M.T., C.F. Anderson, and T.M. Lohman (1978) Thermodynamic analysis of ion effects on binding and conformational equilibria of proteins and nucleic-acids - roles of ion association or release, screening, and ion effects on water activity. *Quarterly Reviews of Biophysics* **11**(2): 103.
[42] Auton, M. and D.W. Bolen (2007) Application of the transfer model to understand how naturally occuring osmolytes affect protein stability. *Osmosensing and Osmosignaling* **428**: 397.
[43] Pierce, V., Kang M, Aburi M, *et al.* (2008) Recent applications of Kirkwood-Buff theory to biological systems. *Cell Biochemistry and Biophysics* **50**(1): 1.
[44] Smith, P.E. (2004) Cosolvent interactions with biomolecules: relating computer simulation data to experimental thermodynamic data. *Journal of Physical Chemistry B* **108**(48): 18716.
[45] Smith, P.E. (2006) Equilibrium dialysis data and the relationships between preferential interaction parameters for biological systems in terms of Kirkwood-Buff integrals. *Journal of Physical Chemistry B* **110**: 2862.
[46] Collins, K.D. and M.W. Washabaugh (1985) The Hofmeister effect and the behavior of water at interfaces. *Quarterly Reviews of Biophysics* **18**(4): 323.
[47] Jungwirth, P. and D.J. Tobias (2002) Ions at the air/water interface. *Journal of Physical Chemistry B* **106**(25): 6361.
[48] Jungwirth, P. and D.J. Tobias (2006) Specific ion effects at the air/water interface. *Chemical Reviews* **106**(4): 1259.
[49] Onorato, R.M., D.E. Otten, and R.J. Saykally (2009) Adsorption of thiocyanate ions to the dodecanol/water interface characterized by UV second harmonic generation. *Proceedings of the National Academy of Sciences of the United States of America* **106**(36): 15176.
[50] Horinek, D., Herz A., Vrbka L., *et al.* (2009) Specific ion adsorption at the air/water interface: the role of hydrophobic solvation. *Chemical Physics Letters* **479**(4/6): 173.
[51] Horinek, D., S.I. Mamatkulov, and R.R. Netz (2009) Rational design of ion force fields based on thermodynamic solvation properties. *Journal of Chemical Physics* **130**(12): 124507.
[52] Bostrom, M., Tavares, F. W., Finet, S., *et al.* (2005) Why forces between proteins follow different Hofmeister series for pH above and below pI. *Biophysical Chemistry* **117**(3): 217.
[53] Bostroem, M., Parsons DF, Salis A, *et al.* (20011) Possible origin of the inverse and direct Hofmeister series for lysozyme at low and high salt concentrations. *Langmuir* **27**(15): 9504.
[54] Parsons, D.F., Boström M, Lo Nostro P, Ninham BW. (2011) Hofmeister effects: interplay of hydration, nonelectrostatic potentials, and ion size. *Physical Chemistry Chemical Physics* **13**(27): 12352.
[55] Lund, M., L. Vrbka, and P. Jungwirth (2008) Specific ion binding to nonpolar surface patches of proteins. *Journal of the American Chemical Society* **130**(35): 11582.
[56] Thomas, A.S. and A.H. Elcock (2011) Molecular dynamics simulations predict a favorable and unique mode of interaction between lithium (Li+) ions and hydrophobic molecules in aqueous solution. *Journal of Chemical Theory and Computation* **7**(4): 818.
[57] Pluharova, E., P.E. Mason, and P. Jungwirth (2013) Ion pairing in aqueous lithium salt solutions with monovalent and divalent counter-anions. *Journal of Physical Chemistry A* **117**(46): 11766.
[58] Collins, K.D. (2006) Ion hydration: implications for cellular function, polyelectrolytes, and protein crystallization. *Biophysical Chemistry* **119**(3): 271.
[59] Fennell, C.J., Bizjak A, Vlachy V, Dill KA. (2009) Ion pairing in molecular simulations of aqueous alkali halide solutions. *Journal of Physical Chemistry B* **113**(19): 6782.

[60] Jagoda-Cwiklik, B., Vacha R, Lund M, *et al.* (2007) Ion pairing as a possible clue for discriminating between sodium and potassium in biological and other complex environments. *Journal of Physical Chemistry B* **111**(51): 14077.
[61] Vrbka, L., Vondrásek J, Jagoda-Cwiklik B, *et al.* (2006) Quantification and rationalization of the higher affinity of sodium over potassium to protein surfaces. *Proceedings of the National Academy of Sciences of the United States of America* **103**(42): 15440.
[62] Heyda, J., Lund M, Oncák M, *et al.* (2010) Reversal of Hofmeister Ordering For Pairing of NH4+ vs alkylated ammonium cations with halide anions in water. *Journal of Physical Chemistry B* **114**(33): 10843.
[63] Heyda, J., T. Hrobarik, and P. Jungwirth (2009) Ion-specific interactions between halides and basic amino acids in water. *Journal of Physical Chemistry A* **113**(10): 1969.
[64] Kherb, J., S.C. Flores, and P.S. Cremer (2012) Role of carboxylate side chains in the cation hofmeister series. *Journal of Physical Chemistry B* **116**(25): 7389.
[65] Uejio, J.S., Schwartz CP, Duffin AM, *et al.* (2008) Characterization of selective binding of alkali cations with carboxylate by X-ray absorption spectroscopy of liquid microjets. *Proceedings of the National Academy of Sciences of the United States of America* **105**(19): 6809.
[66] Aziz, E.F., Ottosson N, Eisebitt S, *et al.* (2008) Cation-specific interactions with carboxylate in amino acid and acetate aqueous solutions: X-ray absorption and ab initio calculations. *Journal of Physical Chemistry B* **112**(40): 12567.
[67] Hess, B. and N.F.A. van der Vegt (2009) Cation specific binding with protein surface charges. *Proceedings of the National Academy of Sciences of the United States of America* **106**(32): 13296.
[68] Leontyev, I.V. and A.A. Stuchebrukhov (2010) Electronic continuum model for molecular dynamics simulations of biological molecules. *Journal of Chemical Theory and Computation* **6**(5): 1498.
[69] Leontyev, I. and A. Stuchebrukhov (2011) Accounting for electronic polarization in non-polarizable force fields. *Physical Chemistry Chemical Physics* **13**(7): 2613.
[70] Mason, P.E., E. Wernersson, and P. Jungwirth (2012) Accurate description of aqueous carbonate ions: an effective polarization model verified by neutron scattering. *Journal of Physical Chemistry B* **116**(28): 8145.
[71] Rembert, K.B., Paterová J., Heyda J., *et al.* (2012) Molecular mechanisms of ion-specific effects on proteins. *Journal of the American Chemical Society* **134**(24): 10039.
[72] Zhang, Y.J., Furyk S., Bergbreiter D.E., Cremer P.S. (2005) Specific ion effects on the water solubility of macromolecules: PNIPAM and the Hofmeister series. *Journal of the American Chemical Society* **127**(41): 14505.
[73] Cho, Y.H., Zhang Y., Christensen T., *et al.* (2008) Effects of Hofmeister anions on the phase transition temperature of elastin-like polypeptides. *Journal of Physical Chemistry B* **112**(44): 13765.
[74] Heyda, J., Vincent J.C., Tobias D.J., *et al.* (2010) Ion specificity at the peptide bond: molecular dynamics simulations of n-methylacetamide in aqueous salt solutions. *Journal of Physical Chemistry B* **114**(2): 1213.
[75] Dzubiella, J. (2008) Salt-specific stability and denaturation of a short salt-bridge-forming alpha-helixChang. *Journal of the American Chemical Society* **130**(42): 14000.
[76] Algaer, E.A. and N.F.A. van der Vegt (2011) Hofmeister ion interactions with model amide compounds. *Journal of Physical Chemistry B* **115**(46): 13781.
[77] Du, H., R. Wickramasinghe, and X. Qian (2010) Effects of salt on the lower critical solution temperature of poly (N-Isopropylacrylamide). *Journal of Physical Chemistry B* **114**(49): 16594.
[78] Du, H., S.R. Wickramasinghe, and X. Qian (2013) Specificity in cationic interaction with poly(N-isopropylacrylamide). *Journal of Physical Chemistry B* **117**(17): 5090.
[79] Okur, H.I., J. Kherb, and P.S. Cremer (2013) Cations bind only weakly to amides in aqueous solutions. *Journal of the American Chemical Society* **135**(13): 5062.
[80] Tome, L.I.N., Pinho S.P., Jorge M., *et al.* (2013) Salting-in with a salting-out agent: explaining the cation specific effects on the aqueous solubility of amino acids. *Journal of Physical Chemistry B* **117**(20): 6116.
[81] Hladilkova, J., Heyda J., Rembert K.B., *et al.* (2013) Effects of end-group termination on salting-out constants for triglycine. *Journal of Physical Chemistry Letters* **4**(23): 4069.

[82] Arakawa, T., Tsumoto K., Nagase K., Ejima D. (2007) The effects of arginine on protein binding and elution in hydrophobic interaction and ion-exchange chromatography. *Protein Expression and Purification* **54**(1): 110.
[83] Arakawa, T., Tsumoto K., Kita Y., *et al.* (2007) Biotechnology applications of amino acids in protein purification and formulations. *Amino Acids* **33**: 587.
[84] Vondrasek, J., Mason P.E., Heyda J., *et al.* (2009) The molecular origin of like-charge arginine-arginine pairing in water. *Journal of Physical Chemistry B* **113**(27): 9041.
[85] Vazdar, M., Vymětal J., Heyda J., *et al.* (2011) Like-charge guanidinium pairing from molecular dynamics and Ab initio calculations. *Journal of Physical Chemistry A* **115**(41): 11193.
[86] Shah, D., Li J., Shaikh A.R., Rajagopalan R. (2012) Arginine-aromatic interactions and their effects on arginine-induced solubilization of aromatic solutes and suppression of protein aggregation. *Biotechnology Progress* **28**(1): 223.
[87] Mason, P.E., Brady JW, Neilson GW, Dempsey CE. (2007) The interaction of guanidinium ions with a model peptide. *Biophysical Journal* **93**(1): L4.
[88] Shukla, D. and B.L. Trout (2011) Understanding the synergistic effect of arginine and glutamic acid mixtures on protein solubility. *Journal of Physical Chemistry* **115**(41): 11831.
[89] Vazdar, M., P. Jungwirth, and P.E. Mason (2013) Aqueous guanidinium-carbonate interactions by molecular dynamics and neutron scattering: relevance to ion-protein interactions. *Journal of Physical Chemistry B* **117**(6): 1844.
[90] Hunter, R.J. (2001) *Foundations of Colloid Science*. Oxford: Oxford University Press.
[91] Arzensek, D., D. Kuzman, and R. Podgornik (2012) Colloidal interactions between monoclonal antibodies in aqueous solutions. *Journal of Colloid and Interface Science* **384**: 207.
[92] Curtis, R.A., J.M. Prausnitz, and H.W. Blanch (1998) Protein-protein and protein-salt interactions in aqueous protein solutions containing concentrated electrolytes. *Biotechnology and Bioengineering* **57**(1): 11.
[93] Muschol, M. and F. Rosenberger (1995) Interactions in undersaturated and supersaturated lysozyme solutions: static and dynamic light scattering results. *Journal of Chemical Physics* **103**(24): 10424.
[94] Eberstein, W., Y. Georgalis, and W. Saenger (1994) Molecular-interactions in crystallizing lysozyme solutions studied by photon-correlation spectroscopy. *Journal of Crystal Growth.* **143**(1–2): 71.
[95] Curtis, R.A. and L. Lue (2005) Electrolytes at spherical dielectric interfaces. *Journal of Chemical Physics* **123**(17) 174702.
[96] Hatlo, M.M., R.A. Curtis, and L. Lue (2008) Electrostatic depletion forces between planar surfaces. *Journal of Chemical Physics* **128**(16): 164717.
[97] Arakawa, T. and S.N. Timasheff (1984) Mechanism of protein salting in and salting out by divalent-cation salts – balance between hydration and salt binding. *Biochemistry* **23**(25): 5912.
[98] Grigsby, J.J., H.W. Blanch, and J.M. Prausnitz (2000) Diffusivities of lysozyme in aqueous MgCl2 solutions from dynamic light-scattering data: effect of protein and salt concentrations. *Journal of Physical Chemistry B* **104**(15): 3645.
[99] Tessier, P.M., A.M. Lenhoff, and S.I. Sandler (2002) Rapid measurement of protein osmotic second virial coefficients by self-interaction chromatography. *Biophysical Journal* **82**(3): 1620.
[100] Ries-Kautt, M.M. and A.F. Ducruix (1989) Relative effectiveness of various ions on the solubility and crystal-growth of lysozyme. *Journal of Biological Chemistry* **264**(2): 745.
[101] Bonnete, F., S. Finet, and A. Tardieu (1999) Second virial coefficient: variations with lysozyme crystallization conditions. *Journal of Crystal Growth* **196**(2–4): 403.
[102] Zhang, Y.J. and P.S. Cremer (2009) The inverse and direct Hofmeister series for lysozyme. *Proceedings of the National Academy of Sciences of the United States of America* **106** (36): 15249.
[103] Schwierz, N., D. Horinek, and R.R. Netz (2010) Reversed anionic Hofmeister series: the interplay of surface charge and surface polarity. *Langmuir* **26**(10): 7370.
[104] Goegelein, C., Wagner D., Cardinaux F., *et al.* (2012) Effect of glycerol and dimethyl sulfoxide on the phase behavior of lysozyme: theory and experiments. *Journal of Chemical Physics* **136**(1): 015102.

[105] Liu, W., Bratko D., Prausnitz J.M., Blanch H.W. (2004) Effect of alcohols on aqueous lysozyme-lysozyme interactions from static light-scattering measurements. *Biophysical Chemistry* **107**(3): 289.
[106] Farnum, M. and C. Zukoski (1999) Effect of glycerol on the interactions and solubility of bovine pancreatic trypsin inhibitor. *Biophysical Journal* **76**(5): 2716.
[107] Javid, N., Vogtt K., Krywka C., *et al.* (2007) Protein-protein interactions in complex cosolvent solutions. *ChemPhysChem* **8**(5): 679.
[108] James, S. and J.J. McManus (2012) Thermal and solution stability of lysozynne in the presence of sucrose, glucose, and trehalose. *Journal of Physical Chemistry B* **116**(34): 10182.
[109] Neal, B.L., D. Asthagiri, and A.M. Lenhoff (1998) Molecular origins of osmotic second virial coefficients of proteins. *Biophysical Journal* **75**(5): 2469.
[110] Asthagiri, D., B.L. Neal, and A.M. Lenhoff (1999) Calculation of short-range interactions between proteins. *Biophysical Chemistry* **78**(3): 219.
[111] Elcock, A.H. and J.A. McCammon (2001) Calculation of weak protein-protein interactions: the pH dependence of the second virial coefficient. *Biophysical Journal* **80**(2): 613.
[112] Pellicane, G., G. Smith, and L. Sarkisov (2008) Molecular dynamics characterization of protein crystal contacts in aqueous solutions. *Physical Reviews Letters* **101**(24): 248102.
[113] Harrold, J.M., M. Ramanathan, and D.E. Mager (2013) Network-based approaches in drug discovery and early development. *Clinical Pharmacology and Therapeutics* **94**(6): 651.
[114] Roncaglioni, A., Toropov A.A., Toropova A.P., Benfenati E. (2013) In silico methods to predict drug toxicity. *Current Opinion in Pharmacology* **13**(5): 802.
[115] Lipinski, C.A. (2000) Drug-like properties and the causes of poor solubility and poor permeability. *Journal of Pharmacological and Toxicological Methods* **44**(1): 235.
[116] Murriel, C.L. and S.F. Dowdy (2006) Influence of protein transduction domains on intracellular delivery of macromolecules. *Expert Opinion on Drug Delivery* **3**(6): 739.
[117] Warwicker, J., S. Charonis, and R.A. Curtis (2013) Lysine and arginine content of proteins: computational analysis suggests a new tool for solubility design. *Molecular Pharmaceutics* **2**(4): 72.
[118] Kuroda, D., Shirai H., Jacobson M.P., Nakamura H. (2012) Computer-aided antibody design. *Protein Engineering Design and Selection* **25**(10): 507.
[119] Mach, H., Gregory S.M., Mackiewicz A., *et al.* (2011) Electrostatic interactions of monoclonal antibodies with subcutaneous tissue. *Therapeutic Delivery* **2**(6): 727.
[120] Fandrich, M., M.A. Fletcher, and C.M. Dobson (2001) Amyloid fibrils from muscle myoglobin. *Nature* **410** (6825): 165.
[121] Sunde, M., Serpell L.C., Bartlam M., *et al.* (1997) Common core structure of amyloid fibrils by synchrotron X-ray diffraction. *Journal of Molecular Biology* **273**(3): 729.
[122] Chiti, F., Stefani M., Taddei N., *et al.* (2003) Rationalization of the effects of mutations on peptide and protein aggregation rates. *Nature* **424**(6950): 805.
[123] Chou, P.Y. and G.D. Fasman (1974) Conformational parameters for amino acids in helical, beta-sheet, and random coil regions calculated from proteins. *Biochemistry* **13**(2): 211.
[124] Richardson, J.S. and D.C. Richardson (2002) Natural beta-sheet proteins use negative design to avoid edge-to-edge aggregation. *Proceedings of the National Academy of Sciences of the United States of America* **99**(5): 2754.
[125] Linding, R., Schymkowitz J., Rousseau F., *et al.* (2004) A comparative study of the relationship between protein structure and beta-aggregation in globular and intrinsically disordered proteins. *Journal of Molecular Biology* **342**(1): 345.
[126] Trovato, A., F. Seno, and S.C. Tosatto (2007) The PASTA server for protein aggregation prediction. *Protein Engineering Design and Selection* **20** (10): 521.
[127] Tartaglia, G.G. and M. Vendruscolo (2008) The Zyggregator method for predicting protein aggregation propensities. *Chemical Society Reviews.* **37** (7): 1395.
[128] Tartaglia, G.G., Pechmann S, Dobson CM, Vendruscolo M. (2009) A relationship between mRNA expression levels and protein solubility in E. coli. *Journal of Molecular Biology* **388** (2): 381.
[129] Zimmerman, S.B. and A.P. Minton (1993) Macromolecular crowding: biochemical, biophysical, and physiological consequences. *Annual Review of Biophysics and Biomolecular Structure* **22**: 27.

[130] de Groot, N.S., Castillo V., Graña Montes R., *et al.* (2012) AGGRESCAN: method, application, and perspectives for drug design. *Methods in Molecular Biology (Clifton)* **819**: 199.
[131] Tsolis, A.C., Papandreou, N.C., Iconomidou, V.A., Hamodrakas, S.J. (2013) A consensus method for the prediction of "aggregation-prone" peptides in globular proteins. *PLoS One* **8**(1): e54175.
[132] Niwa, T., Ying B.W., Saito K., *et al.* (2009) Bimodal protein solubility distribution revealed by an aggregation analysis of the entire ensemble of Escherichia coli proteins. *Proceedings of the National Academy of Sciences of the United States of America* **106**(11): 4201.
[133] Wilkinson, D.L. and R.G. Harrison (1991) Predicting the solubility of recombinant proteins in Escherichia coli. *Biotechnology* **9**(5): 443.
[134] Idicula-Thomas, S. and P.V. Balaji (2007) Correlation between the structural stability and aggregation propensity of proteins. *In Silico Biology* **7**(2): 225.
[135] Smialowski, P., Martin-Galiano A.J., Mikolajka A., *et al.* (2007) Protein solubility: sequence based prediction and experimental verification. *Bioinformatics* **23**(19): 2536.
[136] Magnan, C.N., A. Randall, and P. Baldi (2009) SOLpro: accurate sequence-based prediction of protein solubility. *Bioinformatics* **25**(17): 2200.
[137] Chen, L., Oughtred R., Berman H.M., Westbrook J. (2004) TargetDB: a target registration database for structural genomics projects. *Bioinformatics* **20**(16):2860.
[138] Price, W.N., II, Handelman SK, Everett JK, *et al.* (2011) Large-scale experimental studies show unexpected amino acid effects on protein expression and solubility in vivo in E. coli. *Microbial Informatics and Experimentation* **1**(1): 6.
[139] Singh, G.P. and D. Dash (2013) Electrostatic mis-interactions cause overexpression toxicity of proteins in E. coli. *PLoS One* **8**(5): e64893.
[140] Jones, S., A. Marin, and J.M. Thornton (2000) Protein domain interfaces: characterization and comparison with oligomeric protein interfaces. *Protein Engineering* **13**(2): 77.
[141] Chennamsetty, N., Voynov V., Kayser V., *et al.* (2009) Design of therapeutic proteins with enhanced stability. *Proceedings of the National Academy of Sciences of the United States of America* **106**(29): 11937.
[142] Chennamsetty, N., Voynov V., Kayser V., *et al.* (2010) Prediction of aggregation prone regions of therapeutic proteins. *Journal of Physical Chemistry* **114**(19): 6614.
[143] Kramer, R.M., Shende, V.R., Motl, N., *et al.* (2012) Toward a molecular understanding of protein solubility: increased negative surface charge correlates with increased solubility. *Biophysical Journal* **102**(8): 1907.
[144] Dudgeon, K., Rouet R., Kokmeijer I., *et al.* (2012) General strategy for the generation of human antibody variable domains with increased aggregation resistance. *Proceedings of the National Academy of Sciences of the United States of America* **109**(27): 10879.
[145] Kvam, E., Sierks, M.R., Shoemaker, C.B., *et al.* (2010) Physico-chemical determinants of soluble intrabody expression in mammalian cell cytoplasm. *Protein Engineering Design and Selection* **23**(6):489.
[146] Wayne, N. and D.N. Bolon (2010) Charge-rich regions modulate the anti-aggregation activity of Hsp90. *Journal of Molecular Biology* **401** (5): 931.
[147] Neal, B.L., Asthagiri D., Kaler E.W. (1999) Why is the osmotic second virial coefficient related to protein crystallization? *Journal of Crystal Growth* **196**(2–4): 377.

8

Computational Simulation of Inorganic Nanoparticle Drug Delivery Systems at the Molecular Level

Xiaotian Sun, Zhiwei Feng, Tingjun Hou, and Youyong Li

Institute of Functional Nano and Soft Materials (FUNSOM), Soochow University, China

8.1 Introduction

A primary goal in drug delivery research and drug development in general is to identify agents and/or delivery systems that enhance drug efficacy at the intended site of action while reducing toxicity to healthy tissues [1]. To a large extent, the role of drug delivery in achieving this goal is closely related to the ability of a given drug to permeate cell membranes and the degree to which membrane permeation and drug distribution can be selectively manipulated through candidate selection or chemical modification, administration route, dosage regimen, or delivery system design [1b, 2]. For protein drugs and gene therapy, safe and effective drug delivery approach is critical for the success of these macromolecules [3].

An optimal carrier for targeting drugs should have the following characteristics:

1. Restriction of drug distribution to the target area, organ, tissue, cell, and compartments;
2. Undergoing a capillary-level distribution;
3. Prolonged control of the localization of the drugs;
4. Transport of the drugs mainly to the target cells, for example, tumor cells;
5. Controlled rate of drug release;

Computational Pharmaceutics: Application of Molecular Modeling in Drug Delivery, First Edition.
Edited by Defang Ouyang and Sean C. Smith.

6. Drug release without significantly reducing biological activities;
7. Therapeutical concentration of drugs at the desired molecular target;
8. Minimized leakage of free drug during transit in the bloodstream;
9. Protection from inactivation by plasma enzymes;
10. Biodegradation with subsequent elimination and minimal toxicity;
11. Manifold spectrum of substances (diagnostics and therapeutics);
12. Easy preparation.

The predominant methods to deliver drugs are oral and injection, which has limited the progress of drug development. Most drugs have been formulated to accommodate the oral or injection delivery routes, which are not always the most efficient routes for a particular therapy. New biologic drugs such as proteins and nucleic acids require novel delivery technologies [4], which could minimize side effects and lead to better patient compliance.

Market forces are also driving the need for new, effective drug delivery methods. It is estimated that drug delivery will account for more than 40% of all pharmaceutical sales by 2013 [5]. Meanwhile, upcoming patent expirations are driving pharmaceutical companies to reformulate their products.

New drug delivery methods may enable pharmaceutical companies to develop new formulations of off-patent and soon to be off-patent drugs. Reformulating old drugs can reduce side effects and increase patient compliance, thus saving money on health care delivery. Furthermore, drug candidates that did not pass through the trials phases may be reformulated to be used with new drug delivery systems.

Innovative drug delivery systems may make it possible to use certain chemical entities or biologics that were previously impractical because of toxicities or because they were impossible to administer. For example, drug targeting is enabling the delivery of chemotherapy agents directly to tumors, reducing systemic side effects [6].

Researchers are continually investigating new ways to deliver macromolecules that will facilitate the development of new biologic products such as bioblood proteins and biovaccines. Similarly, the success of DNA and RNA therapies will depend on innovative drug delivery techniques [7]. Many times, the success of a drug is dependent on the delivery method. This importance is exemplified by the presence of more than 300 companies based in the United States involved with developing drug delivery platforms.

In addition to the commonly used oral and injection routes, drugs can also be administered through other means, including transdermal, transmucosal, ocular, pulmonary, and implantation. The mechanisms used to achieve alternative drug delivery typically incorporate one or more of the following materials: biologics, polymers, silicon-based materials [8], carbon-based materials [9], or metals [10]. These materials are structured in microscale and, more recently, nanoscale formats [11].

The efficiency of drug delivery to various parts of the body is directly affected by particle size [12]. Nanostructure-mediated drug delivery, a key technology for the realization of nanomedicine, has the potential to enhance drug bioavailability, improve the timed release of drug molecules, and enable precision drug targeting. Nanoscale drug delivery systems can be implemented within pulmonary therapies, as gene delivery vectors, and in the stabilization of drug molecules that would otherwise degrade too rapidly. Additional benefits of using targeted nanoscale drug carriers are reduced drug toxicity and more efficient drug distribution.

Anatomic features such as the blood–brain barrier, the branching pathways of the pulmonary system, and the tight epithelial junctions of the skin make it difficult for drugs to reach many desired physiologic targets. Nanostructured drug carriers will help to penetrate or overcome these barriers of drug delivery. Courrier *et al.* have shown that the greatest efficiency for delivery into the pulmonary system is achieved for particle diameters of <100 nm [13]. Greater uptake efficiency has been shown for gastrointestinal absorption and transcutaneous permeation, with particle sizes around 100 and 50 nm [14], respectively. However, such small particles traveling in the pulmonary tract may also have a greater chance of being exhaled. Larger, compartmental or multilayered drug carrier architectures may help with delivery to the pulmonary extremities [15]. For instance, the outer layers of the carrier architecture may be formulated to biodegrade as the carrier travels through the pulmonary tract. As the drug carrier penetrates further into the lung, additional shedding will allow the encapsulated drug to be released. Biodegradable nanoparticles of gelatin and human serum albumin show promise as pulmonary drug carriers [16].

Advantages of nanostructure-mediated drug delivery include the ability to deliver drug molecules directly into cells and the capacity to target tumors within healthy tissue. For example, DNA and RNA that is packaged within a nanoscale delivery system can be transported into the cell to fix genetic mutations or alter gene expression profiles [7]. The mechanisms of cellular uptake of external particulates include clathrin- and caveoli-mediated endocytosis, pinocytosis, and phagocytosis [17]. However, phagocytosis may not play a role in the uptake of nanoscale particles because of the small size of such particles.

Nanoscale drug delivery architectures are able to penetrate tumors due to the discontinuous, or bleaky, nature of the tumor microvasculature, which typically contains pores ranging from 100 to 1000 nm in diameter. The microvasculature of healthy tissue varies by tissue type, but in most tissues including the heart, brain, and lung, there are tight intercellular junctions less than 10 nm [17]. Therefore, tumors within these tissue types can be selectively targeted by creating drug delivery nanostructures greater than the intercellular gap of the healthy tissue but smaller than the pores found within the tumor vasculature.

Computational methods to predict drug delivery prior to experiments are increasingly desirable to minimize the investment in drug design and development. Significant progress in molecular dynamics (MD) simulation methodologies has been made to study drug delivery [18]. Such methods are directly applicable to the design and optimization of drug delivery systems. MD simulations are particularly valuable in addressing issues that are difficult to be explored in laboratory experiments for drug delivery. This review is intended to review recent progress resulting from computational simulations of drug delivery systems including our latest MD study of drug delivery, which provides us a theoretical view of drug delivery process.

There are several categories for drug delivery systems. One of them is carbon nanotubes (CNTs) as shown in Figure 8.1a. Another category includes graphene and graphene oxide (GO), as shown in Figure 8.1b. Figure 8.1c illustrates drug delivery by silica nanoparticles (SNPs). Gold (Au) nanoparticle is another possible route for drug delivery, as shown in Figure 8.1d. In the following sections, we will review computational simulations in these four categories separately.

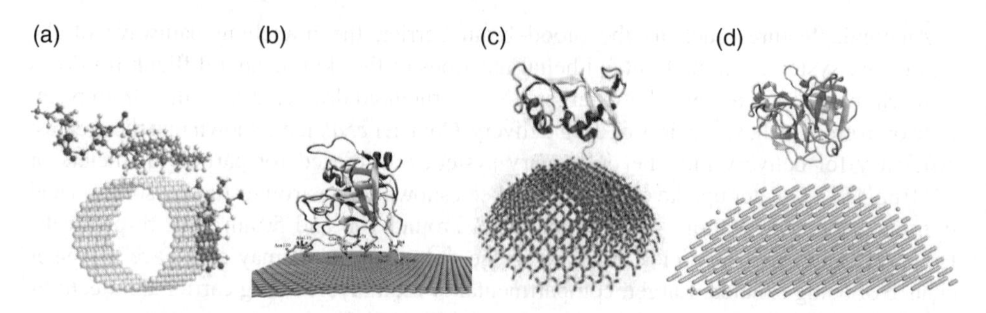

Figure 8.1 Drug delivery strategies with: (a) carbon nanotubes, (b) graphene/graphene oxide, (c) silica nanoparticles, and (d) Au nanoparticles (see colour plate section).

8.2 Materials and Methods

8.2.1 Prepared Structures

Crystal structures are retrieved from the Protein Data Bank (http://www.pdb.org/pdb/) and other structures can be constructed by homology modeling. Then the structures are preparing by software, such as Discovery Studio (including residues repair, loops repair, and energy minimization).

Histidine residue is the only one which ionizes within the physiological pH range (~7.4). To determine the protonation states for histidines and other residues, Discovery Studio and other software are used to predict protein ionization and residue pK values.

Nanomaterials can be collected from Material Studio or VMD, which should provide a sufficient surface for simulations. The nanomaterial modes reflect a typical outcome of a standard oxidation process.

Then the models are prepared by Material Studio. Geometry optimization is calculated by using the Forcite module. The Dreiding force field, Gasteiger (maximum iteration is set at 50000, convergence limit is set at 5.0×10^{-6}) and ultra-fine quality are using for the energy calculation. The output structures are regarded as the initial structures for the following simulations.

For drug molecules (or groups) conjugated to nanomaterials through covalent attachment,the strategies to link a protein to the nanomaterials have taken four main approaches: (i) electrostatic adsorption, (ii) conjugation to the ligand on the nanomaterials surface, (iii) conjugation to a small cofactor molecule that the protein can recognize and bind, and (iv) direct conjugation to the NP surface. Other strategies are described in the review of Medintz *et al.* [19].

8.2.2 MD Simulations

Atoms of nanomaterials are uncharged in accordance with Hummer *et al.* [60].

In simulations without adsorbed proteins, the nanomaterials are free, while in simulations with adsorbed proteins, the nanomaterials are fixed during the simulation. The initial minimum (MIN) distance between nanomaterial surface and adsorbed protein is 2.0 nm; that is, the protein and nanomaterials are not in contact at the beginning of the simulations.

Then the free nanomaterials (or the nanomaterials with adsorbed proteins) are embedded in periodic boundary conditions in a rectangular water box (TIP3P water model) with an efficient size, and the waters within 5/2 Å of the protein/nanomaterials are eliminated.

The system is first equilibrated for 200 ps to 1 ns with the protein/nanomaterials fixed. Then the protein is released and another 200 ps to 1 ns equilibration is performed.

Starting from the last frame of the equilibration, 100 or more nanoseconds of MD simulations are performed. The MD simulations are performed by software, such as GROMACS, AMBER, or NAMD. Trajectory analyses can be carried out with VMD.

8.2.3 Computational Simulation of Drug Delivery Strategies with CNTs

CNTs are low-dimensional sp^2 carbon nanomaterials exhibiting many unique physical and chemical properties that are interesting in a wide range of areas, including nanomedicine. Since 2004, CNTs have been extensively explored as drug delivery carriers for the intracellular transport of chemotherapy drugs [20], proteins [21], and genes [22]. In vivo cancer treatment with CNTs has been demonstrated in animal experiments by several different groups.

Open-ended single-walled carbon nanotubes (SWCNTs) consist of a seamless and hollow cylindrical structure with carbon atoms as backbone [23]. Consequently, they can provide internal cavities that are capable of accommodating biomolecules, yielding an immense potential in drug delivery systems and biotechnology. They are found experimentally to be noncytotoxic in biological environments, and therefore, they are capable of transporting various molecules across cellular membranes via the endocytosis pathway without causing cell damage or death. It has been reported that many molecules [e.g., fullerenes (C60), nonfullerene molecules, DNA oligonucleotides, peptides, protein fragments, and amylase] can be encapsulated into CNTs [18]. The surface of CNTs can play an important role for drug delivery applications too, as shown in Figure 8.2.

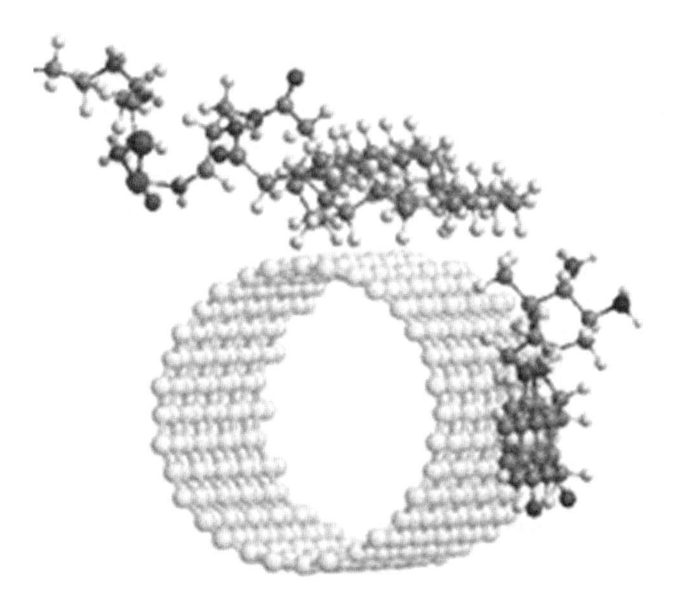

Figure 8.2 Drug delivery strategies with carbon nanotubes. The hydrophobic drug molecules can be attached to the surface or the inner part of carbon nanotubes directly by van der Waals interaction. The drug molecules can also be attached to carbon nanotubes by its hydrophobic tail (see colour plate section).

Liu and Wang reported MD simulation results on the insertion of the peptide drug Zadaxin into CNTs in a water solvent. Zadaxin is the synthetic version of thymosin, a substance that is demonstrated by many experimental tests to be a safe, mild, and effective treatment for chronic hepatitis B in clinical use. The authors showed that Zadaxin is capable of being encapsulated into CNTs with a suitable size, length, and in a water solution. The CNT-Zadaxin attractive interaction and the solvation effect of water play dominant roles in the process of this insertion. The diameter of the (14, 14) CNT was suggested to be the critical size for the Zadaxin insertion [24].

Liu's group put further efforts into studying the interaction between Zadaxin and CNTs by using MD and a series of steered molecular dynamics (SMD) simulations. They found that Zadaxin is encapsulated inside the nanotube after its spontaneous insertion and oscillates around the center of the tube, where the van der Waals interaction energy is observed to be a minimum [23]. SMD results show a maximum value of the pulling force, which demonstrates the ability of the CNTs to trap protein/peptide drugs. They also observed a spontaneous encapsulation of a globular protein into the CNT through MD simulations. And they found that the free energy of the system decreases after encapsulation. The enthalpy decrease was found to make a dominant contribution to the free energy change. During the insertion, the protein makes a stepwise conformational change to maximize its affinity to the CNT walls as well as the protein-CNT interactions.

Water self-diffusion through CNTs is another attractive topic for computer simulation of CNTs for drug delivery applications. Striolo performed a series of MD simulations to understand the extent to which the presence of a few oxygenated active sites, modeled as carbonyls, affects the transport properties of confined water [25]. At low hydration levels, the author found little water diffusion. The diffusion, which appears to be of the Fickian type for sufficiently large hydration levels, becomes faster as the number of confined water molecules increases, reaches a maximum, and slows as water fills the CNTs. The author explained the findings on the basis of two collective motion mechanisms observed from the analysis of sequences of simulation snapshots: "cluster breakage" and "cluster libration" mechanisms. The cluster breakage mechanism produces longer displacements for the confined water molecules than the cluster libration one, but deactivates as water fills the CNT.

Cheng *et al.* reported MD simulation of CNT-based drug delivery and release systems. They show that a peptide encapsulated inside or attached to the outer surface of a CNT can be released by another nanotube through a competitive replacement process [18]. Energy analysis shows that the van der Waals interaction plays the key role in the process and the potential well between two nanotubes drives the competitive replacement.

Noncovalent adsorption of proteins onto CNTs is important to understand the environmental and biological activity of CNTs as well as their potential applications in nanostructure fabrication. Shen *et al.* performed the adsorption dynamics and features of a model protein (the A sub-domain of human serum albumin) onto the surfaces of CNTs with different diameters by MD simulation [26]. The adsorption behaviors were observed by both trajectory and quantitative analyses. During the adsorption process, the secondary structures of α-helices in the model protein were slightly affected. However, the random coils connecting these α-helices were strongly affected and this changed the tertiary structure of the protein. The conformation and orientation selection of the protein were induced by the properties and texture of the surfaces, as indicated by the interaction curve. In addition, the stepwise adsorption dynamics of these processes were found.

Zuo *et al.* used large-scale MD simulations to study the interaction between several proteins (WW domains) and CNTs (one form of hydrophobic nanoparticles) [27]. They found that the CNT can plug into the hydrophobic core of proteins to form stable complexes. This plugging of nanotubes disrupts and blocks the active sites of WW domains from binding to the corresponding ligands, thus leading to the loss of the original function of the proteins. The key to this observation is the hydrophobic interaction between the nanoparticles and the hydrophobic residues, particularly tryptophans, in the core of the domain, which might provide a novel route to nanoparticle toxicity on the molecular level for the hydrophobic nanoparticles.

The influential factors, namely chirality, temperature, radius, and surface chemical modification, of the interaction energy for a polyethylene (PE) molecule encapsulated into single-walled CNTs (SWNTs) was investigated by molecular mechanics (MM) and MD simulation by Li *et al.* [28]. The results showed that all these factors influence the interaction energy between PE and SWNTs. The interaction energy between PE molecule and armchair SWNTs is largest among eight kinds of chiral SWNTs. The interaction energy decreases with increasing temperature or SWNT radius. Methyl, phenyl, hydroxyl, carboxyl, –F, and amino groups were introduced onto the surface of the SWNTs by simulation software and the influence of SWNT chemical modification was also investigated [29]. The interaction energy between PE and chemically modified SWNTs is larger than that between PE and pristine SWNTs, and increases monotonously with increasing concentration of modified groups. In addition, the group electronegativity and van der Waals force affect the interaction energy between PE and chemically modified SWNTs greatly, which can be attributed to the electronic structures of the chemically modified groups [28]. This study can provide some useful suggestions for composite material design and drug transport.

Recently, Cheng *et al.* explored the diameter selectivity of dynamic self-assembly for single-strand DNA (ssDNA) encapsulation in double-walled nanotubes (DWNTs) via MD simulation [30]. Moreover, the pulling out process was carried out by SMD simulations. Considering π–π stacking and solvent accessibility together, base-CNT binding should be strongest on a graphene sheet and weakest on the inner CNT surface. When pulling the ssDNA out of a single-walled carbon nanotube (SWNT), the force exhibits characteristic fluctuations around a plateau about 300 pN. Each fluctuation force pulse to pull ssDNA corresponds to the exit of one base. In addition, the solvents used for the system are also of significant interest. Water plays an important role in the encapsulation process but not in the pulling out process.

8.2.4 Computational Simulation of Drug Delivery Strategies with Graphene/GO

In the past few years, graphene and its water-soluble derivative, GO, have attracted huge attention owing to their interesting physical and chemical properties and shown wide applications in various fields, including biotechnology and biomedicine. GO, in particular, possesses a single-layered, two-dimensional, sp^2 hybrid structure with sufficient surface groups, offering a unique double-sided, easily accessible substrate for multivalent functionalization and efficient loading of molecules from small organic ones

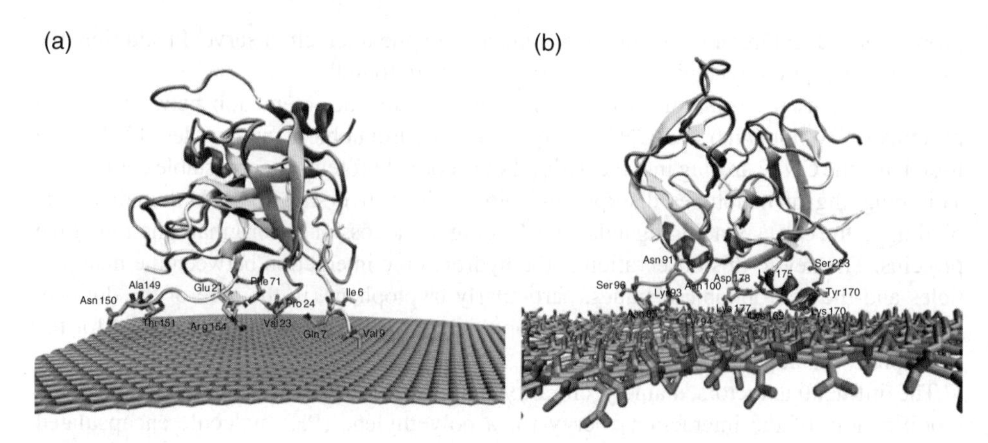

Figure 8.3 *Drug delivery strategies with: (a) graphene and (b) graphene oxide (GO). The hydrophobic drug molecules can be attached to the surface of graphene/GO directly (see colour plate section).*

to biomacromolecules [31]. The potential of functionalized GO lies in gene and drug delivery, cellular imaging, cancer therapeutics, biosensing, as well as an antibacterial agent [32]. For instance, GO nanosheets modified with PE glycol have been employed as aqueous compatible carriers for water-insoluble drug delivery. The intrinsic oxygen-containing functional groups were used as initial sites for the deposition of metal nanoparticles and organic macromolecules, such as porphyrin, on the GO sheets, which opened up a novel route to multifunctional nanometer-scaled catalytic magnetic and optoelectronic materials (Figure 8.3).

MD simulations have been performed to investigate the mechanical properties of hydrogen functionalized graphene for H-coverages spanning the entire range from graphene (H-0%) to graphane (H-100%) [33]. Pei *et al.* find that the Young's modulus, tensile strength, and fracture strain of the functionalized graphene deteriorate drastically with increasing H-coverage up to about 30% [33]. Beyond this limit the mechanical properties remain insensitive to H-coverage. While the Young's modulus of graphane is smaller than that of graphene by 30%, the tensile strength and fracture strain show a much larger drop of about 65%. They show that this drastic deterioration in mechanical strength arises both from the conversion of sp^2 to sp^3 bonding and due to easy rotation of unsupported sp^3 bonds. Their results suggest that the coverage-dependent deterioration of the mechanical properties must be taken into account when analyzing the performance characteristics of nanodevices fabricated from functionalized graphene sheets.

Moreover, MD simulations have been performed to study the mechanical properties of methyl (CH_3) functionalized graphene [34]. It is found that the mechanical properties of functionalized graphene greatly depend on the location, distribution, and coverage of CH_3 radicals on graphene. Surface functionalization exhibits a much stronger influence on the mechanical properties than edge functionalization. For patterned functionalization on graphene surfaces, the radicals arranged in lines perpendicular to the tensile direction lead to larger strength deterioration than those parallel to the tensile direction [35]. For random functionalization, the elastic modulus of graphene decreases gradually with increasing CH_3

coverage, while both the strength and fracture strain show a sharp drop at low coverage. When CH_3 coverage reaches saturation, the elastic modulus, strength, and fracture strain of graphene drop by as much as 18, 43, and 47%, respectively [34].

MD simulations were performed to study interaction between the graphene nanoribbon (GNR) and SWCNT by the group of Liu [36]. The GNR enters the SWCNT spontaneously to display a helical configuration which is quite similar to the chloroplast in a *Spirogyra* cell. This unique phenomenon results from the combined action of the van der Waals potential well and the π–π stacking interaction. The size of SWCNT and GNR should satisfy some certain conditions in the helical encapsulation process. A DNA-like double helix would be formed inside the SWCNT with the encapsulation of two GNRs [37]. A water cluster enclosed in the SWCNT affects the formation of the GNR helix in the tube significantly. Furthermore, they also studied the possibility that the spontaneous encapsulation of GNR is used for substance delivery. The expected outcome of these properties is to provide novel strategies to design nanoscale carriers and reaction devices.

Titov *et al.* demonstrate by MD simulations that graphene sheets could be hosted in the hydrophobic interior of biological membranes formed by amphiphilic phospholipid molecules [38]. Their simulation shows that these hybrid graphene–membrane superstructures might be prepared by forming hydrated micelles of individual graphene flakes covered by phospholipids, which can be then fused with the membrane. Since the phospholipid layers of the membrane electrically isolate the embedded graphene from the external solution, the composite system might be used in the development of biosensors and bioelectronic materials.

In order to produce water-dispersible nanocrystals, including upconversion nanoparticles (UCNPs) which are the new generation fluorophores and magnetic nanoparticles (Fe_3O_4), a polyethylenimine-modified graphene oxide (PEI-GO) was used as a nanocarrier of nanocrystals, and PEI-GO-nanocrystal hybrids were prepared by transferring hydrophobic nanocrystals from an organic phase to water [39]. Nanocrystals were anchored onto the hydrophobic plane of PEI-GO, which was confirmed by atomic force microscopy (AFM) and electron microscopy. MD simulation further showed that hydrophobic interaction between PEI-GO and oleic acid molecules coated on the surface of the nanocrystals was the major driving force in the transfer process. The hybrids show high stability in both water and physiological solutions, and they combine the functionalities of the nanocrystals and PEI-GO, such as luminescence, super-paramagnetism, and drug delivery capability. Through π–π stacking interaction between PEI-GO-UCNP and an aromatic drug, PEI-GO-UCNP was able to load a water-insoluble anticancer drug, doxorubicin (DOX), with a superior loading capacity of 100 wt%. In addition, PEI-GO-UCNP did not exhibit toxicity on human endothelial cells and PEI-GO-UCNP-DOX showed a high potency of killing cancer cells *in vitro*.

Poly(vinyl alcohol)/graphene oxide (PVA/GO) composites were studied by MM and MD methods to analyze the effect of GO sheet addition on PVA material by the group of Lu [40]. The properties of polymer/GO composites with different oxidation degrees and dispersion states of GO sheets in a PVA matrix were compared. The interfacial binding characteristics, mechanical properties, and glass transition temperature of PVA/GO composites were obtained. It was found that the oxidation degree of the GO sheet influenced the strength of the interfacial binding characteristics between the polymer and GO sheet. A high oxidation degree of GO enhanced the interaction between the GO sheet and PVA matrix, thus improving the properties of PVA/GO composites. By reinforcing pure PVA with GO to give PVA/GO composites, its Young's modulus, bulk modulus, and shear modulus, as well as the glass transition temperature of the PVA/GO composites, were obviously enhanced.

Understanding the pH-dependent behavior of GO aqueous solutions is important to the production of assembled GO or reduced GO films for electronic, optical, and biological applications. Shih *et al.* carried out a comparative experimental and MD simulation study to uncover the mechanisms behind the aggregation and the surface activity of GO at different pH values [41]. At low pH, the carboxyl groups are protonated such that the GO sheets become less hydrophilic and form aggregates. MD simulations further suggest that the aggregates exhibit a GO-water-GO sandwich-like structure and as a result are stable in water instead of precipitating. However, at high pH, the deprotonated carboxyl groups are hydrophilic such that individual GO sheets prefer to dissolve in bulk water like a regular salt. The GO aggregates formed at low pH are found to be surface active and do not exhibit characteristic features of surfactant micelles. Their findings suggest that GO does not behave like conventional surfactants in aqueous solutions at pH 1 and 14. The molecular-level understanding of the solution behavior of GO presented here can facilitate and improve the experimental techniques used to synthesize and sort large, uniform GO dispersions in a solution phase.

Li's group investigated the interactions between serine proteases (chymotrypsin and trypsin) and graphene/GO by using MD simulation. First, their results show that chymotrypsin is adsorbed onto the surface of both graphene and GO with different surface curvature and area. Moreover, the active site of the S1 specificity pocket in chymotrypsin is far away from the graphene surface, while it is adsorbed onto the GO surface by its cationic residues and hydrophilic residues, which strongly inhibits enzymatic activity. By carefully examination, they found that the position and conformation of the active site of the S1 specificity pocket are the reason that GO exhibited the highest inhibition dose response for chymotrypsin inhibition compared with all other reported artificial inhibitors: adsorption of the active site of the S1 specificity pocket onto GO is not available for ligands with large conformational changes, while it is available for ligands with small conformational changes adsorbed on the graphene. Their study shows that the position and the conformation of the active site of the S1 specificity pocket are important for enzymatic activity. These findings might provide a novel route to the interaction of nanomaterials and enzymes.

8.2.5 Computational Simulation of Drug Delivery Strategies with Silicon Nanomaterials

Silicon nanomaterials are a type of important nanomaterials with attractive properties, including excellent electronic/mechanical properties, favorable biocompatibility, huge surface to volume ratios, surface tailorability, improved multifunctionality, and compatibility with conventional silicon technology [41]. Consequently, there has been great interest in developing functional silicon nanomaterials for various applications ranging from electronics to biology (Figure 8.4).

MD simulations were used to study the effect of passivating ligands of varying lengths grafted to a nanoparticle and placed in various alkane solvents. Average height and density profiles for methyl-terminated alkoxylsilane ligands [-O-Si$(OH)_2(CH_2)$nCH_3, with

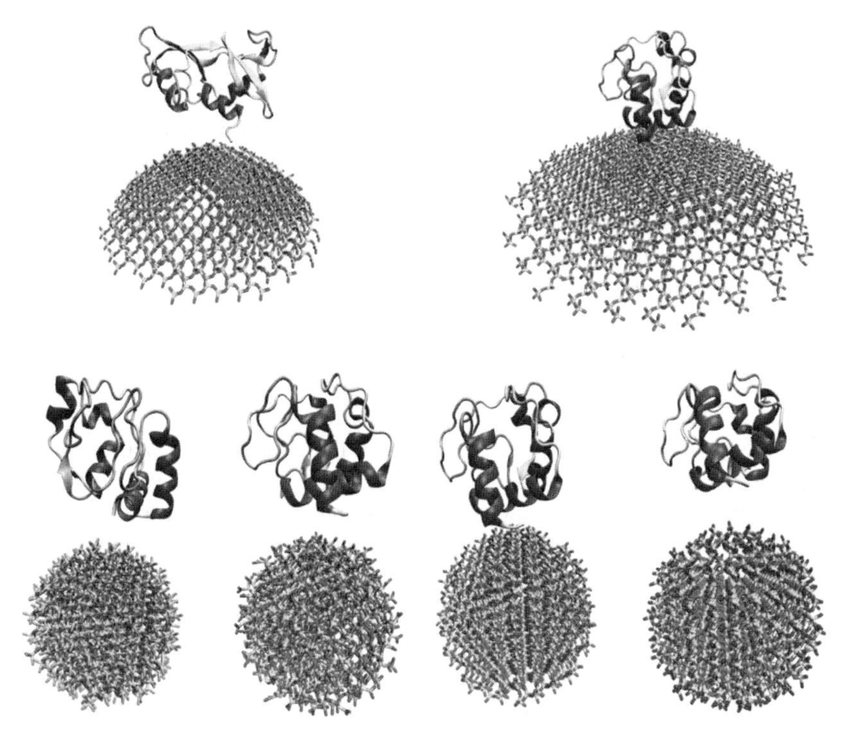

Figure 8.4 *Drug delivery strategies with silicon nanomaterials. The hydrophobic drug molecules can be attached to the surface of silicon nanomaterials directly.*

$n=9$, 17, and 35] attached to a 5-nm diameter amorphous silica nanoparticle with coverages of between 1.0 and 3.0 chains/nm^2 were presented for explicitly modeled, short-chain hydrocarbon solvents and for implicit good and poor solvents. Three linear solvents, $C_{10}H_{22}$ (decane), $C_{24}H_{50}$, and $C_{48}H_{96}$, and a branched solvent, squalene, were studied by Peters *et al.* [42]. An implicit poor solvent captured the effect of the longest chain length solvent at lower temperatures, while its temperature dependence was similar to that of the branched solvent squalene. In contrast, an implicit good solvent produced coating structures that were far more extended than those found in any of the explicit solvents tested and showed little dependence on temperature. Coatings equilibrated in explicit solvents were more compact in longer chain solvents because of autophobic dewetting. Changes in the coating density profiles were more pronounced as the solvent chain length was increased from decane to $C_{24}H_{50}$ than from $C_{24}H_{50}$ to $C_{48}H_{98}$ for all coatings. The response of coatings in squalene was not significantly different from that of a linear chain of equal mass. Significant interpenetration of the solvent chains with the brush coating was observed only for the shortest grafted chains in decane. In all cases, the methyl terminal group was not confined to the coating edge but was found throughout the entire coating volume, from the core to the outermost shell. Increasing the temperature from 300 to 500 K led to greater average brush heights, but the dependence was weak.

Control over selective recognition of biomolecules on inorganic nanoparticles is a major challenge for the synthesis of new catalysts, functional carriers for therapeutics, and the assembly of renewable biobased materials. Perry's group found low sequence similarity among sequences of peptides strongly attracted to amorphous SNPs of various size (15–450 nm) using combinatorial phage display methods [43]. Characterization of the surface by acid base titrations and zeta potential measurements revealed that the acidity of the silica particles increased with larger particle size, corresponding to between 5 and 20% ionization of silanol groups at pH 7. The wide range of surface ionization results in the attraction of increasingly basic peptides to increasingly acidic nanoparticles, along with major changes in the aqueous interfacial layer, as seen in MD simulation. They identified the mechanism of peptide adsorption using binding assays, zeta potential measurements, IR spectra, and molecular simulations of the purified peptides (without phage) in contact with uniformly sized silica particles. Positively charged peptides are strongly attracted to anionic silica surfaces by ion pairing of protonated N-termini, Lys side chains, and Arg side chains with negatively charged siloxide groups. Further, attraction of the peptides to the surface involves hydrogen bonds between polar groups in the peptide with silanol and siloxide groups on the silica surface, as well as ion–dipole, dipole–dipole, and van der Waals interactions. Electrostatic attraction between peptides and particle surfaces is supported by neutralization of zeta potentials, an inverse correlation between the required peptide concentration for measurable adsorption and the peptide pI, and the proximity of cationic groups to the surface in the computation [44]. The importance of hydrogen bonds and polar interactions is supported by adsorption of noncationic peptides containing Ser, His, and Asp residues, including the formation of multilayers. They also demonstrate the tuning of interfacial interactions using mutant peptides with an excellent correlation between adsorption measurements, zeta potentials, computed adsorption energies, and the proposed binding mechanism [45]. Follow-on questions about the relation between peptide adsorption on SNPs and mineralization of silica from peptide-stabilized precursors were raised.

Atomistic MD simulations of a composite consisting of an ungrafted or a grafted spherical silica nanoparticle embedded in a melt of 20-monomer atactic polystyrene (PS) chains were performed Ndoro *et al.* [45, 46]. The structural properties of the polymer in the vicinity of a nanoparticle were studied. The nanoparticle modifies the polymer structure in its neighborhood. These changes increase for higher grafting densities and larger particle diameters. Mass and number density profiles show layering of the polymer chains around the nanoparticle, which extends to 2 nm. In contrast, the increase in the polymer's radius of gyration and other induced ordering (alignment of the chains parallel to the surface and orientation of backbone segments) are shorter-ranged [47]. The infiltration of free PS chains into the grafted chains region is reduced with increasing grafting density. Therefore, the interpenetration of grafted and free chains at high grafting densities, which is responsible for the mechanical anchoring of nanoparticles in the PS matrix, is less than what would be desirable for a well-reinforced composite [47].

Fang *et al.* investigated the mechanical properties of cubic silicon nanoparticles with side lengths ranging from 2.7 to 16.3 nm using MD simulation with a parallel computing technique [48]. The results revealed that the surface energy of the particles increases significantly as the particle size decreases. Furthermore, having passed the point of maximum compressive load, the phase transformation region of the particles gradually transfers from the core to the surface. The small volume of the current nanoparticles suppresses the nucleation of dislocations, and as a result, the maximum strength and Young's modulus values of all but the smallest of the current nanoparticles are greater than the corresponding values in bulk silicon. Finally, it is found that the silicon nanoparticles with a side length of 10.86 nm exhibit the greatest maximum strength (24 GPa). In nanoparticles with shorter side lengths, the maximum strength decreases significantly as the volume of the nanoparticle is reduced.

MD simulation was applied in analyzing the material removal mechanism of silicon substrate under the impact of a large porous silica cluster with different pore diameters by the group of Liang [49]. As the pore diameter of the porous cluster increases, the number of atoms removed from the impact silicon surface first increases and then decreases until the cluster is adhered to the substrate, which is due to the combinational effects of the cluster plough, adhesion between the cluster and the substrate, and permeation of the substrate atoms through the pore of the cluster. Among these three effects, adhesion is the most significant. Meanwhile, the damage of the impact substrate becomes weaker due to the decreasing penetration depth with increasing pore diameter. In addition, it is found that the effect of an enlarged real contact area between the cluster and the substrate is more significant than that of deeper penetration of the cluster in order to enhance the material removal rate (MRR) during the impact process. These findings are instructive in optimizing the process parameters to obtain lower surface roughness and higher MRR during the chemical mechanical polishing process.

Self-assembled monolayers (SAMs) of alkylsilanes have been considered as wear-reducing layers in tribological applications, particularly to reduce friction and wear in microelectromechanical systems (MEMSs) devices. Though these films successfully reduce interfacial forces, they are easily damaged during impact and shear. Surface roughness at the nanoscale is believed to play an important role in the failure of these films because it affects both the formation and quality of SAMs, and it focuses interfacial contact forces onto very small areas, magnifying the locally applied pressure and shear on the lubricant film. To complement prior studies employing Fourier transform infrared spectroscopy (FTIR) and AFM experiments in which SNPs are used to simulate nanoasperities and to refine our analysis of these films to a molecular level, Ewers and Batteas [50] employed classical MD simulations to understand the impact of nanoscopic surface curvature on the properties of alkylsilane SAMs. Amorphous SNPs of various radii were prepared to simulate single asperities on a rough MEMS device surface, or AFM tips, which were then functionalized with alkylsilane SAMs of varying chain lengths. Factors related to the tribological performance of the film, including gauche defect density and exposed silica surface area, were examined to understand the impact of surface curvature on the film. Additionally, because the packing density of the films has been found to be relatively low for alkylsilane SAMs on surfaces with nanoscopic curvature, packing density studies were performed on simulated silica surfaces lacking curvature to understand the relative impact of these two important factors. It was found that both curvature and packing density affect the film quality; however, packing density was found to show the strongest correlation to film quality, demonstrating that greater priority should be given to the reduction of free volume within the films to improve their structural rigidity, to better passivate the underlying surfaces of the devices, and to improve the extent and accessibility of nondestructive dissipation pathways. All of these lead to improved friction and wear resistance. While focused on silica nanoasperities, these MD simulations afford general approaches for studies of ligand effects on a range of surfaces with nanoscopic curvature such as metal oxide nanoparticles and quantum dots [50].

Recently, Xiaotian Sun *et al.* [51] investigated the orientation and adsorption between several enzymes (cytochrome c/RNase A/lysozyme) and SNPs of 4 and 11 nm by using MD simulation. First, their results show the small SNPs induce greater structural stabilization: their results show three enzymes are adsorbed onto the surfaces of both 4 and 11 nm SNPs. Moreover, the active site of cytochrome c is far away from the surface of 4 nm SNPs, while it is adsorbed onto the surface of 11 nm SNPs. Importantly, the active site of cytochrome c adsorption onto 11 nm SNPs is not available for ligands with large conformational changes, while it is available for ligands with small conformational changes adsorbing onto the 4 nm SNPs. They also explored the influences of different groups (-OH, -COOH, $-NH_2$, and CH_3) coated onto SNPs. Their results show different groups induce huge differences on the structure of enzymes: the active site of cytochrome c adsorbed onto 2 nm SNPs coated with different groups is far away from the surface of SNPs, but it has a different degree of conformational changes. The active site of cytochrome c endures few conformational changes when SNPs were coated with -COOH. However, it has large conformational changes when SNPs were coated with $-NH_2$ and -OH.

8.2.6 Computational Simulation of Drug Delivery Strategies with Au Nanomaterials

With the advent of nanotechnology in medicine and biology, there is an increasing interest in understanding and controlling the interactions of nanomaterials with proteins. Among all nanostructured materials, gold nanoparticles (GNPs) have attracted particular interest because of their chemical stability, biocompatibility, surface Plasmon resonance effect, and unique catalytic activities. Gold–protein conjugates have a broad application in biomedical fields, such as DNA analysis, protein microarrays, pregnancy testing in humans, diverse biosensors, and cancer cell imaging [52] (Figure 8.5).

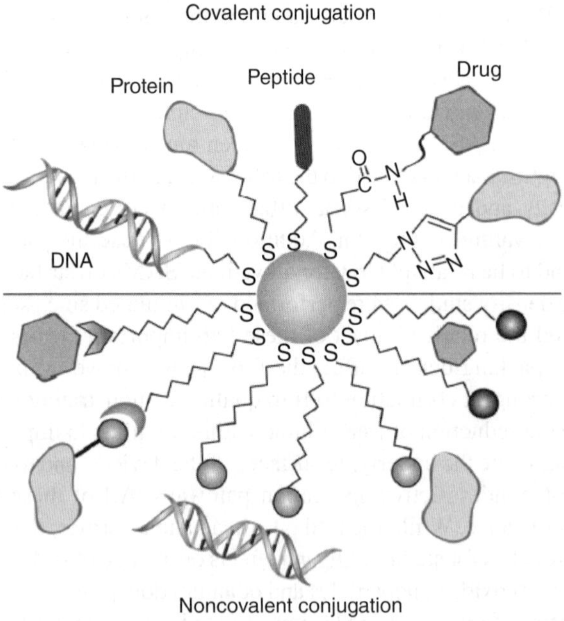

Figure 8.5 *Drug delivery strategies with Au nanomaterials. The hydrophobic drug molecules can be attached to the surface of Au nanomaterials directly.*

Classical MD simulations were employed to investigate the structural and dynamical properties of water near an Au nanoparticle at room temperature by Yang and Weng [53]. The simulation results showed that a well-defined multi-layered structure of water was formed close to the surface of the Au nanoparticle and the orientation of water molecules in the interfacial region changed gradually from a random arrangement to an ordered arrangement with a reduction in the radial distance. By analyzing the mean square displacement and occupation time distribution in different water layers, they found that water molecules in the first and second layers display very low diffusivity, whereas water molecules in the third and fourth layers can migrate from the interfacial region to the bulk region at short time. Additionally, the average number of hydrogen bonds per water molecule in the interfacial region is higher than that in the bulk phase.

The thermal stability of unsupported gold (Au) nanoparticles, containing 140–6708 atoms, was investigated using MD simulation in combination with the modified embedded-atom method. Shim *et al.* found that the melting temperature of the Au nanoparticles decreases drastically with decreasing particle size [54]. The melting temperatures calculated in the present study are in excellent agreement with previous experimental data. It is further confirmed that the calculated equilibrium shape of the Au nanoparticles is a truncated octahedron bounded by eight (1 1 1) and six (1 0 0) facets, which can be explained by the anisotropy of the surface energy of Au. On heating, the premelting phenomenon of the surface atoms is apparently observed prior to the melting of the whole particle.

Hao *et al.* used MD simulations to characterize the structure and dynamics for several peptides and the effect of conjugating them to a gold nanoparticle [55]. Peptide structure and dynamics were compared for two cases: unbound peptides in water and peptides bound to the gold nanoparticle surface in water. The results show that conjugating the peptides to the gold nanoparticle usually decreases conformational entropy, but sometimes increases entropy. Conjugating the peptides can also result in more extended structures or more compact structures depending on the amino acid sequence of the peptide. The results also suggest that if one wishes to use peptide–nanoparticle conjugates for drug delivery it is important that the peptides contain a secondary structure in solution because in our simulations peptides with little to no secondary structure adsorbed onto the nanoparticle surface.

Lee *et al.* used MD methods to study a 2 nm gold nanoparticle functionalized with four ssDNAs at the atomistic level [56]. The DNA strands, which were attached to the faces of a 201 atom truncated octahedral gold particle with a -$S(CH_2)_6$- linker, were found to be perpendicular to the surface of the particle, with the alkane chain lying on the surface. There was no significant hydrogen bonding interactions between the adsorbed ssDNAs during the simulation. Even though the expected radius would be 49 Å (3.4 Å per base) for a Watson–Crick DNA structure, the simulation with 0.5 M salt showed a radius of about 29 Å (2.2 Å per base), which is a result consistent with recent experimental reports. It was also found that the sodium concentration within 30 Å of the gold particle is about 20% higher than the bulk concentration. This is consistent with an observed increase in the melting temperature of DNA when many functionalized gold particles are hybridized together.

Atomistic MD simulations of self-assembled alkanethiol monolayers were performed to investigate the ligand shell organization of homoligand surfactants on spherical gold nanoparticle surfaces as a function of temperature, nanoparticle size, and ligand tail length. At high temperature, Pradip *et al.* showed that the ligands orient randomly with respect to the surface normal with a small tilt angle [57]. As the temperature decreases, the molecules order and adopt a larger tilt angle. The effects of alkanethiol tail length and nanoparticle size on the tilt structure are also significant. At low temperature, the equilibrium conformation of alkanethiols obeys the crystallographic model, whereas at high temperature the continuous model is valid. The dependence of tilt angle on different parameters and comparison with SAMs on flat surfaces was also discussed.

MD simulations have been performed to obtain detailed all-atom models of the interface between PS and GNPs [58]. Considering their relevance in memory technology, systems containing gold nanoparticles included in PS polymer melts also in the presence of 8-hydroxyquinoline (8-HQ) molecules have been studied. Four different systems, including a coated or a noncoated nanoparticle, were compared. Calculated radial density profiles showed that the presence of noncoated nanoparticles in a polymer melt causes an ordering of polymer chains. A similar ordering behavior is found for the 8-HQ molecule. In the presence of a coated gold nanoparticle, calculated radial density profiles show much less order. When 8-HQ is present, this molecule is closer to the nanoparticle surface, and when in contact with a coated nanoparticle it shows a partial penetration into the thiols layer. The molecular description obtained from simulations supports some of the hypothesis made on the basis of the experimental behavior of nonvolatile memory devices.

Yu *et al.* presented a MD study of the binding process of peptide A3 (AYSSGAPPMPPF) and other similar peptides onto gold surfaces, and identified the functions of many amino acids [59]. Their results provide a clear picture of the separate regimes present in the binding process: diffusion, anchoring, crawling, and binding. Moreover, they explored the roles of individual residues and found that tyrosine, methionine, and phenylalanine are strong binding residues; serine serves as an effective anchoring residue; proline acts as a dynamic anchoring point; while glycine and alanine give flexibility to the peptide backbone. They then showed that their findings apply to unrelated phage-derived sequences that were reported recently to facilitate AuNP synthesis. This new knowledge may aid in the design of new peptides for the synthesis of gold nanostructures with novel morphologies.

8.3 Summary

Significant progress in the field of drug delivery has been achieved recently. This is not only due to the development of various strategies in drug delivery, but also benefits from the computational methods and applications in drug delivery systems. Computational methods to predict the binding and dynamics between a drug molecule and its carrier are increasingly desirable to minimize investment in drug design and development. In this review, we summarize the computational studies performed in four categories: drug delivery strategies with CNTs, graphene/GO, SNPs, and GNPs. Significant progress in computational simulation is making it possible to understand the mechanism of drug delivery under different categories. Computational simulations are particularly valuable in addressing issues that are difficult to explore in laboratory experiments, such as diffusion and dynamics. Despite rapid progress in the molecular modeling of drug delivery, plenty of important problems still remain. The accuracy for predicting the release rate for a drug delivery system is one of the important issues remaining. The strategy to improve the description of various environmental factors for simulation, such as pH value, solvent effect, and temperature, are critically important, since the drug delivery system responds differently under different environments. The prediction of loading capacity and the distribution of drug molecules are also very important, since they are directly related to the rational design and optimization of drug delivery system.

Acknowledgements

This work is supported by the National Basic Research Program of China (973 Program, Grant No. 2012CB932400), the National Natural Science Foundation of China (Grant No. 91233115, 21273158, and 91227201), a project funded by the Priority Academic Program Development of Jiangsu Higher Education Institutions (PAPD). This is also a project supported by the Fund for Innovative Research Teams of Jiangsu Higher Education Institutions, the Jiangsu Key Laboratory for Carbon-Based Functional Materials and Devices, and the Collaborative Innovation Center of Suzhou Nano Science and Technology.

References

[1] (a) Kreuter, J., Nanoparticle-based drug delivery systems. Journal of Controlled Release 1991, 16 (1/2), 169–176; (b) Orive, G.; Hernandez, R. M.; Gascon, A. R.; Dominguez-Gil, A.; Pedraz, J. L., Drug delivery in biotechnology: present and future. *Current Opinion in Biotechnology* 2003, **14** (6), 659–664.

[2] Abe, Y.; Shibata, H.; Kamada, H.; Tsunoda, S.-I.; Tsutsumi, Y.; Nakagawa, S., Promotion of optimized protein therapy by bioconjugation as a polymeric DDS. Anti-Cancer Agents in Medicinal Chemistry 2006, 6 (3), 251–258.

[3] (a) Rolland, A. P., From genes to gene medicines: recent advances in nonviral gene delivery. Critical Reviews in Therapeutic Drug Carrier Systems 1998, 15 (2), 143–198; (b) Takeshita, F.; Minakuchi, Y.; Nagahara, S.; Honma, K.; Sasaki, H.; Hirai, K.; Teratani, T.; Namatame, N.; Yamamoto, Y.; Hanai, K.; Kato, T.; Sano, A.; Ochiya, T., Efficient delivery of small interfering RNA to bone-metastatic tumors by using atelocollagen in vivo. Proceedings of the National Academy of Sciences of the United States of America 2005, 102 (34), 12177–12182.

[4] Liu, B. R.; Liou, J.-S.; Chen, Y.-J.; Huang, Y.-W.; Lee, H.-J., Delivery of nucleic acids, proteins, and nanoparticles by arginine-rich cell-penetrating peptides in rotifers. Marine Biotechnology (New York) 2013, 15 (5), 584–595.

[5] Harris, M. S.; Lichtenstein, G. R., Review article: delivery and efficacy of topical 5-aminosalicylic acid (mesalazine) therapy in the treatment of ulcerative colitis. Alimentary Pharmacology and Therapeutics 2011, 33 (9), 996–1009.

[6] Bai, F.; Wang, C.; Lu, Q.; Zhao, M.; Ban, F.-Q.; Yu, D.-H.; Guan, Y.-Y.; Luan, X.; Liu, Y.-R.; Chen, H.-Z.; Fang, C., Nanoparticle-mediated drug delivery to tumor neovasculature to combat P-gp expressing multidrug resistant cancer. Biomaterials 2013, 34 (26), 6163–6174.

[7] Zhang, L.; Wang, Z.; Lu, Z.; Shen, H.; Huang, J.; Zhao, Q.; Liu, M.; He, N.; Zhang, Z., PEGylated reduced graphene oxide as a superior ssRNA delivery system. Journal of Materials Chemistry B 2013, 1 (6), 749.

[8] Peng, F.; Su, Y.; Wei, X.; Lu, Y.; Zhou, Y.; Zhong, Y.; Lee, S. T.; He, Y., Silicon-nanowire-based nanocarriers with ultrahigh drug-loading capacity for in vitro and in vivo cancer therapy. Angewandte Chemie 2013, 52 (5), 1457–1461.

[9] Wu, C. H.; Cao, C.; Kim, J. H.; Hsu, C. H.; Wanebo, H. J.; Bowen, W. D.; Xu, J.; Marshall, J., Trojan-horse nanotube on-command intracellular drug delivery. Nano Letters 2012, 12 (11), 5475–5480.

[10] Wate, P. S.; Banerjee, S. S.; Jalota-Badhwar, A.; Mascarenhas, R. R.; Zope, K. R.; Khandare, J.; Misra, R. D., Cellular imaging using biocompatible dendrimer-functionalized graphene oxide-based fluorescent probe anchored with magnetic nanoparticles. Nanotechnology 2012, 23 (41), 415101.

[11] Sanchez, V. C.; Jachak, A.; Hurt, R. H.; Kane, A. B., Biological interactions of graphene-family nanomaterials: an interdisciplinary review. Chemical Research in Toxicology 2012, 25 (1), 15–34.
[12] Parra, J.; Abad-Somovilla, A.; Mercader, J. V.; Taton, T. A.; Abad-Fuentes, A., Carbon nanotube-protein carriers enhance size-dependent self-adjuvant antibody response to haptens. Journal of Controlled Release 2013, 170 (2), 242–251.
[13] Ferreira, A. J.; Cemlyn-Jones, J.; Robalo Cordeiro, C., Nanoparticles, nanotechnology and pulmonary nanotoxicology. Revista Portuguesa de Pneumologia 2013, 19 (1), 28–37.
[14] Gaumet, M.; Gurny, R.; Delie, F., Localization and quantification of biodegradable particles in an intestinal cell model: The influence of particle size. European Journal of Pharmaceutical Sciences 2009, 36 (4/5), 465–473.
[15] Beck-Broichsitter, M.; Merkel, O. M.; Kissel, T., Controlled pulmonary drug and gene delivery using polymeric nano-carriers. Journal of Controlled Release 2012, 161 (2), 214–224.
[16] Brzoska, M.; Langer, K.; Coester, C.; Loitsch, S.; Wagner, T.; Mallinckrodt, C., Incorporation of biodegradable nanoparticles into human airway epithelium cells—in vitro study of the suitability as a vehicle for drug or gene delivery in pulmonary diseases. Biochemical and Biophysical Research Communications 2004, 318 (2), 562–570.
[17] Hughes, G. A., Nanostructure-mediated drug delivery. Nanomedicine: Nanotechnology, Biology and Medicine 2005, 1 (1), 22–30.
[18] Li, Y.; Hou, T., Computational simulation of drug delivery at molecular level. Current Medicinal Chemistry 2010, 17 (36), 4482–4491.
[19] Medintz, I. L.; Mattoussi, H.; Clapp, A. R., Potential clinical applications of quantum dots. International Journal of Nanomedicine 2008, 3 (2), 151–167.
[20] Bianco, A., Carbon nanotubes for the delivery of therapeutic molecules. Expert Opinion on Drug Delivery 2004, 1 (1), 57–65.
[21] Shi Kam, N. W.; Jessop, T. C.; Wender, P. A.; Dai, H., Nanotube molecular transporters: internalization of carbon nanotube-protein conjugates into Mammalian cells. Journal of the American Chemical Society 2004, 126 (22), 6850–6851.
[22] Pantarotto, D.; Singh, R.; McCarthy, D.; Erhardt, M.; Briand, J. P.; Prato, M.; Kostarelos, K.; Bianco, A., Functionalized carbon nanotubes for plasmid DNA gene delivery. Angewandte Chemie International Edition 2004, 43 (39), 5242–5246.
[23] Chen, Q.; Wang, Q.; Liu, Y.-C.; Wu, T.; Kang, Y.; Moore, J. D.; Gubbins, K. E., Energetics investigation on encapsulation of protein/peptide drugs in carbon nanotubes. The Journal of Chemical Physics 2009, 131, 015101.
[24] Liu, Y.-C.; Wang, Q., Dynamic behaviors on zadaxin getting into carbon nanotubes. The Journal of Chemical Physics 2007, 126, 124901.
[25] Striolo, A., Water self-diffusion through narrow oxygenated carbon nanotubes. Nanotechnology 2007, 18 (47), 475704.
[26] Shen, J.-W.; Wu, T.; Wang, Q.; Kang, Y., Induced stepwise conformational change of human serum albumin on carbon nanotube surfaces. Biomaterials 2008, 29 (28), 3847–3855.
[27] Zuo, G.; Huang, Q.; Wei, G.; Zhou, R.; Fang, H., Plugging into proteins: poisoning protein function by a hydrophobic nanoparticle. ACS Nano 2010, 4 (12), 7508–7514.
[28] Li, Q.; He, G.; Zhao, R.; Li, Y., Investigation of the influence factors of polyethylene molecule encapsulated into carbon nanotubes by molecular dynamics simulation. Applied Surface Science 2011, 257 (23), 10022–10030.
[29] Xie, J.; Xue, Q.; Chen, H.; Xia, D.; Lv, C.; Ma, M., Influence of solid surface and functional group on the collapse of carbon nanotubes. Journal of Physical Chemistry C 2010, 114 (5), 2100–2107.
[30] Cheng, C.-L.; Zhao, G.-J., Steered molecular dynamics simulation study on dynamic self-assembly of single-stranded DNA with double-walled carbon nanotube and graphene. Nanoscale 2012, 4 (7), 2301–2305.
[31] Jin, L.; Yang, K.; Yao, K.; Zhang, S.; Tao, H.; Lee, S.-T.; Liu, Z.; Peng, R., Functionalized graphene oxide in enzyme engineering: a selective modulator for enzyme activity and thermostability. ACS Nano 2012, 6 (6), 4864–4875.
[32] Feng, L.; Liu, Z., Graphene in biomedicine: opportunities and challenges. Nanomedicine 2011, 6 (2), 317–324.

[33] Pei, Q.; Zhang, Y.; Shenoy, V., A molecular dynamics study of the mechanical properties of hydrogen functionalized graphene. Carbon 2010, 48 (3), 898–904.
[34] Pei, Q.-X.; Zhang, Y.-W.; Shenoy, V. B., Mechanical properties of methyl functionalized graphene: a molecular dynamics study. Nanotechnology 2010, 21 (11), 115709.
[35] Ueta, A.; Tanimura, Y.; Prezhdo, O. V., Infrared spectral signatures of surface-fluorinated graphene: a molecular dynamics study. Journal of Physical Chemistry Letters 2012, 3 (2), 246–250.
[36] Yu, D.; Liu, F., Synthesis of carbon nanotubes by rolling up patterned graphene nanoribbons using selective atomic adsorption. Nano Letters 2007, 7 (10), 3046–3050.
[37] Jiang, Y.; Li, H.; Li, Y.; Yu, H.; Liew, K. M.; He, Y.; Liu, X., Helical encapsulation of graphene nanoribbon into carbon nanotube. ACS Nano 2011, 5 (3), 2126–2133.
[38] Titov, A. V.; Král, P.; Pearson, R., Sandwiched graphene – membrane superstructures. ACS Nano 2009, 4 (1), 229–234.
[39] Yan, L.; Chang, Y.-N.; Zhao, L.; Gu, Z.; Liu, X.; Tian, G.; Zhou, L.; Ren, W.; Jin, S.; Yin, W., The use of polyethylenimine-modified graphene oxide as a nanocarrier for transferring hydrophobic nanocrystals into water to produce water-dispersible hybrids for use in drug delivery. Carbon 2013, 216 (3), 35–41.
[40] Ding, N.; Chen, X.; Wu, C.-M. L.; Lu, X., Computational investigation on the effect of graphene oxide sheets as nanofillers in poly (vinyl alcohol)/graphene oxide composites. Journal of Physical Chemistry C 2012, 116 (42), 22532–22538.
[41] Shih, C.-J.; Lin, S.; Sharma, R.; Strano, M. S.; Blankschtein, D., Understanding the pH-dependent behavior of graphene oxide aqueous solutions: a comparative experimental and molecular dynamics simulation study. Langmuir: The ACS Journal of Surfaces and Colloids 2011, 28 (1), 235–241.
[42] Peters, B. L.; Lane, J. M. D.; Ismail, A. E.; Grest, G. S., Fully atomistic simulations of the response of silica nanoparticle coatings to alkane solvents. Langmuir: The ACS Journal of Surfaces and Colloids 2012, 28 (50), 17443–17449.
[43] Puddu, V.; Perry, C. C., Peptide adsorption on silica nanoparticles: evidence of hydrophobic interactions. ACS Nano 2012, 6 (7), 6356–6363.
[44] Hartvig, R. A.; van de Weert, M.; Østergaard, J.; Jorgensen, L.; Jensen, H., Protein adsorption at charged surfaces: the role of electrostatic interactions and interfacial charge regulation. Langmuir: The ACS Journal of Surfaces and Colloids 2011, 27 (6), 2634–2643.
[45] Patwardhan, S. V.; Emami, F. S.; Berry, R. J.; Jones, S. E.; Naik, R. R.; Deschaume, O.; Heinz, H.; Perry, C. C., Chemistry of aqueous silica nanoparticle surfaces and the mechanism of selective peptide adsorption. Journal of the American Chemical Society 2012, 134 (14), 6244–6256.
[46] (a) Ghanbari, A.; Ndoro, T. V.; Leroy, F. D. R.; Rahimi, M.; Böhm, M. C.; Müller-Plathe, F., Interphase structure in silica–polystyrene nanocomposites: a coarse-grained molecular dynamics study. Macromolecules 2011, 45 (1), 572–584; (b) Ndoro, T. V.; Böhm, M. C.; Müller-Plathe, F., Interface and interphase dynamics of polystyrene chains near grafted and ungrafted silica nanoparticles. Macromolecules 2011, 45 (1), 171–179.
[47] Ndoro, T. V.; Voyiatzis, E.; Ghanbari, A.; Theodorou, D. N.; Böhm, M. C.; Müller-Plathe, F., Interface of grafted and ungrafted silica nanoparticles with a polystyrene matrix: atomistic molecular dynamics simulations. Macromolecules 2011, 44 (7), 2316–2327.
[48] Fang, K.-C.; Weng, C.-I.; Ju, S.-P., An investigation into the mechanical properties of silicon nanoparticles using molecular dynamics simulations with parallel computing. Journal of Nanoparticle Research 2009, 11 (3), 581–588.
[49] Chen, R.; Jiang, R.; Lei, H.; Liang, M., Material removal mechanism during porous silica cluster impact on crystal silicon substrate studied by molecular dynamics simulation. Applied Surface Science 2012, 264, 148–156.
[50] Ewers, B. W.; Batteas, J. D., Molecular dynamics simulations of alkylsilane monolayers on silica nanoasperities: impact of surface curvature on monolayer structure and pathways for energy dissipation in tribological contacts. Journal of Physical Chemistry C 2012, 116 (48), 25165–25177.
[51] Sun, X., Feng, Z., Zhang, L., Hou, T., Li, Y., The selective interaction between silica nanoparticles and enzymes from molecular dynamics simulations. PLoS One 2014, 9 (9), e107696.

[52] (a) Pissuwan, D.; Cortie, C.; Valenzuela, S.; Cortie, M., Gold nanosphere-antibody conjugates for hyperthermal therapeutic applications. Gold Bulletin 2007, 40 (2), 121–129; (b) Lee, J. H.; Yigit, M. V.; Mazumdar, D.; Lu, Y., Molecular diagnostic and drug delivery agents based on aptamer-nanomaterial conjugates. Advanced Drug Delivery Reviews 2010, 62 (6), 592–605; (c) Yun, Y.-H.; Eteshola, E.; Bhattacharya, A.; Dong, Z.; Shim, J.-S.; Conforti, L.; Kim, D.; Schulz, M. J.; Ahn, C. H.; Watts, N., Tiny medicine: nanomaterial-based biosensors. Sensors 2009, 9 (11), 9275–9299.

[53] Yang, A.-C.; Weng, C.-I., Structural and dynamic properties of water near monolayer-protected gold clusters with various alkanethiol tail groups. Journal of Physical Chemistry C 2010, 114 (19), 8697–8709.

[54] Shim, J.-H.; Lee, B.-J.; Cho, Y. W., Thermal stability of unsupported gold nanoparticle: a molecular dynamics study. Surface Science 2002, 512 (3), 262–268.

[55] Wei, H.; Hao, F.; Huang, Y.; Wang, W.; Nordlander, P.; Xu, H., Polarization dependence of surface-enhanced Raman scattering in gold nanoparticle – nanowire systems. Nano Letters 2008, 8 (8), 2497–2502.

[56] Lee, O.-S.; Schatz, G. C., Molecular dynamics simulation of DNA-functionalized gold nanoparticles. Journal of Physical Chemistry C 2009, 113 (6), 2316–2321.

[57] Ghorai, P. K.; Glotzer, S. C., Molecular dynamics simulation study of self-assembled monolayers of alkanethiol surfactants on spherical gold nanoparticles. Journal of Physical Chemistry C 2007, 111 (43), 15857–15862.

[58] Milano, G.; Santangelo, G.; Ragone, F.; Cavallo, L.; Di Matteo, A., Gold nanoparticle/polymer interfaces: all atom structures from molecular dynamics simulations. Journal of Physical Chemistry C 2011, 115 (31), 15154–15163.

[59] Yu, J.; Becker, M. L.; Carri, G. A., The influence of amino acid sequence and functionality on the binding process of peptides onto gold surfaces. Langmuir: The ACS Journal of Surfaces and Colloids 2011, 28 (2), 1408–1417.

[60] Hummer, G.; Rasaiah, J. C.; Noworyta, J. P., Water conduction through the hydrophobic channel of a carbon nanotube. Nature 2001, 414 (6860), 188–190.

9

Molecular and Analytical Modeling of Nanodiamond for Drug Delivery Applications

Lin Lai and Amanda S. Barnard

CSIRO Virtual Nanoscience Laboratory, Australia

9.1 Introduction

Nanodiamonds (NDs) are nanoscale diamond particles, which were firstly discovered as a byproduct of detonation in 1963 in the USSR [1]. However, the synthesis of NDs remained largely unknown to the rest of the world until the end of the 1980s, when similar synthesis techniques were developed in the United States [2]. Subsequently, NDs gradually attracted increasing attention worldwide, especially following a number of breakthroughs in synthesis, purification, and isolation techniques achieved in the late 1990s [3, 4].

Today, NDs can be synthesized in large quantities at a relatively low cost. Synthesis techniques reported include detonation, laser ablation, high-energy ball milling of high-pressure high-temperature (HPHT) diamond microcrystals, plasma-assisted chemical vapor deposition (CVD), autoclave synthesis from supercritical fluids, chlorination of carbides, ion irradiation of graphite, electron irradiation of carbon onions, and ultrasound cavitation [3, 4]. NDs synthesized via detonation have already been commercialized and are most frequently employed in biomedical experiments at the laboratory level.

Since it was established that NDs are nontoxic [4], the door was opened for a range of biomedical applications [5–9], such as drug delivery [7–9] and biolabeling [5, 6], due to their excellent biocompatibility and superior optical properties. Each of these applications,

Computational Pharmaceutics: Application of Molecular Modeling in Drug Delivery, First Edition.
Edited by Defang Ouyang and Sean C. Smith.

however, largely depends on the structure and properties of the ND surface, and future applications demand large quantities of monodispersed single-digit diamond nanoparticles. To promote stability and efficiency (particularly when seeking to guarantee safety in medical applications), postsynthesis treatments such as surface functionalization are essential to improve the homogeneity of the ND surface and moderate the interactions with molecules such as antibodies and drugs.

Following detonation synthesis, the primary diamond particles are tightly bound to each other, forming large aggregates (agglutinates) which are difficult to separate using conventional ultrasonic treatments. De-aggregation is necessary to obtain the monodispersed NDs required by most biomedical applications [10–14]. Therefore, in addition to structural integrity and functional surface chemistry, control of the interactions between individual particles and surrounding molecules and nanostructures is imperative in drug delivery applications. In the following sections, we will see how computational modeling is a powerful way to interrogate this complex problem and provide both practical guidance and a detailed understanding of the underlying mechanisms that make ND an ideal platform for future technologies.

9.2 Structure of Individual NDs

It is well established that a bare ND surface is highly unstable, chemically reactive, and readily forms covalent terminations with available atoms and molecules such as hydrogen and oxygen. Consequently, there exist a large number of dangle bonds on the surface of a single ND, and in the absence of surface passivation, sp^3 hybridization of the carbon atoms (characteristic of the diamond lattice) cannot be maintained at the surface. The lattice at the surface tends to distort and bonding changes to increase the $2p\pi$–$2p\pi$ overlap on the adjacent carbon in order to reduce the number of dangling bonds and lower the energy. Although reconstruction on various diamond surfaces has been well studied for many years, it is different on NDs due to the unique (finite) structure, which combines several different surface orientations within a single particle.

Raty *et al.* used computer simulations based on the density functional theory (DFT) to study small diamond nanoparticles, and predicted that a single ND will have a core–shell structure (with a diamond core covered by a fullerene-like shell) [15, 16]. Further, Barnard *et al.* pointed out that the ND morphology plays an important role in the structural stability of diamond nanoparticles [17, 18]. They used the density functional based tight-binding with self-consistent charges method (SCC-DFTB) [19, 20] to study a series of diamond nanoparticles with various sizes (ranging from 1.0 to 3.3 nm) and shapes. The results show the graphitization of the (111) facets and the 2×1 reconstruction on the (100) facets, as illustrated in Figure 9.1. Other simulations have also predicted that, under specific conditions, multilayers of fullerene-like structure can be formed, resulting in a carbon onion structure [22–26].

The morphology and surface structure of NDs have also been extensively characterized by various experimental techniques, including high-resolution transmission electron microscopy (HRTEM) [27–30], X-ray photoelectron spectroscopy (XPS) [31–33], X-ray diffraction (XRD) [28, 32, 34], small-angle X-ray scattering (SAXS) [31, 35], Raman [36–39], infrared (IR) [40], NMR [34], and thermodesorption [41]. In general, these measurements agree

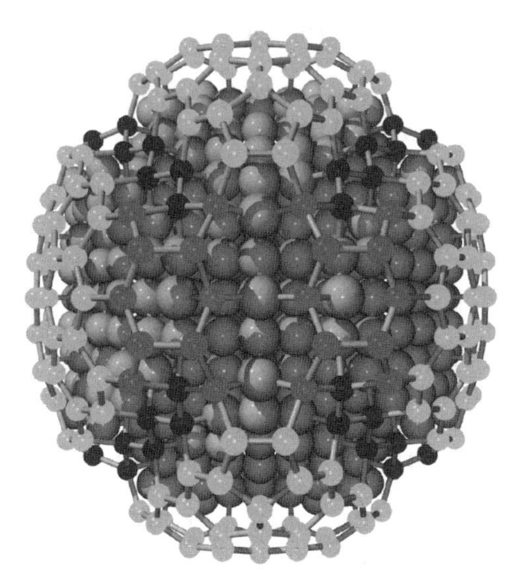

Figure 9.1 *Fully relaxed structure of a bare C_{705} ND. The outmost C atoms on the {100}, {110}, and {111} facets are highlighted in red, blue, and green, respectively [21] (see color plate section). Adapted from [21] with permission from The Royal Society of Chemistry.*

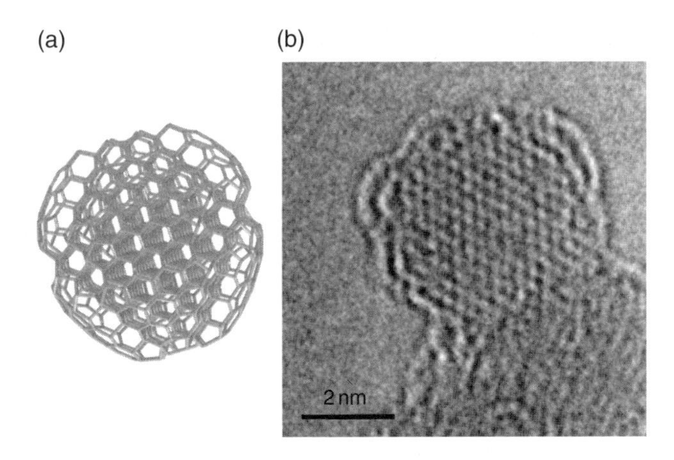

Figure 9.2 *Illustration of the core–shell structure of a single ND by: (a) computer simulations and (b) high-resolution TEM of a 4-nm ND (taken at 80 kV), showing the graphitized facets in the <111> directions. Courtesy of Shery Chang, Forschungszentrum Juelich GmbH).*

with the theoretical predictions of the polyhedral and core–shell structure of NDs (Figure 9.2). The diamond core is found to be completely covered by graphitic carbon in ND powders, and partially encapsulated by a thin layer of graphene following purification [3]. The graphitic shells can be completely removed by oxidation in air [3], which is attributed to the termination of dangling bonds at the surface by oxygen or other adsorbates [42–44].

9.3 Surface Chemistry and Interactions

As stated above, the surface chemistry of NDs is of significant importance for a variety of biomedical applications. However, deliberate modification and/or functionalization can be complicated. Samples produced using different synthesis techniques often have distinct sizes and characteristic surface structures [45, 46], which lead to significant differences in chemical reactivity and affinity to specific adsorbates. Purification treatments always introduce additional chemical terminations to the ND surface, often leaving them far from pure. For instance, cooling during detonation with water or ice inevitably leads to the addition of hydroxyl groups. Purification using oxidizing mineral acids and/or air oxidation could result in the formation of carboxyl and carbonyl groups on the surface [39, 41, 47, 48]. Therefore, characterization of the surface chemistry reported in the literature often varies and appears inconsistent. In general, the identified covalent terminations on the ND surfaces include H, O, COOH, CNH, OH, CO, NH_2, CH_2, SH, alcohol functions, ether, anhydrides, and lactones [8, 37, 40, 49–52].

The precise control of the surface chemistry of NDs requires homogeneity of the surface passivation and its stability under a range of desirable (and relevant) environmental conditions. The homogeneity and thermal stability of the surface modification can be significantly affected by the surrounding environmental conditions, and this can be detrimental to performance. While the static modification of ND surfaces has been extensively investigated, far less attention has been devoted to the thermal stability of the surface functionalization under different pressures and temperatures. Further experiments are certainly needed in this area.

Thermodynamic stability of the surface passivation/functionalization of NDs can be theoretically and computationally examined using the formation enthalpy, ΔH [53]. Assuming a single ND is in equilibrium with a reservoir of adsorbates, we can define ΔH as

$$\Delta H = \Delta E + n_a \times \mu_a, \tag{9.1}$$

where

$$\Delta E = E(\text{host}, a) - E(\text{host}). \tag{9.2}$$

Here, $E(\text{host}, a)$ and $E(\text{host})$ are the respective total energies of passivated/functionalized and bare (i.e., no adsorbate) ND with an identical size. It should be noted that throughout this discussion the bare ND (without surface termination, but with characteristic reconstructions) provides the reference state (i.e., zero free enthalpy). This implies the chemical potential of carbon (μ_C) is taken as an average energy per atom of a bare ND. Finally, μ_a is the chemical potential of the adsorbate a, and n_a is the number of the adsorbate on ND surfaces.

Assuming an ideal gas reservoir of N particles/molecules at constant pressure P and temperature T, the chemical potential in Equation 9.1 can be described by

$$\mu_a = E_a + E_a^{\text{ZPE}} + \Delta\mu_a\left(T, P^0\right) + k_B T \ln\left(P / P^0\right), \tag{9.3}$$

where

$$\Delta\mu_a\left(T, P^0\right) = k_B T \ln\left[\frac{P^0}{k_B T}\left(\frac{h^2}{2\pi m_a k_B T}\right)^{3/2}\right] + \mu^{\text{rot}} + \mu^{\text{vib}} + \mu^{\text{elec}} + \mu^{\text{nucl}}. \tag{9.4}$$

In Equation 9.3, E_a is the calculated total energy of a particle/molecule a; E_a^{ZPE} is its zero-point vibrational energy,

$$E_a^{\text{ZPE}} = \sum_{i=1}^{l} \bar{h}\omega_i / 2,$$

where ω_i is ith of all l fundamental modes. Throughout this discussion, molecular vibrational frequencies from previous experiments are used (as they have been in the previous works we discuss), but they can also be directly calculated at the same theory level as for total energies. $\Delta\mu_a(T,P^0)$ is deliberately separated from the first two terms as it is the primary contribution to the free energy that depends on the temperature at standard pressure $P^0 = 1$ atm.

In Equation 9.4, the first term is the translational free energy, while the following terms are the rotational free energy (μ^{rot}), vibrational free energy without zero-point energy (μ^{vib}), electronic free energy depending on the spin degeneracy (μ^{elec}), and nuclear free energy apart from total energy of the ground state (μ^{nucl}), respectively. Their detailed expression can be found elsewhere [54, 55]. In practice, the last four terms can be safely neglected, since they are generally two orders of magnitude smaller than the other terms [54, 55]. The last term (μ^{nucl}) is actually neglected in the DFT by the Born–Oppenheimer approximation. Finally, m_a denotes the mass of the adsorbate a, while k_B and h are Boltzmann and Planck's constants, respectively.

The total energy of each configuration can be calculated with any method with sufficient sensitivity to distinguish different bonding configurations and electronic states, and those presented here have been calculated with the SCC-DFTB method implemented in DFTB+ code [56, 57]. The two models for NDs shown in Figure 9.3 are C_{705} and C_{837} with a rhombitruncated cuboctahedral and a truncated octahedral structure, respectively. For each type of passivation or functionalization, seven different configurations (three for C_{837}) have previously been reported, providing one of the most detailed studies of ND surfaces in the literature, denoted as (100):X, (110):X, (111):X, (110)/(111):X, (100)/(111):X, (100)/(111):X, and full:X, respectively (Figure 9.4). In each case, X can be substituted by any passivating or functional group.

9.3.1 Surface Passivation and Environmental Stability

9.3.1.1 H Passivation

Now, let us start by examining hydrogen passivation of NDs, assuming a direct chemical reaction between bare NDs and hydrogen gas. This assumption is reasonable and coincides with the actual reaction in experiments, although it is commonly preceded by a decarbonylation and/or carboxylation process [58–60]. This reaction is demonstrated by the disappearance of C=O signals and the emergence of new (or enhanced) C-H signals in IR spectra (as shown in Figure 9.5) for H-treated carboxylated ND [60]. This scenario is also true for hydrogenation via hydrogen plasma treatment [59].

According to Equations 9.3 and 9.4, the chemical potential of hydrogen gas is

$$\mu_{H_2}(P,T) = E_{H_2} + E_{H_2}^{\text{ZPE}} + k_B T \ln\left[\frac{P^0}{k_B T}\left(\frac{h^2}{2\pi m_{H_2} k_B T}\right)^{3/2}\right] + k_B T \ln\left(P / P^0\right).$$

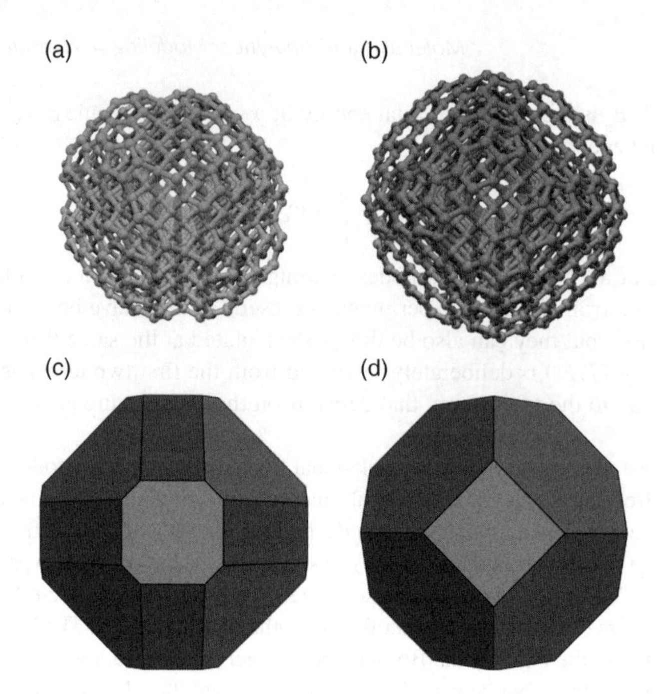

Figure 9.3 *(a, b) Fully relaxed structures of: (a) C_{705} and (b) C_{837} NDs, in ball and stick model, in which gray balls denote the C atoms. (c, d) Geometric morphology for the corresponding schematic representations: (c) C_{705} and (d) C_{837} NDs. In (c) and (d) the {111}, {100}, {110} facets are highlighted in dark gray, pale gray, and midgray, respectively. Reprinted with permission from Ref. [43]. Copyright ©2009, American Chemical Society.*

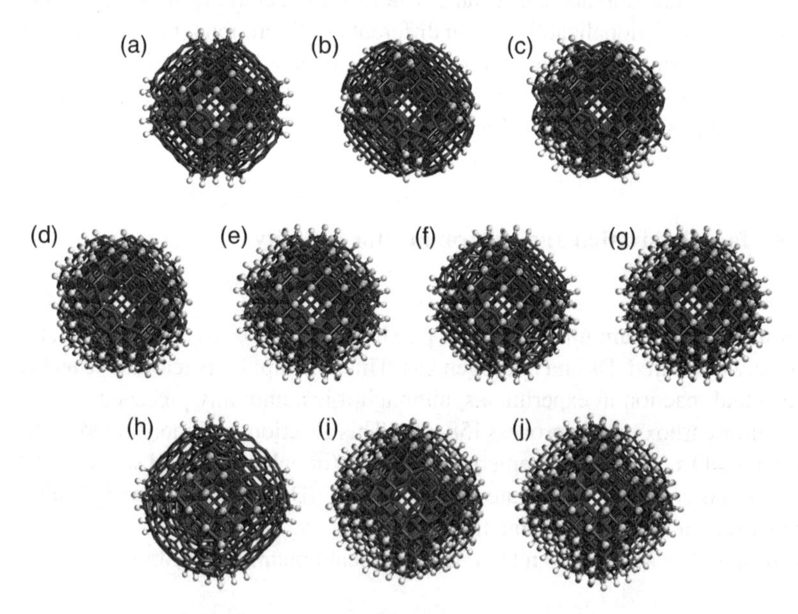

Figure 9.4 *Illustration of the studied configurations: (a) (100):X, (b) (110):X, (c) (111):X, (d) (110)/(111):X, (e) (100)/(111):X, (f) (100)/(110):X, (g) full:X for X-passived/functionalized C705 ND, (h) (100):X, (i) (111):X, and (j) full:X for X-passivated/functionalized C837 ND. White and gray balls represent the adsorbate (X: atoms or groups) and carbon atoms, respectively. Reproduced from Ref. [44] with permission from The Royal Society of Chemistry.*

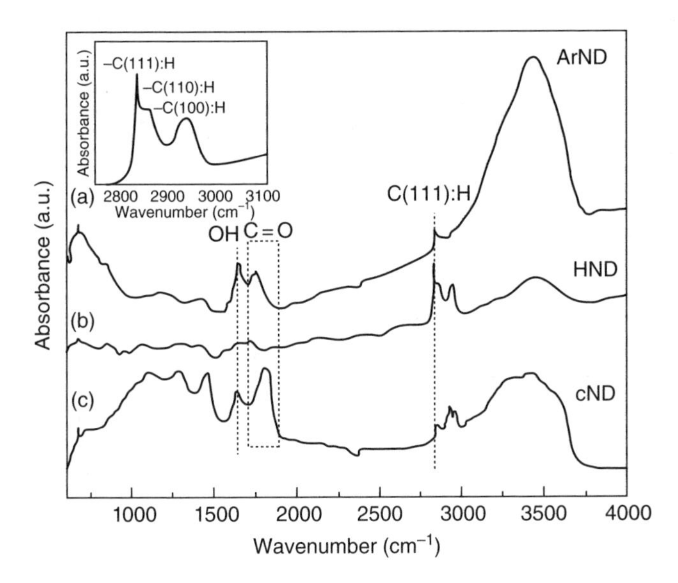

Figure 9.5 *IR spectra for 100 nm NDs: (i) hydrogen-treated nanodiamond (HND), (ii) argon-treated nanodiamond (ArND), and (iii) pure carboxylated nanodiamond (cND). Inset shows the magnified C-H peak with different crystal facets for the hydrogenated ND. Reprinted from Ref. [60] with permission from Elsevier.*

Considering decomposition of H_2, the chemical potential of H (μ_H) is equal to $\mu_{H_2}/2$. In order to guarantee covalent C-H bonds and prevent the formation of H_2 on the ND surfaces, the upper limit of μ_H is taken as $\left(E_{H_2} + E_{H_2}^{ZPE}\right)/2$. The reservoir pressure P is set to be standard pressure ($P = P^0 = 1\,\text{atm}$) as it is in experiments.

Figure 9.6 presents the relationship between thermal stability of hydrogenated NDs and temperature. It is clear that full/homogeneous H passivation is energetically more favorable than partial passivation and bare surfaces, under these conditions. These results also show that the configuration with clean {*hlk*} facets (where *hlk* includes the low-index forms: 100, 110, and 111) than the configuration with only the same facets passivated by H.

Experimentally, it is difficult to achieve hydrogenation of oxidized diamond surfaces, especially for NDs smaller than ~ 100 nm. High temperatures (~850 to 900 °C) or active hydrogen atoms in plasma are required to achieve sufficient reactivity. The results in Figure 9.6 indicate that high temperatures are actually used to overcome the reaction barrier due to the dissociation of H_2 and desorption of carbonyl and/or carboxyl groups. Once homogeneous hydrogenation is achieved, it is rather stable, and desorption of H is not preferred.

9.3.1.2 O Passivation

Detonation-synthesized NDs contains large amounts of nondiamond carbon surrounding their diamond cores. These sp^2-bonded carbon atoms can be reduced or removed by air [39, 61] or ozone [62–64] treatment. As demonstrated by Osswald *et al.* [39], the sp^3/sp^2 ratio in detonation NDs is increased by two orders after isothermal oxidation for 5 h in ambient air at atmospheric pressure. This oxidation process has been suggested to completely remove CH_2 and CH_3 groups and shift the signals of C=O vibrations by 20–40 cm^{-1}, as

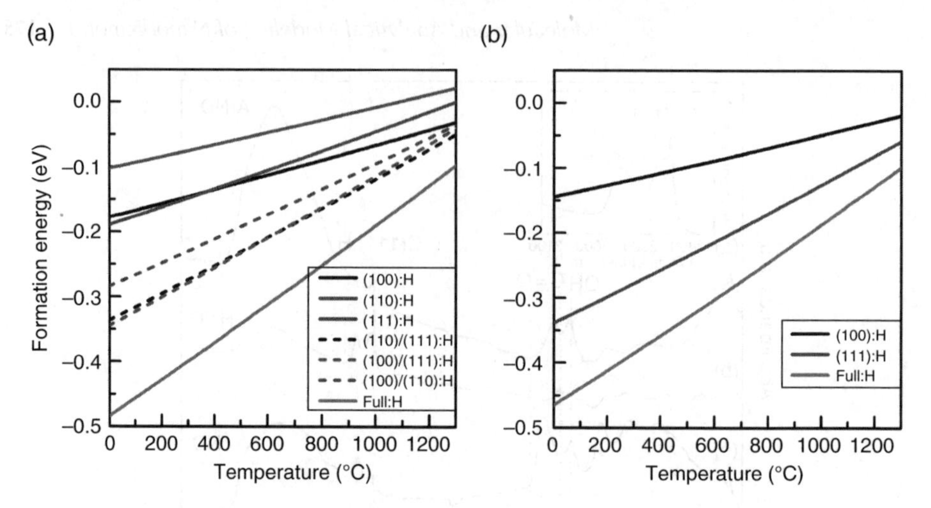

Figure 9.6 *Formation energies as a function of temperature for hydrogenated C_{705} (a) and C_{837} (b) NDs in a hydrogen gas reservoir [44]. Adapted from [44] with permission from The Royal Society of Chemistry.*

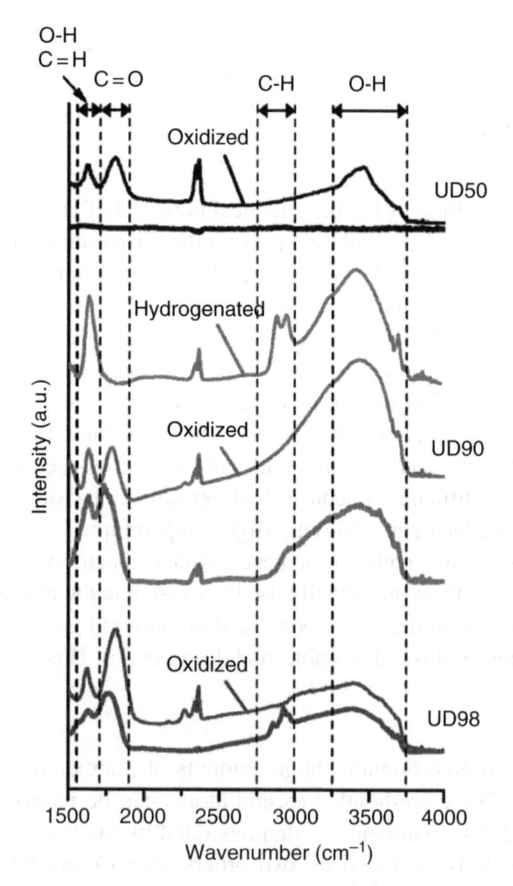

Figure 9.7 *FTIR spectra of UD50, UD90, and UD98 samples before and after oxidation for 5 h at 425 °C in air. A spectrum of oxidized UD90 that was annealed for 2 h at 800 °C in hydrogen (20 ml/min) is shown to demonstrate the possibility to control the surface chemistry of ND after oxidation. Reprinted with permission from Ref. [39]. Copyright ©2006, American Chemical Society.*

shown in the Fourier transform infrared spectroscopy (FTIR) spectra (Figure 9.7) of three different samples: UD50, UD90, and UD98. Sample UD50 is the raw detonation soot containing nondiamond carbons. Samples UD90 and UD98 were NDs treated by different multistage acidic purifications using nitric and sulfuric acids (see details in Ref. [62]). An optimal temperature range of 400–430 °C was suggested to achieve no or minimal loss of diamond carbon [39], and a reduction of the average particle size [61] has been reported. Therefore, it is of interest to study the relationship between stability and oxidizing conditions.

Previously reported studies of the stability of the O passivation of NDs assume bare NDs react with O_2 gas, to produce ketone passivated (C=O) NDs in equilibrium with a reservoir of O_2. Following Equations 9.3 and 9.4, the chemical potential of O_2 is defined as

$$\mu_{O_2}(P,T) = E_{O_2} + E_{O_2}^{ZPE} + k_B T \ln\left[\frac{P^0}{k_B T}\left(\frac{h^2}{2\pi m_{O_2} k_B T}\right)^{3/2}\right] + k_B T \ln\left(P / P^0\right).$$

Similarly, the chemical potential of O (μ_O) is equal to a half of μ_{O_2}, if considering the decomposition of O_2. The pressure P is again set to be 1 atm in accordance with reaction conditions in existing experiments.

Figure 9.8 shows the thermal stability, with respect to increasing temperature, of each O-passivated configuration for C_{705} and C_{837} NDs. With the lowest formation energy, full O passivation is energetically favorable up to 907 and 1157 °C for C_{705} and C_{837}, respectively. Further increase in temperature will change the relative stability of the configurations. The {110}/{100}:O configuration, followed by the {100}:O, becomes the most energetically stable. This indicates that heat treatment exceeding the critical temperatures above can lead to the desorption of adsorbed O on the {110} and {111} facets. These results agree well with a recent study on the thermal stability of detonation NDs using XPS [65]. In this work, Zeppilli and coworkers reported there are detectable oxygen signals up to 1050 °C, and the fraction of C=O bonds signals drops from $15 \pm 1\%$ at 750 °C to $12 \pm 1\%$ at 850 °C, and even to $9 \pm 1\%$ at 970 °C.

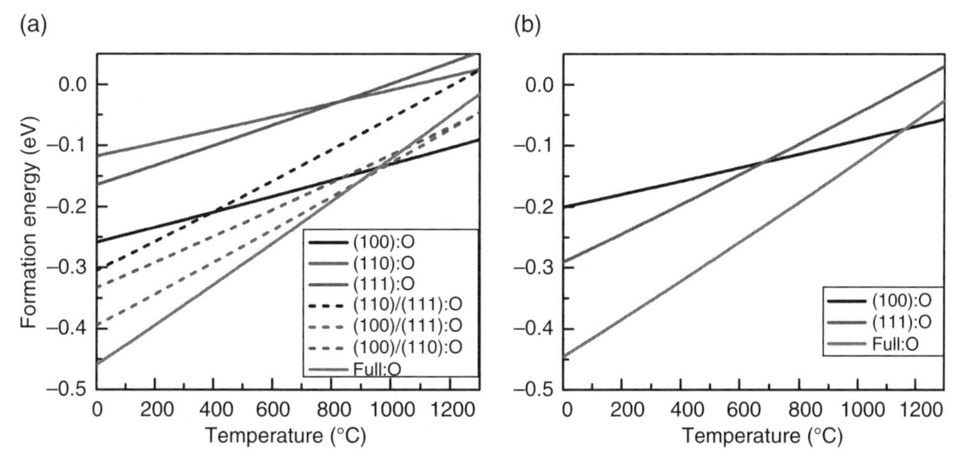

Figure 9.8 *Formation energies, E_f, in an O_2 reservoir (in units of eV/atom) for O-passivated NDs: (a) C_{705} and (b) C_{837} [42]. Adapted from [42] with permission from The Royal Society of Chemistry.*

9.3.2 Surface Functionalization

9.3.2.1 OH Functionalization

Hydroxylation of NDs is useful because it makes NDs hydrophilic and easily dispersed in aqueous solution and polar solvents. Hydroxylation of ND can be achieved with several approaches, including treatments by borane, $LiAlH_4$, fenton reagent, mechanochemical treatment in water, and photochemical reaction with water vapor. In addition, OH groups may also be introduced at the surface during the hydrogenation of oxidized NDs. Therefore, it is important to understand the hydroxylation stability of nanodiamonds.

As discussed above, OH is introduced from different precursors under distinct conditions. Different chemical reservoirs are required in order to examine the effect. Mainly, two reservoirs are considered: a mixture of H_2 and O_2 gases and water vapor. The chemical potentials of H_2 and O_2 were derived in previous sections, and that of H_2O (μ_{H_2O}) can be written as

$$\mu_{H_2O}(P,T) = E_{H_2O} + E_{H_2O}^{ZPE} + k_B T \ln\left[\frac{P^0}{k_B T}\left(\frac{h^2}{2\pi m_{H_2O} k_B T}\right)^{3/2}\right] + k_B T \ln\left(P / P^0\right).$$

Accordingly, the chemical potential of OH (μ_{OH}) is equal to $\left(\mu_{H_2} + \mu_{O_2}\right)/2$ and $\mu_{H_2O} - \mu_{H_2}/2$, respectively.

Figure 9.9 shows thermal stability of OH-functionalized NDs with respect to the increasing temperature at the atmospheric pressure. In general, the stability of the OH-functionalization significantly depends on the chemical reservoir. Within a mixture of H_2 and O_2 gases, all of the OH-functionalized configurations examined for two models show a negative formation energy, even at high temperatures, indicating that OH functionalization is energetically preferable to bare reconstructed surfaces. However, both of the OH-functionalized NDs have higher formation energies in a water vapor reservoir, quickly exceeding zero at ~277 °C. This implies that in a water vapor reservoir desorption of OH will occur as the temperature increases, and OH groups can be effectively removed with a heat treatment. However, homogeneous OH functionalization is energetically favored in both reservoirs at room temperature, which guarantees the biomedical application of OH-functionalized NDs.

9.3.2.2 NH_x Functionalization

Amino functionalization is important for NDs to bind a large variety of functional molecules, such as bioactive compounds and polymer building blocks, through amide formation, reductive amination, nucleophilic attack, or direct condensation reactions. However, grafting of amino onto ND surfaces by direct amination reactions is difficult and has not yet been achieved with great success. Theoretical and computational methods are therefore very enlightening and useful in explaining the thermal stability and reaction conditions of NH_x functionalization.

According to Equations 9.3 and 9.4, the chemical potential of NH_3 can be derived as

$$\mu_{NH_3}(P,T) = E_{NH_3} + E_{NH_3}^{ZPE} + k_B T \ln\left[\frac{P^0}{k_B T}\left(\frac{h^2}{2\pi m_{NH_3} k_B T}\right)^{3/2}\right] + k_B T \ln\left(P / P^0\right).$$

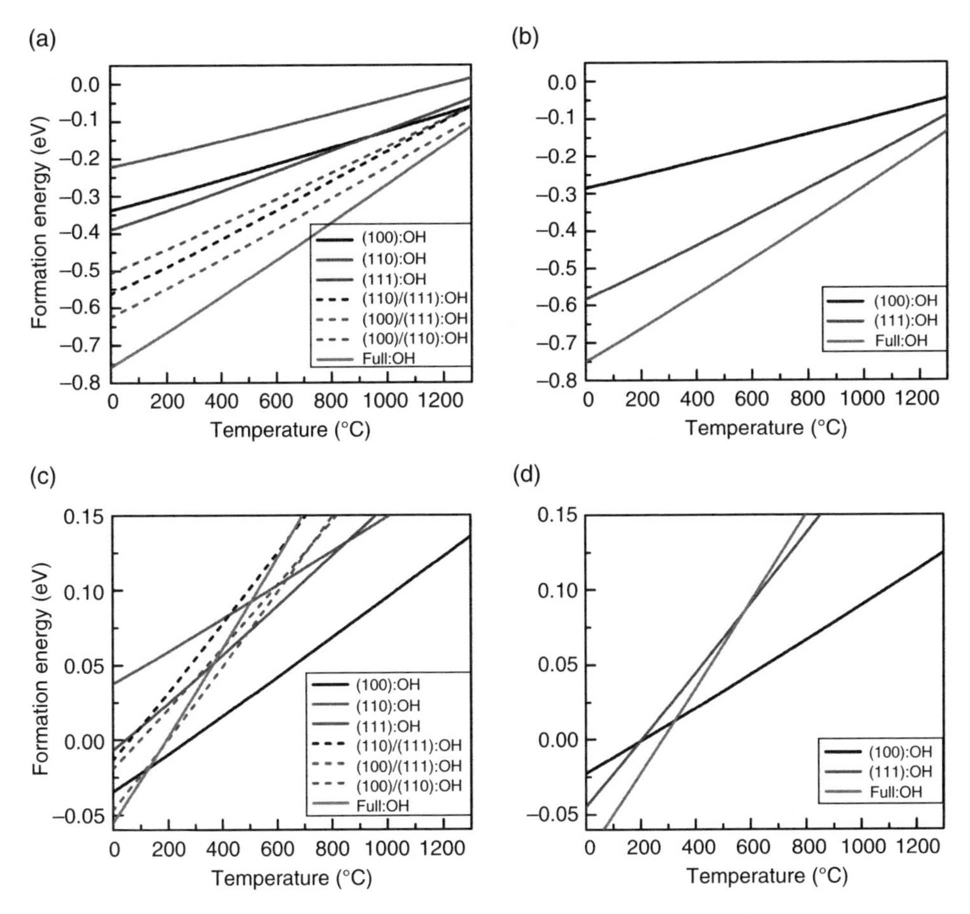

Figure 9.9 *Formation energies, E_f, in units of eV/atom for OH-functionalized C_{705} and C_{837} NDs in: (a,b) a mixture of H_2 and O_2 gases and (c,d) a water vapor reservoir [42]. Adapted from [42] with permission from The Royal Society of Chemistry.*

Assuming direct amination $\left(\mathrm{ND}+\mathrm{NH}_3 \rightarrow \mathrm{ND}-\mathrm{NH}_x+\frac{3-x}{2}\mathrm{H}_2\right)$ the chemical potentials of NH and NH_2 are equal to and $\mu NH_3 - \mu H_2$ and $\mu_{\mathrm{NH}_3}-\frac{1}{2}\mu_{\mathrm{H}_2}$, respectively.

The formation energies of NH- and NH_2-functionalized NDs as a functional of temperature at atmospheric pressure are presented in Figure 9.10. The results show that NH functionalization of NDs is energetically unfavorable in both cases. Compared with NH, NH_2-functionalized NDs have relatively smaller formation energies and are more stable than the bare surface at low temperature. However, in both cases homogeneous functionalization is either unfavorable (NH) or only stable at temperature lower than room temperature, where NH_2 starts to be desorbed from the ND surface. Clearly, this explains why direct amination of NDs is hard to achieve in experiments.

The thermal stability of NH and NH_2 functionalization can be enhanced by changing the chemical reservoir, for instance, a mixture of N_2 and H_2 gases (Figure 9.11). Experimentally,

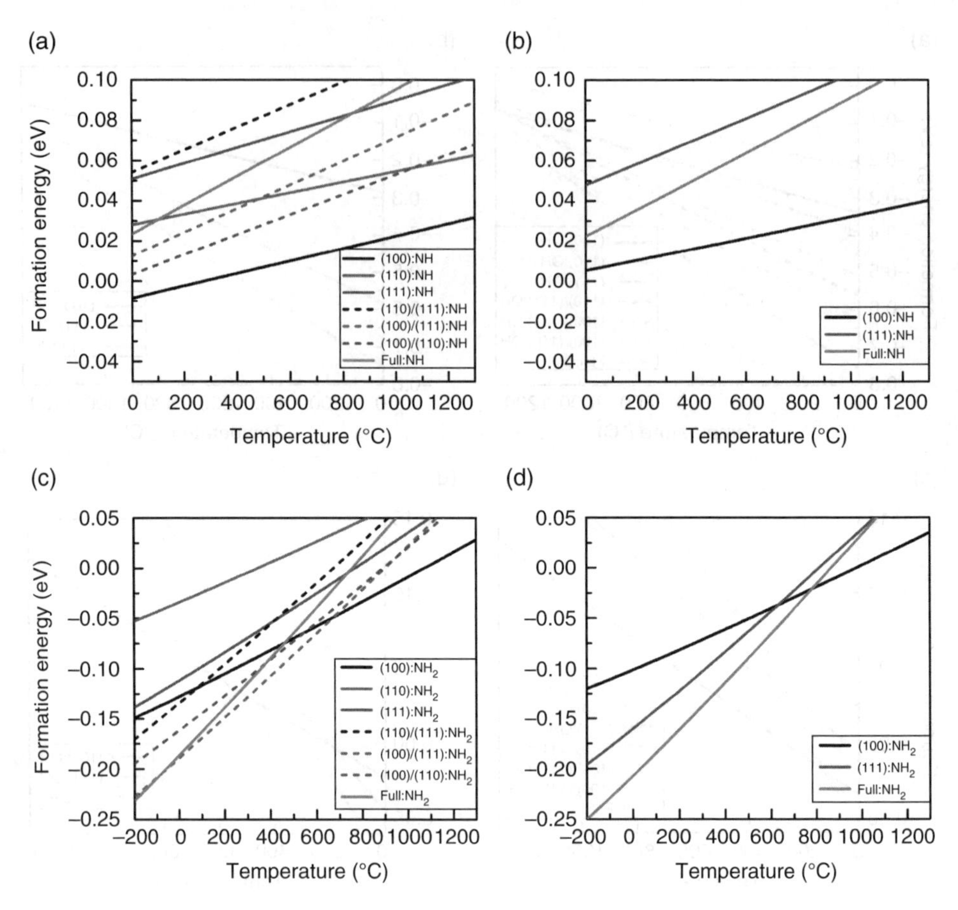

Figure 9.10 *Formation energies, E_f, in a NH_3 reservoir (in units of eV/atom) for: (a,b) NH- and (c,d) NH_2-functionalized C_{705} and C_{837} NDs. Modified with permission from Ref. [43]. Copyright 2009, American Chemical Society.*

amination of NDs is achieved using aminated silanes or aminated aromatic moieties starting from hydroxylated NDs at elevated temperature.

9.3.3 Consequences for Interactions and Self-Assembly

The surface electrostatic potential leads to strong long-ranged Coulomb interactions between two NDs. It has also been suggested to be the cause of abnormally persistent hydration on the surface of monodispersed NDs [66]. More importantly, it helps to reveal the unsolved mechanism underlying the inactivated transportation and slow release of drugs observed in experiments, which will significantly advance the development of ND-based drug delivery platforms.

Denotation NDs have been found to form tightly bound aggregates with a large size (~100 nm), and conventional techniques like ultrasound treatment alone are unable to disperse the NDs aggregates. Several approaches such as beading milling, graphitization/oxidation, or burning in air were suggested to reduce the aggregates into primary particles. Moreover, surface

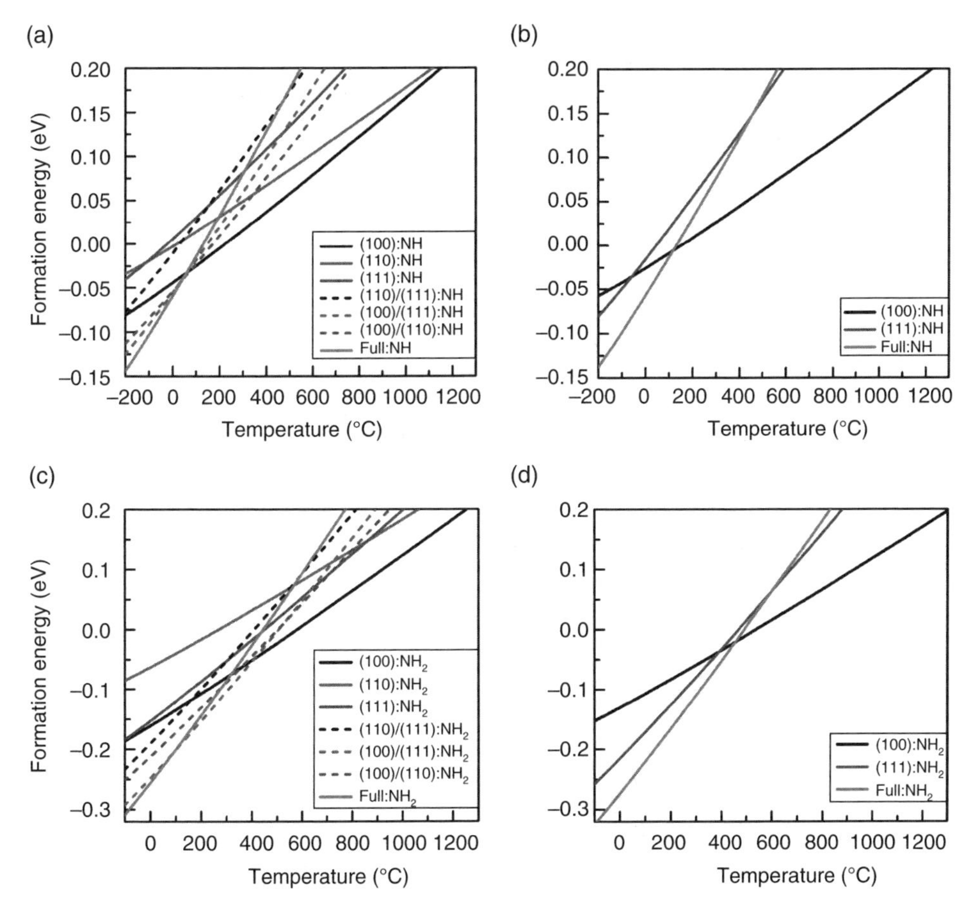

Figure 9.11 *Formation energies, E_f, in a reservoir of a mixture of N_2 and H_2 gases (in units of eV/atom) for: (a,b) NH- and (c,d) NH_2-functionalized C_{705} and C_{837} NDs. Modified with permission from Ref. [43]. Copyright 2009, American Chemical Society.*

passivation/functionalization was found to be an effective way of achieving the monodispersion of NDs. For this reason, the mechanism of the interparticle interaction of the ND within aggregates has also received attention in recent years.

The interparticle interaction has been previously simulated by placing two identical NDs together with their centers aligned linearly and with interacting facets parallel to one another. The total energies of the systems as a function of a series of interfacet separation distances were calculated and reported, so that the relationship between the relative binding energies and the separation distance, and the interparticle interaction are revealed.

9.3.3.1 Interactions between Bare NDs

Figure 9.12 shows the total binding energies with respect to the separation distance for different configurations of the C_{837} ND. The results above also indicate that NDs prefers to form aggregates through the $(111)_a$-$(111)_b$ configuration. The mechanism becomes clear

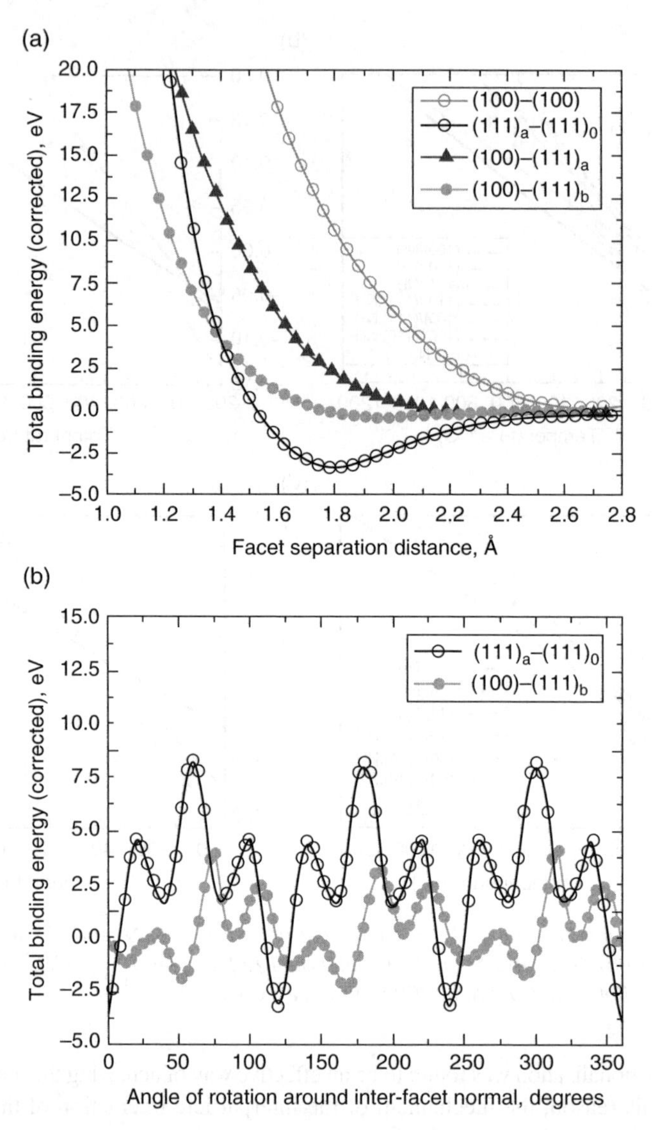

Figure 9.12 *The (corrected) total binding as a function of: (a) facet–facet separation distance in the (100)–(100), $(111)_a$–$(111)_b$, (100)–$(111)_a$, and (100)–$(111)_b$ configurations; (b) the angle of rotation around the common interfacet normal vector, for particles separated by 1.78Å in the $(111)_a$–$(111)_b$ and (100)–$(111)_b$ configurations. The results are corrected with respect to two C_{837} nanoparticles at ±∞ (noninteracting) [67]. Reproduced from Ref. [67] by permission of The Royal Society of Chemistry.*

when we examine the surface electrostatic potentials of NDs as shown in Figure 9.13. Here we can see the (100) facets and the edges between the (100) and (111) facets exhibit a strong positive potential. In contrast, half of the graphitized (111) facets shows a negative potential [denoted as type $(111)_a$ in Figures 9.12 and 9.13], and the remaining (111) surfaces [denote as type $(111)_b$] exhibit more variation in potential depending on the position relative to the

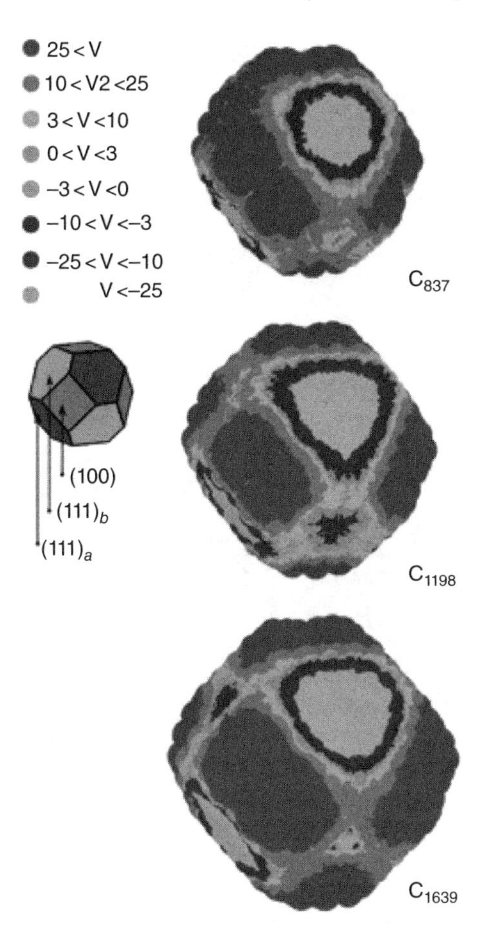

Figure 9.13 *Normalized surface electrostatic potential for the relaxed structures of truncated octahedral C_{837}, C_{1169}, and C_{1639} structures. The model below the legend (left) denotes the relative orientation of the (100), $(111)_a$, and $(111)_b$ facets [67]. Reproduced from Ref. [67] by permission of The Royal Society of Chemistry.*

facet center. As a result, only the (100)–$(111)_b$ and $(111)_a$–$(111)_b$ configurations have an exothermic minima due to the existence of attractive Coulomb interactions.

In general, the (100)–(100) and (100)–$(111)_a$ interactions are thermodynamically unfavorable at all separation distances. The curve of the $(111)_a$–$(111)_b$ interaction configuration has a exothermic minimum of −3.28 eV at 1.78 Å, while the (100)–$(111)_b$ interaction configuration only exhibits a small minimum of −0.38 eV at 1.90 Å. Since the van der Waals interactions were not included within the calculations, the exothermic minima are attributed to the Coulomb interactions. The results explain why ND aggregates are bound more tightly than other aggregates of nanoparticles, based on van der Waals interactions, although the aggregates can still be dispersed in other ways.

Once the equilibrium separation distance was determined, additional calculations were conducted by fixing one of the pair of NDs in each configuration and rotating the other one around the line connecting their centers. By fixing the separation to the equilibrium distance, these calculations presented the coherent interparticle interactions and revealed the

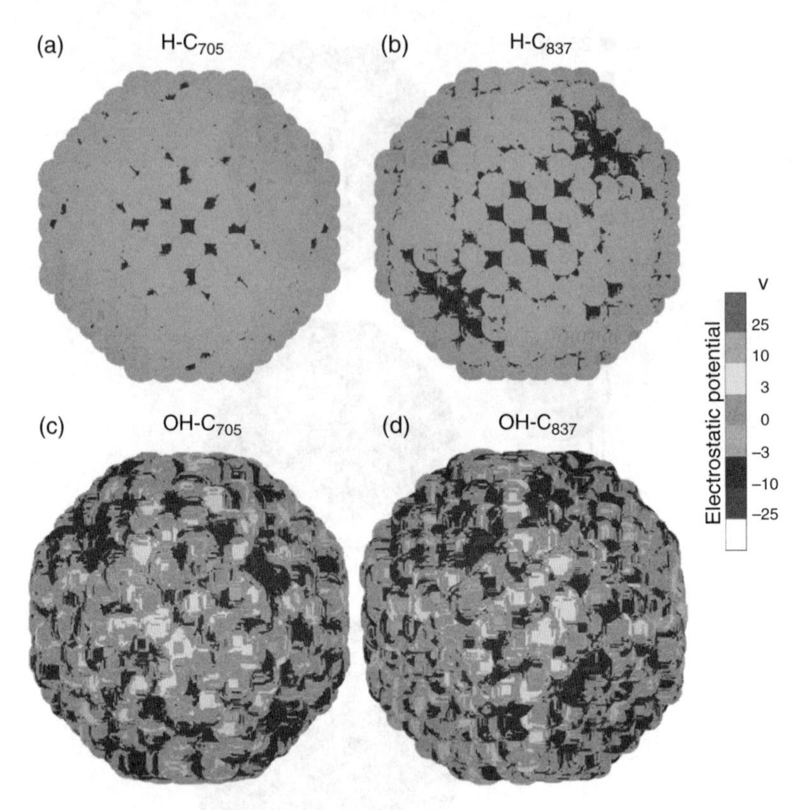

Figure 9.14 Surface electrostatic potential of (a,b) hydrogen- and (c,d) hydroxyl-passivated C_{705} and C_{837} NDs. This potential is calculated using Coulomb's law considering the charge associated with each atom. The legend shows the static assignment of colors using a dielectric constant value of 10. Reprinted with permission from Ref. [69]. Copyright 2012, American Chemical Society.

relationship between the binding energy and rotation angles for each configuration, as presented in Figure 9.12b.[1] Recent experiments by high-resolution aberration-corrected electron microscopy has confirmed the electrostatic interaction mechanism in the formation of agglutinates and agglomerates of NDs [68].

9.3.3.2 *Interactions between Passivated/Functionalized NDs*

Once the ND surfaces are passivated or functionalized, the electrostatic potential changes significantly (as shown in Figure 9.14). For instance, the surface electrostatic potentials of the H-passivated surfaces of C_{705} and C_{837} NDs were found to be homogeneous and slightly negative, especially for C_{705}. There is a vague anisotropy in the hydrogenated C_{837} ND, where the $\{111\}_a$ facets show more negative potential than the $\{111\}_b$ surfaces. In contrast, hydroxylated NDs have a complicated distribution of surface electrostatic potential. Generally, the $\{100\}$ and $\{110\}$ facets have a largely positive potential, whereas the $\{111\}_b$ facets have a largely

[1] In practice, the binding energy as a function of rotating angles can also be determined before calculating the equilibrium distances, as done in the next section. There is no difference in the final result by these two routes.

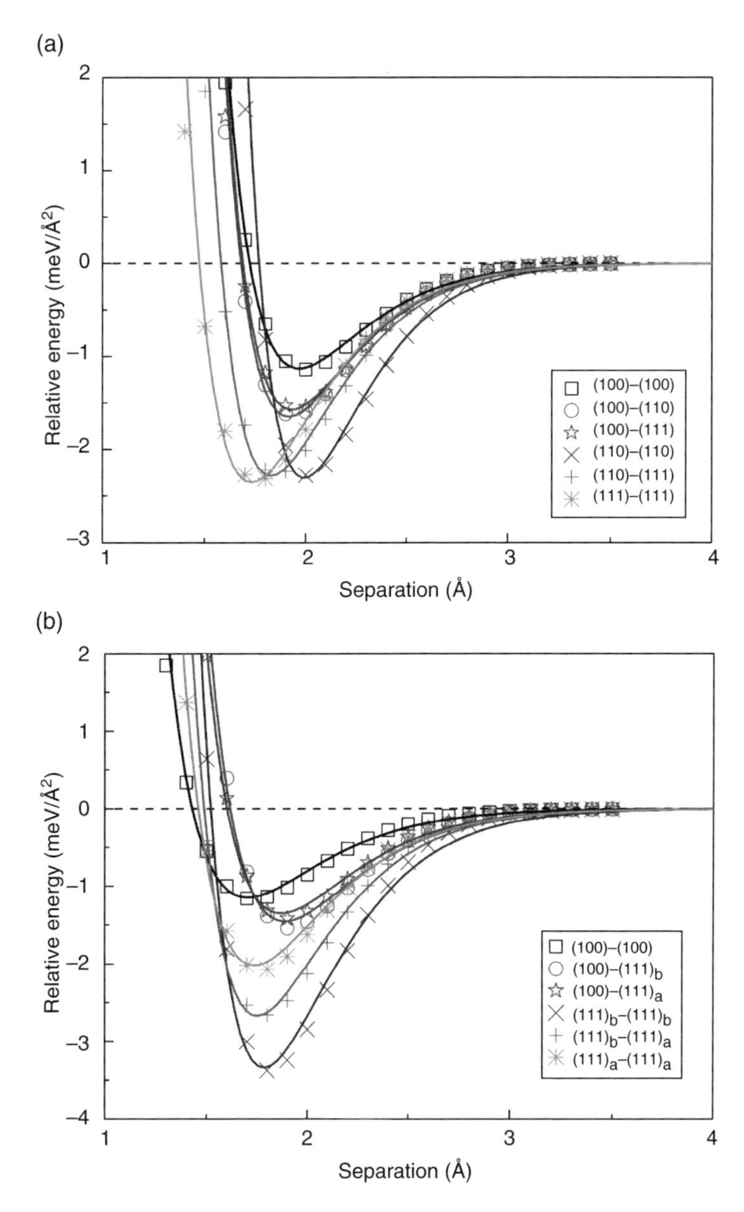

Figure 9.15 *Interparticle interactions as a function of separation distance between two hydrogenated (a)* C_{705} *and (b)* C_{837} *NDs. Reprinted with permission from Ref. [69]. Copyright 2012, American Chemical Society.*

negative potential. While the trend is qualitatively consistent with bare NDs, the degree of facet-dependent anisotropy is significantly lower. Moreover, since the density of surface charges reduces, the value of the surface electrostatic potential has also decreased significantly.

Consequently, the interparticle interactions in passivated or functionalized NDs will be altered due to the changes in the surface electrostatic potential (see Figures 9.15 and 9.16) [69].

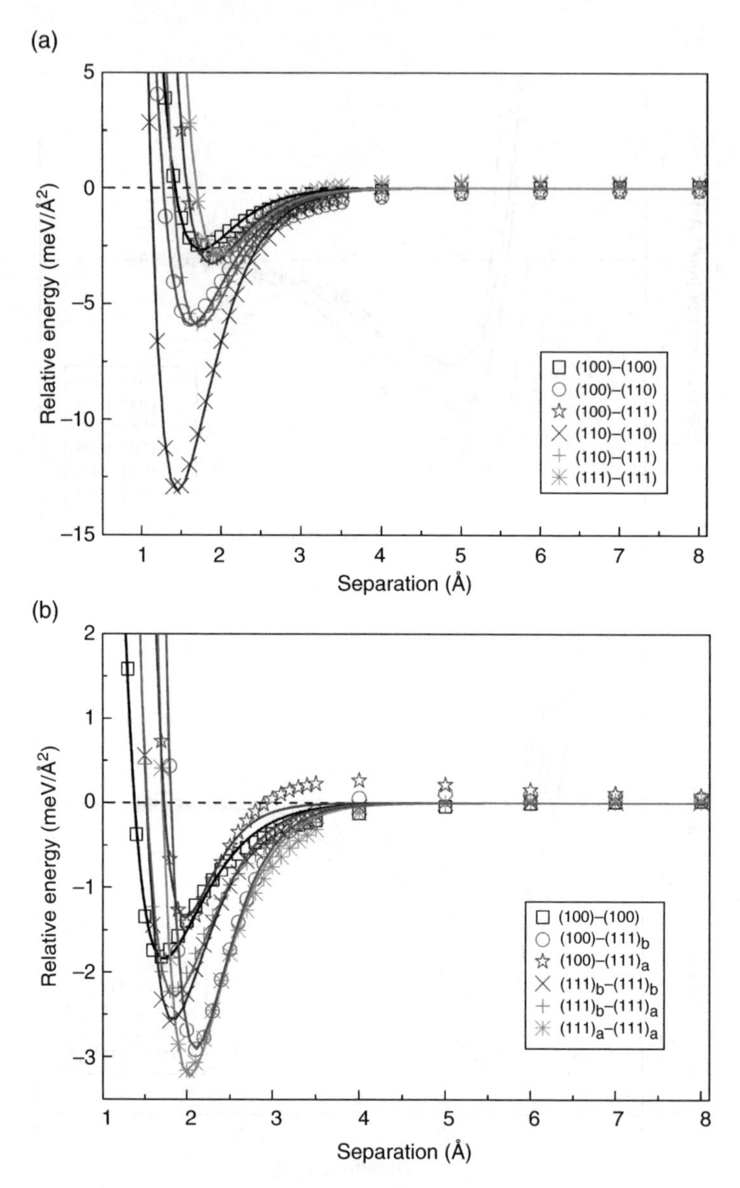

Figure 9.16 *Interparticle interactions as a function of separation distance between two hydroxylated (a) C_{705} and (b) C_{837} NDs. Reprinted with permission from Ref. [69]. Copyright 2012, American Chemical Society.*

All interaction configurations are thermodynamically favorable and exhibit a exothermic minimum. These results indicates that H-passivated and OH-functionalized NDs are weakly bound due to Coulomb interactions. However, some interacting configurations [i.e., (111)–(111), (110)–(111), and (110)–(110) for H-C_{705}; $(111)_b$–$(111)_b$ for H-C_{837}; (110)–(110) for OH-C_{705}; $(111)_b$–$(111)_a$ and $(111)_b$–$(111)_b$ for OH-C_{837}], have much lower binding energies, particularly at the equilibrium separation distances (i.e., the energy minima). This result implies that these configurations are more probable than others to be present in the ND aggregates.

9.4 NDs as a Therapeutic Platform

NDs have attracted increasing research interests worldwide in recent years, since they were suggested to be an ideal platform to carry and deliver therapeutic drugs to treat cancer. The advantages of the ND-based platform include their excellent *in vivo* and *in vitro* biocompatibilities and the large surface area that can be readily conjugated by a wide range of chemicals and drugs. It has been reported that NDs covered by OH and COOH groups are able to reversibly load and release Doxorubicin hydrochloride (DOX; a apoptosis-inducing chemotherapeutic widely used to treat cancer) [70]. The complex of NDs and DOX were found to be able to overcome drug efflux and to significantly increase apoptosis and tumor growth inhibition beyond conventional DOX treatment, when treating liver and mammary cancer in mouse models [71]. Moreover, *in vivo* toxicity was found to be significantly reduced compared with standard DOX treatment.

9.4.1 Simulations with Doxorubicin

The interaction between DOX and NDs has been reported to be weak under ambient conditions due to low aqueous solubility. To maintain the complex of NDs and DOX, salts or an increase in the solvent basicity is required. Presently, it is still a challenge to understand the underlying mechanism of drug adsorption and release. Therefore, atomistic simulations with the constant pH molecular dynamics (CpHMD) method has been used to study the interactions between NDs and DOX drugs in solvents (Figure 9.17) [72].

The DOX molecules were found to only bind to the ND surfaces at high pH after a simulation of 100 ps (see Figure 9.18). This is because the electrostatic interactions are enhanced at high pH, and more DOX molecules will be bound. It was also shown (Figure 9.19) that the binding of DOX molecules requires at least ~10% of the ND surface to be fully titrated (charged). These results demonstrate that pH is a critical factor to effect the interaction between DOX and NDs. However, discrepancies exist between experiments and simulations (Figure 9.19), which is attributed to the complicated surface chemistry of NDs used in the experiments. A recent study combining experimental and computational techniques quantitatively characterized the surface functional groups of NDs and reported approximately 22 000 phenols, 7000 pyrones, and 9000 sulfonic acids in the average 50 nm diameter ND aggregates, with at least 2000 fixed positive charges stabilized within pyrones and/or chromenes [73]. Therefore, further work is required, especially on the interactions between DOX and NDs with mixed types of passivation or functionalization.

9.4.2 Experimental Progress

Since the early attempts in ND-mediated delivery of DOX for cancer therapy, significant progresses have been achieved in the development of ND-based therapeutic platform [7, 9, 70, 71, 74–82]. NDs have also been used as a vector for other types of drugs [74, 80, 82–88].

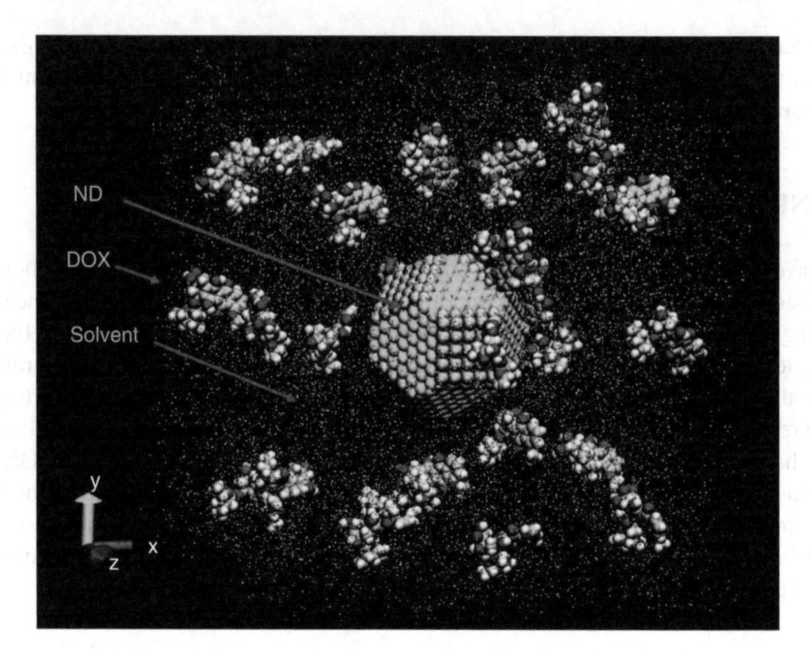

Figure 9.17 Initial structure of atomistic model of DOX-ND-solvent (pH buffer) system simulated via molecular dynamics (MD) to study the pH dependent DOX-ND binding. In the solvent media, 26 DOX molecules are allowed to interact with one truncated octahedral ND particle that is functionalized (30% of total ND surface area). The surface electrostatics on the ND surface depends on the pH of the solvent. Here the green and cyan colors on ND refer to the atoms belonging to (111) and (100) facets, respectively. Reprinted with permission from Ref. [72]. Copyright 2011, American Chemical Society.

Apart from chemotherapeutics, gene and protein delivery have also been demonstrated. Gene therapy has been delivered using polymer-functionalized NDs [82] and fenton-treated hydroxylated NDs [86] as the vectors. Protein drug delivery has also been achieved using N,O-carboxy-methyl chitosan-modified NDs [88]. Similarly, bovine insulin has been carried via noncovalent interaction with ND surfaces coated with -OH and -COOH groups, which ensure the successful delivery and pH-dependent release of adsorbed insulins [80]. NDs have also been shown to covalently bind and carry drugs inside cells, while still maintaining their medical function [82, 85].

NDs with hydrophilic functionalization are water soluble. As a carrier, they enable the suspension of various water-insoluble therapeutics with preserved functionality, which can offer a great therapeutic advantage. For instance, Chen *et al.* have demonstrated the enhanced water solubility of several therapeutics using NDs, including Purvalanol A (a drug for liver cancer treatment), 4-hydroxytamoxifen (4-OHT, a drug for breast cancer treatment), and dexamethasone (an anti-inflammatory drug for the treatment of blood and brain cancers and rheumatic and renal disorders) using NDs [7].

In addition, new therapeutic techniques have emerged based on the complex of NDs and drugs. For instance, Lam *et al.* have developed hybrid parylene-ND-based microfilms, which offer a platform for flexible, robust, and slow drug release, as shown in Figure 9.20 [78]; and Loh *et al.* have demonstrated that functionalized NDs (i.e., a complex of NDs

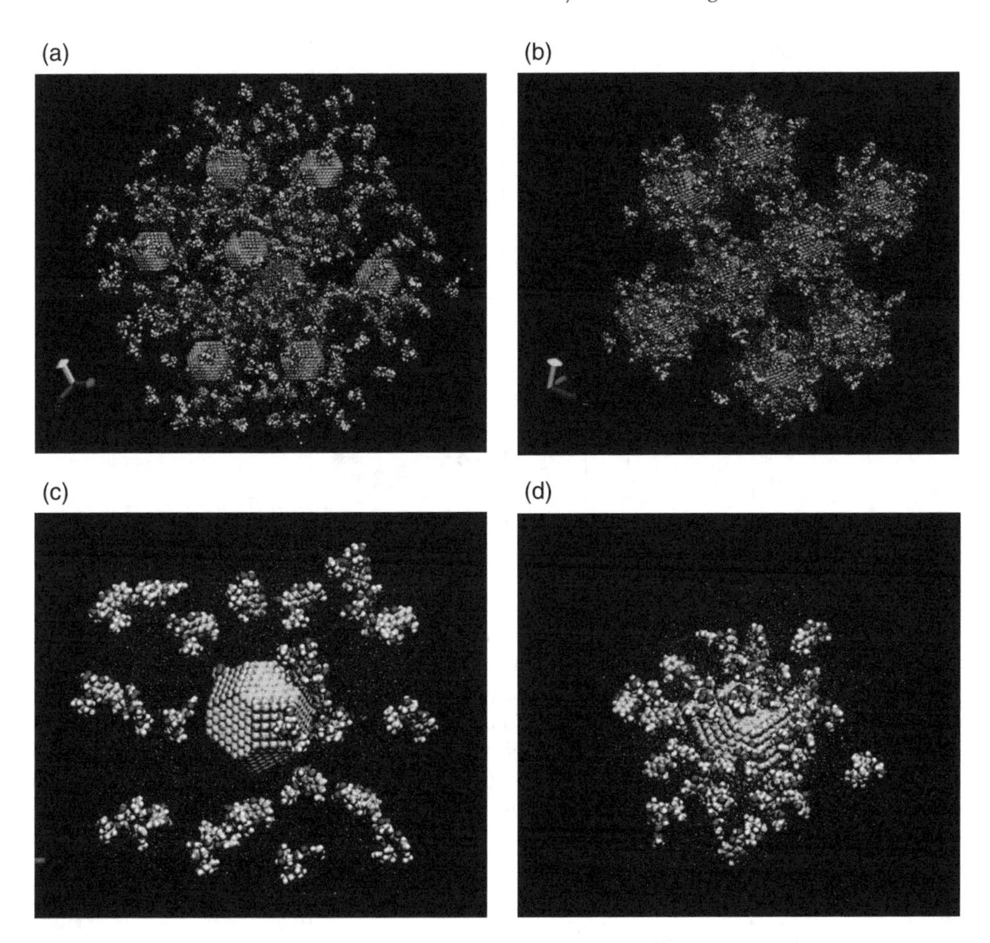

Figure 9.18 Representative final MD snapshots for the DOXND interaction simulation at different pH level: (a,b) large model with 8 NDs and 208 DOX in the solvent and (c,d) small model with 1 ND and 26 DOX in the solvent. The coverage of functional group is 30% of the ND surface. (a) and (b) pH equal to 7 and 11, respectively. (c) and (d) pH equal to 6.5 and 10.5, respectively. For clarity the water molecules are shown in point form whereas DOX and ND are shown in VDW sphere form. Different ND facets are shown in different colors. Note that at lower pH ($pH<8$) only a few DOX molecules bind to ND, but at higher pH level ($pH>8$), most are bound. Reprinted with permission from Ref. [72]. Copyright 2011, American Chemical Society.

and DOX drugs) can be delivered and directly injected into a single cell using the nanofountain probe [79].

9.5 Outlook

Diamond nanoparticles have been the topic of computational modeling for approximately two decades, and during this time we have learned a lot. We have a reasonable understanding of their size-dependent phase relationships with other nanocarbons, the structure

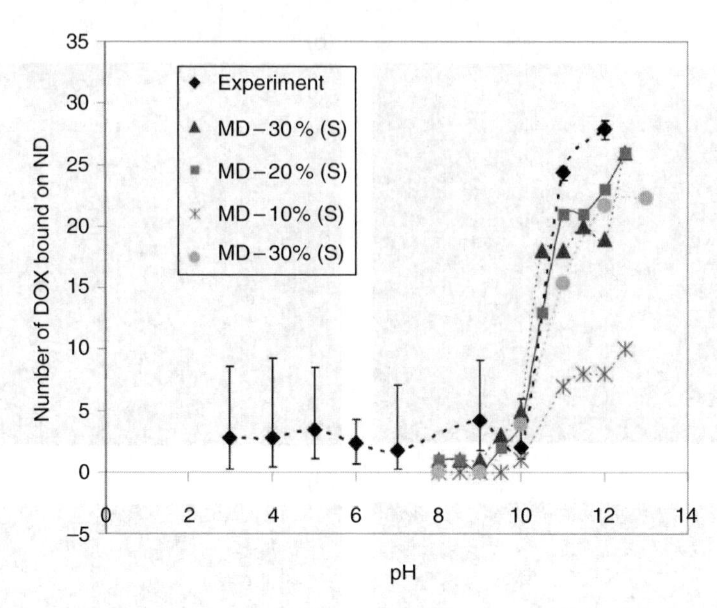

Figure 9.19 pH-dependent DOX-ND binding capacity. At lower pH, few DOX molecules are adsorbed because of the limited availability of titrated ND sites. At higher pH, enhanced adsorption is attributed to a larger availability of titrated ND sites. For the smaller system (shown as S in the chart legend) MD simulations are performed for three different percentages of functional groups on the ND surface, namely, 30, 20, and 10%. For the larger system (shown as L in the chart legend), only the case of the 30% functional group is studied. Reprinted with permission from Ref. [72]. Copyright 2011, American Chemical Society.

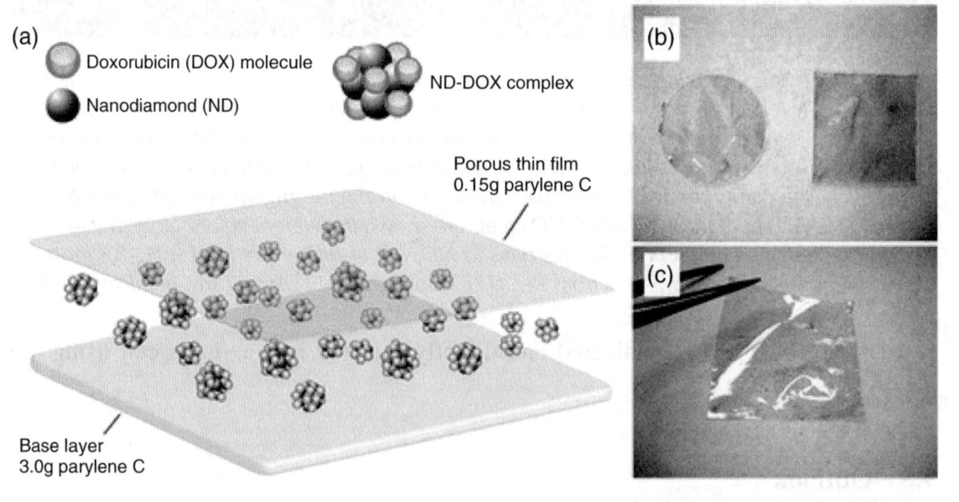

Figure 9.20 (a) Illustrated schematic of hybrid film patch. NDs and DOX molecules bound through physical interactions in various configurations are deposited atop a base layer of parylene. A final layer of parylene film is then deposited for additional elution control. (b) Hybrid films with a 10g base layer of varied size and shapes. (c) The patch exhibits innate flexibility and a thin physical profile. Reprinted with permission from Ref. [78]. Copyright 2008, American Chemical Society.

of their surfaces, and the impact that deliberate or spontaneous structural changes have on their affinity for different types of interactions. As computational techniques developed, improvements to our understanding followed, and simulations were able to systematically embrace a greater degree of structural and chemical complexity. In more recent years we have developed an appreciation for surface functionalization, and many groups have successfully simulated the surface structure and chemistry of NDs under a range conditions.

However, far fewer studies have addressed functionalization and interactions with medically relevant molecules, such as drugs. Since drug delivery is currently one of the hottest topics in ND research, and laboratory experiments show such promise, this is surprising. NDs have been used in a variety of *in vitro* and *in vivo* studies, and with the current computational resources and technology there are numerous opportunities of complement these experiments with targeted studies *in silico*. For example, an important part of any drug delivery platform is the stable adsorption of the drug (to carry the payload to site of interest) and the selective desorption of the drug (delivery of the payload). Both adsorption and desorption can now be simulated with great accuracy.

There are, however, a number of challenges that remain: accurate representation of solvation, the inclusion of weak interactions at an acceptable level of theory, and the dynamics of drug/particle and particle/particle interactions. In the past most studies aggressively focused on thermodynamic considerations and largely ignored the kinetics. There was also a focus on covalent bonding at surfaces, but this is not how drug molecules and NDs interact. If it were, the particles would (almost) never deliver their payload. A greater emphasis on weak and long-ranged interactions between NDs and other molecules (and particles) is required to take the work outlined in this chapter to the next level, to move from simulations of this (very promising) drug carrier to simulations of ND drug delivery.

References

[1] V. V. Danilenko. On the history of the discovery of nanodiamond synthesis. *Physics of the Solid State*, **46**(4):595–599, 2004.

[2] N. R. Greiner, D. S. Phillips, J. D. Johnson, and F. Volk. Diamonds in detonation soot. *Nature*, **333**(6172):440–442, 1988.

[3] V.N. Mochalin, O. Shenderova, D. Ho, and Y. Gogotsi (2012) The properties and applications of nanodiamonds. *Nature Nanotechnology*, **7** (1):11–23.

[4] A.M. Schrand, S.A. Ciftan Hens, and O.A. Shenderova (2009) Nanodiamond particles: properties and perspectives for bioapplications. *Critical Reviews in Solid State and Materials Sciences*, **34**(1–2):18–74.

[5] A.S. Barnard. Diamond standard in diagnostics: nanodiamond biolabels make their mark. *Analyst*, **134**(9):1751–1764, 2009.

[6] J.-I. Chao, E. Perevedentseva, P.-H. Chung, K.-K. Liu, C.-Y. Cheng, C.-C. Chang, and C.-L. Cheng. Nanometer-sized diamond particle as a probe for biolabeling. *Biophysical Journal*, **93**(6):2199–2208, 2007.

[7] M. Chen, E. D. Pierstorff, R. Lam, S.-Y. Li, H. Huang, E. Osawa, and D. Ho. Nanodiamond-mediated delivery of water-insoluble therapeutics. *ACS Nano*, **3**(7):2016–2022, 2009.

[8] D. Ho, ed. *Nanodiamonds: Applications in Biology and Nanoscale Medicine*. Springer Science+Business Media, New York, 2009.

[9] X.-Q. Zhang, M. Chen, R. Lam, Xu X., E. Osawa, and D. Ho. Polymer-functionalized nanodiamond platforms as vehicles for gene delivery. *ACS Nano*, **3**(9):2609–2616, 2009.

[10] N. Gibson, O. Shenderova, T. J. M. Luo, S. Moseenkov, V. Bondar, A. Puzyr, K. Purtov, Z. Fitzgerald, and D. W. Brenner. Colloidal stability of modified nanodiamond particles. *Diamond and Related Materials*, **18**(4):620–626, 2009.

[11] A. Krueger, J. Stegk, Y. Liang, Lu L., and G. Jarre. Biotinylated nanodiamond: simple and efficient functionalization of detonation diamond. *Langmuir*, **24**(8):4200–4204, 2008.

[12] A. Kruger, F. Kataoka, M. Ozawa, T. Fujino, Y. Suzuki, A. E. Aleksenskii, A. Y. Vul, and E. Osawa (2005) Unusually tight aggregation in detonation nanodiamond: identification and disintegration. *Carbon*, **43**(8):1722–1730.

[13] M. Ozawa, M. Inaguma, M. Takahashi, F. Kataoka, A. Krueger, and E. Osawa (2007) Preparation and behavior of brownish, clear nanodiamond colloids. *Advanced Materials*, **19**(9):1201–1206.

[14] Kang X. and Qunji X. (2007) Deaggregation of ultradispersed diamond from explosive detonation by a graphitization-oxidation method and by hydriodic acid treatment. *Diamond and Related Materials*, **16**(2):277–282.

[15] J. Y. Raty and G. Galli. Ultradispersity of diamond at the nanoscale. *Nature Materials*, **2**(12):792–795, 2003.

[16] J. Y. Raty, G. Galli, C. Bostedt, T. W. van Buuren, and L. J. Terminello (2003) Quantum confinement and fullerenelike surface reconstructions in nanodiamonds. *Physical Review Letters*, **90**(3):037401.

[17] A. S. Barnard, S. P. Russo, and I. K. Snook. Visualization of hybridization in nanocarbon systems. *Journal of Computational and Theoretical Nanoscience*, **2**(1):68–74, 2005.

[18] A.S. Barnard and M. Sternberg. Crystallinity and surface electrostatics of diamond nanocrystals. *Journal of Materials Chemistry*, **17**(45):4811–4819, 2007.

[19] Th. Frauenheim, G. Seifert, M. Elstner, Th. Niehaus, C. Köhler, M. Amkreutz, M. Sternberg, Z. Hajnal, A. Di Carlo, and S. Suhai. Atomistic simulations of complex materials: ground-state and excited-state properties. *Journal of Physics: Condensed Matter*, **14**:3015, 2002.

[20] D. Porezag, Th. Frauenheim, Th. Köhler, G. Seifert, and R. Kaschner. Construction of tight-binding-like potentials on the basis of density-functional theory: application to carbon. *Physical Review B*, **51**:12947–12957, 1995.

[21] L. Lai and A.S. Barnard (2013) Diamond nanoparticles as a new platform for the sequestration of waste carbon. *Physical Chemistry Chemical Physics*, **15**(23):9156–9162.

[22] A. S. Barnard and M. Sternberg. Vacancy induced structural changes in diamond nanoparticles. *Journal of Computational and Theoretical Nanoscience*, **5**(11):2089–2095, 2008.

[23] A. Brodka, L. Hawelek, A. Burian, S. Tomita, and V. Honkimaki. Molecular dynamics study of structure and graphitization process of nanodiamonds. *Journal of Molecular Structure*, **887**(1–3):34–40, 2008.

[24] A. Brodka, T. W. Zerda, and A. Burian. Graphitization of small diamond cluster – molecular dynamics simulation. *Diamond and Related Materials*, **15**(11–12):1818–1821, 2006.

[25] J. M. Leyssale and G. L. Vignoles. Molecular dynamics evidences of the full graphitization of a nanodiamond annealed at 1500 k. *Chemical Physics Letters*, **454**(4–6):299–304, 2008.

[26] Qian X. and Xiang Z. (2012) Bucky-diamond versus onion-like carbon: end of graphitization. *Physical Review B*, **86**(15):155417.

[27] K. Iakoubovskii, K. Mitsuishi, and K. Furuya (2008) High-resolution electron microscopy of detonation nanodiamond. *Nanotechnology*, **19**(15): 155705.

[28] O. O. Mykhaylyk, Y. M. Solonin, D. N. Batchelder, and R. Brydson (2005) Transformation of nanodiamond into carbon onions: a comparative study by high-resolution transmission electron microscopy, electron energy-loss spectroscopy, x-ray diffraction, small-angle x-ray scattering, and ultraviolet raman spectroscopy. *Journal of Applied Physics*, **97**(7):074302.

[29] A. I. Shames, A. M. Panich, W. Kempinski, A. E. Alexenskii, M. V. Baidakova, A. T. Dideikin, V. Y. Osipov, V. I. Siklitski, E. Osawa, M. Ozawa, and A. Y. Vul (2002) Defects and impurities in nanodiamonds: EPR, NMR and TEM study. *Journal of Physics and Chemistry of Solids*, **63**(11):1993–2001.

[30] S. Turner, O. I. Lebedev, O. Shenderova, I. I. Vlasov, J. Verbeeck, and G. Van Tendeloo (2009) Determination of size, morphology, and nitrogen impurity location in treated detonation nanodiamond by transmission electron microscopy. *Advanced Functional Materials*, **19** (13): 2116–2124.

[31] A. E. Aleksenskii, V. Y. Osipov, A. Y. Vul, B. Y. Ber, A. B. Smirnov, V. G. Melekhin, G. J. Adriaenssens, and K. Iakoubovskii. Optical properties of nanodiamond layers. *Physics of the Solid State*, **43**(1):145–150, 2001.

[32] P. W. Chen, Y. S. Ding, Q. Chen, F. L. Huang, and S. R. Yun. Spherical nanometer-sized diamond obtained from detonation. *Diamond and Related Materials*, **9**(9–10):1722–1725, 2000.

[33] F. Y. Xie, W. G. Xie, L. Gong, W. H. Zhang, S. H. Chen, Q. Z. Zhang, and J. Chen. Surface characterization on graphitization of nanodiamond powder annealed in nitrogen ambient. *Surface and Interface Analysis*, **42**(9):1514–1518, 2010.

[34] X. W. Fang, J.D. Mao, E. M. Levin, and K. Schmidt-Rohr (2009) Nonaromatic core-shell structure of nanodiamond from solid-state NMR spectroscopy. *Journal of the American Chemical Society*, **131**(4):1426–1435.

[35] A. E. Aleksenskii, M. V. Baidakova, A. Y. Vul, and V. I. Siklitskii. The structure of diamond nanoclusters. *Physics of the Solid State*, **41**(4):668–671, 1999.

[36] J. Cebik, J.K. McDonough, F. Peerally, R. Medrano, I. Neitzel, Y. Gogotsi, and S. Osswald. Raman spectroscopy study of the nanodiamond-to-carbon onion transformation. *Nanotechnology*, **24**(20):205703, 2013.

[37] V. Mochalin, S. Osswald, and Y. Gogotsi (2009) Contribution of functional groups to the raman spectrum of nanodiamond powders. *Chemistry of Materials*, **21**(2):273–279.

[38] S. Osswald, V. N. Mochalin, M. Havel, G. Yushin, and Y. Gogotsi (2009) Phonon confinement effects in the raman spectrum of nanodiamond. *Physical Review B*, **80**(7):075419.

[39] S. Osswald, G. Yushin, V. Mochalin, S. O. Kucheyev, and Y. Gogotsi (2006) Control of sp2/sp3 carbon ratio and surface chemistry of nanodiamond powders by selective oxidation in air. *Journal of the American Chemical Society*, **128**(35): 11635–11642.

[40] T. Jiang and K. Xu. Ftir study of ultradispersed diamond powder synthesized by explosive detonation. *Carbon*, **33**(12):1663–1671, 1995.

[41] Y. V. Butenko, V. L. Kuznetsov, E. A. Paukshtis, A. I. Stadnichenko, I. N. Mazov, S. I. Moseenkov, A. I. Boronin, and S. V. Kosheev (2006) The thermal stability of nanodiamond surface groups and onset of nanodiamond graphitization, *Fullerenes, Nanotubes, and Carbon Nanostructures*, **14** (2–3): 557–564.

[42] Lin Lai and A. S. Barnard. Modeling the thermostability of surface functionalisation by oxygen, hydroxyl, and water on nanodiamonds. *Nanoscale*, **3**(6):2566–2575, 2011.

[43] L. Lai and A. S. Barnard (2011) Stability of nanodiamond surfaces exposed to N, NH, and NH2. *Journal of Physical Chemistry C*, **115**(14):6218–6228.

[44] L. Lai and A. S. Barnard. Nanodiamond for hydrogen storage: temperature-dependent hydrogenation and charge-induced dehydrogenation. *Nanoscale*, **4**(4):1130–1137, 2012.

[45] C. L. Cheng, C. F. Chen, W. C. Shaio, D. S. Tsai, and K. H. Chen. The CH stretching features on diamonds of different origins. *Diamond and Related Materials*, **14**(9):1455–1462, 2005.

[46] J. B. Donnet, C. Lemoigne, T. K. Wang, C. M. Peng, M. Samirant, and A. Eckhardt. Detonation and shock synthesis of nanodiamonds. *Bulletin de la Société Chimique de France*, **134**(10–11):875–890, 1997.

[47] D. Mitev, R. Dimitrova, M. Spassova, Ch. Minchev, and S. Stavrev. Surface peculiarities of detonation nanodiamonds in dependence of fabrication and purification methods. *Diamond and Related Materials*, **16**(4–7):776–780, 2007.

[48] V. Pichot, M. Comet, E. Fousson, C. Baras, A. Senger, F. Le Normand, and D. Spitzer. An efficient purification method for detonation nanodiamonds. *Diamond and Related Materials*, **17**(1):13–22, 2008.

[49] M. Comet, V. Pichot, B. Siegert, F. Britz, and D. Spitzer (2010) Detonation nanodiamonds for doping kevlar. *Journal for Nanoscience and Nanotechnology*, **10** (7): 4286–4292.

[50] A. Krueger (2008) New carbon materials: biological applications of functionalized nanodiamond materials. *Chemistry A European Journal*, **14**(5):1382–1390.

[51] J. S. Tu, E. Perevedentseva, P.H. Chung, and C.L. Cheng (2006) Size-dependent surface CO stretching frequency investigations on nanodiamond particles. *Journal of Chemical Physics*, **125** (17): 174713.

[52] Q. Zou, M. Z. Wang, and Y. G. Li. Analysis of the nanodiamond particle fabricated by detonation. *Journal of Experimental Nanoscience*, **5**(4):319–328, 2010.

[53] SB Zhang, SH Wei, and A Zunger Intrinsic n-type versus p-type doping asymmetry and the defect physics of ZnO. *Physical Review B*, **63**(7):075205, 2001.

[54] K. Reuter and M. Scheffler (2002) Composition, structure, and stability of RuO_2(110) as a function of oxygen pressure. *Physical Review B*, **65**(3):035406.
[55] K. Reuter and M. Scheffler (2003) Composition and structure of the RuO_2(110) surface in an O-2 and CO environment: implications for the catalytic formation of CO_2. *Physical Review B*, **68**(4):045407.
[56] B. Aradi, B. Hourahine, and Th. Frauenheim. DFTB+, a sparse matrix-based implementation of the DFTB method. *Journal of Physical Chemistry A*, **111**(26):5678–5684, 2007.
[57] M. Elstner, D. Porezag, G. Jungnickel, J. Elsner, M. Haugk, T. Frauenheim, S. Suhai, and G. Seifert. Self-consistent-charge density-functional tight-binding method for simulations of complex materials properties. *Physical Review B*, **58**(11):7260–7268, 1998.
[58] H. A. Girard, J. C. Arnault, S. Perruchas, S. Saada, T. Gacoin, J.P. Boilot, and P. Bergonzo. Hydrogenation of nanodiamonds using MPCVD: a new route toward organic functionalization. *Diamond and Related Materials*, **19**(7–9):1117–1123, 2010.
[59] A. Krueger and D. Lang (2012) Functionality is key: recent progress in the surface modification of nanodiamond. *Advanced Functional Materials*, **22**(5):890–906.
[60] J. Mona, J.-S. Tu, T.-Y. Kang, C.-Y. Tsai, E. Perevedentseva, and C.-L. Cheng (2012) Surface modification of nanodiamond: photoluminescence and raman studies. *Diamond and Related Materials*, **24**:134–138.
[61] T. Gaebel, C. Bradac, J. Chen, J.M. Say, L. Brown, P. Hemmer, and J. R. Rabeau. Size-reduction of nanodiamonds via air oxidation. *Diamond and Related Materials*, **21**:28–32, 2012.
[62] G. Cunningham, A. M. Panich, A. I. Shames, I. Petrov, and O. Shenderova. Ozone-modified detonation nanodiamonds. *Diamond and Related Materials*, **17**(4–5):650–654, 2008.
[63] I. Petrov, O. Shenderova, V. Grishko, V. Grichko, T. Tyler, G. Cunningham, and G. McGuire. Detonation nanodiamonds simultaneously purified and modified by gas treatment. *Diamond and Related Materials*, **16**(12):2098–2103, 2007.
[64] O. Shenderova, A. Koscheev, N. Zaripov, I. Petrov, Y. Skryabin, P. Detkov, S. Turner, and G. Van Tendeloo. Surface chemistry and properties of ozone-purified detonation nanodiamonds. *Journal of Physical Chemistry C*, **115**(20):9827–9837, 2011.
[65] S. Zeppilli, J. C. Arnault, C. Gesset, P. Bergonzo, and R. Polini. Thermal stability and surface modifications of detonation diamond nanoparticles studied with x-ray photoelectron spectroscopy. *Diamond and Related Materials*, **19**(7–9, SI): 846–853, 2010.
[66] E. Osawa, D. Ho, H. Huang, M. V. Korobov, and N.N. Rozhkova (2009) Consequences of strong and diverse electrostatic potential fields on the surface of detonation nanodiamond particles. *Diamond and Related Materials*, **18** (5): 904–909.
[67] Barnard A.S. Self-assembly in nanodiamond agglutinates. *Journal of Materials Chemistry*, **18**(34):4038–4041, 2008.
[68] L.-Y. Chang, E. Osawa, and A. S. Barnard. Confirmation of the electrostatic self-assembly of nanodiamonds. *Nanoscale*, **3**(3):958–962, 2011.
[69] L. Lai and A.S. Barnard (2012) Interparticle interactions and self-assembly of functionalized nanodiamonds. *Journal of Physical Chemistry Letters*, **3**(7):896–901.
[70] H. Huang, E. Pierstorff, E. Osawa, and D. Ho (2007) Active nanodiamond hydrogels for chemotherapeutic delivery. *Nano Letters*, **7**(11):3305–3314.
[71] E.K. Chow, X.-Q. Zhang, M. Chen, R. Lam, E. Robinson, H. Huang, D. Schaffer, E. Osawa, A. Goga, and D. Ho (2011) Nanodiamond therapeutic delivery agents mediate enhanced chemoresistant tumor treatment. *Science Translational Medicine*, **3**(73): 73ra21.
[72] A. Adnan, R. Lam, H. Chen, J. Lee, D. J. Schaffer, A.S. Barnard, G.C. Schatz, D. Ho, and W. K. Liu (2011) Atomistic simulation and measurement of ph dependent cancer therapeutic interactions with nanodiamond carrier. *Molecular Pharmaceutics*, **8**(2):368–374.
[73] J.T. Paci, H. B. Man, B. Saha, D. Ho, and G. C. Schatz (2013) Understanding the surfaces of nanodiamonds. *Journal of Physical Chemistry C*, **117**(33): 17256–17267.
[74] M. Chen, X.-Q. Zhang, H. B. Man, R. Lam, E. K. Chow, and D. Ho (2010) Nanodiamond vectors functionalized with polyethylenimine for sirna delivery. *Journal of Physical Chemistry Letters*, **1**(21): 3167–3171.
[75] D. Ho. Beyond the sparkle: the impact of nanodiamonds as biolabeling and therapeutic agents. *ACS Nano*, **3**(12):3825–3829, 2009.

[76] H. Huang, M. Chen, P. Bruno, R. Lam, E. Robinson, D. Gruen, and D. Ho (2009) Ultrananocrystalline diamond thin films functionalized with therapeutically active collagen networks. *Journal of Physical Chemistry B*, **113**(10):2966–2971.

[77] H. Kim, H. B. Man, B. Saha, A. M. Kopacz, O.-S. Lee, G. C. Schatz, D. Ho, and W. K. Liu (2012) Multiscale simulation as a framework for the enhanced design of nanodiamond-polyethylenimine-based gene delivery. *Journal of Physical Chemistry Letters*, **3**(24): 3791–3797.

[78] R. Lam, M. Chen, E. Pierstorff, H. Huang, E. Osawa, and D. Ho. Nanodiamond-embedded microfilm devices for localized chemotherapeutic elution. *ACS Nano*, **2**(10):2095–2102, 2008.

[79] O. Loh, R. Lam, M. Chen, N. Moldovan, H. Huang, D. Ho, and H. D. Espinosa. Nanofountain-probe-based high-resolution patterning and single-cell injection of functionalized nanodiamonds. *Small*, **5**(14):1667–1674, 2009.

[80] R.A. Shimkunas, E. Robinson, R. Lam, S. Lu, X. Xu, X.-Q. Zhang, H. Huang, E. Osawa, and D. Ho (2009) Nanodiamond-insulin complexes as ph-dependent protein delivery vehicles. *Biomaterials*, **30** (29):5720–5728.

[81] A.H. Smith, E.M. Robinson, X.-Q. Zhang, E. K. Chow, Y. Lin, E. Osawa, J. Xi, and D. Ho. Triggered release of therapeutic antibodies from nanodiamond complexes. *Nanoscale*, **3**(7):2844–2848, 2011.

[82] X.-Q. Zhang, R. Lam, Xu X., E. K. Chow, H.-J. Kim, and D. Ho (2011) Multimodal nanodiamond drug delivery carriers for selective targeting, imaging, and enhanced chemotherapeutic efficacy. *Advanced Materials*, **23** (41):4770–4775.

[83] A. Alhaddad, M.-P. Adam, J. Botsoa, G. Dantelle, S. Perruchas, T. Gacoin, C. Mansuy, S. Lavielle, C. Malvy, F. Treussart, and J.-R. Bertrand. Nanodiamond as a vector for sirna delivery to ewing sarcoma cells. *Small*, **7**(21):3087–3095, 2011.

[84] J. Li, Y. Zhu, W. Li, X. Zhang, Y. Peng, and Q. Huang. Nanodiamonds as intracellular transporters of chemotherapeutic drug. *Biomaterials*, **31**(32):8410–8418, 2010.

[85] K.-K. Liu, W.-W. Zheng, C.-C. Wang, Y.-C. Chiu, C.-L. Cheng, Y.-S. Lo, C. Chen, and J.-I Chao. Covalent linkage of nanodiamond-paclitaxel for drug delivery and cancer therapy. *Nanotechnology*, **21**(31):315106, 2010.

[86] R. Martin, M. Alvaro, J. R. Herance, and H. Garcia. Fenton-treated functionalized diamond nanoparticles as gene delivery system. *ACS Nano*, **4**(1):65–74, 2010.

[87] K. V. Purtov, A. I. Petunin, A. E. Burov, A. P. Puzyr, and V. S. Bondar. Nanodiamonds as carriers for address delivery of biologically active substances. *Nanoscale Research Letters*, **5**(3):631–636, 2010.

[88] H.-D. Wang, Q. Yang, and C.H. Niu (2010) Functionalization of nanodiamond particles with N,O-carboxymethyl chitosan. *Diamond and Related Materials*, **19**(5/6):441–444.

10

Molecular Modeling of Layered Double Hydroxide Nanoparticles for Drug Delivery

Vinuthaa Murthy[1], *Zhi Ping Xu*[2], *and Sean C. Smith*[3]

[1] *School of Psychological and Clinical Sciences, Charles Darwin University, Australia*
[2] *Australian Institute for Bioengineering and Nanotechnology, University of Queensland, Australia*
[3] *School of Chemical Engineering, University of New South Wales, Australia*

10.1 Introduction

Layered double hydroxides (LDHs), also known as hydrotalcite-like compounds, are anionic clay materials. The interest in LDHs has increased rapidly in recent years due to their role in a wide range of applications in catalysis [1, 2], photocatalysis [3], heat stabilizers [4], ion exchangers [5], biosensors [6], halogen scavengers, medicine [7, 8], and environmental remediation [9, 10].

As LDHs exhibit permanent positive charges with a large surface area, a variety of anionic organic and inorganic structures can be intercalated into the LDH interlayer. The anions in the interlayer gallery are generally exchangeable. Many different kinds of anions have been successfully intercalated into LDH, including almost all of the common inorganic anions [11, 12]. Many organic and biomolecular anions, including carboxylates [13], benzoates [14], sulfonates [15], amino acids, and peptides [16, 17], as well as nucleotide phosphates and DNA chains [18, 19], can be intercalated within the interlayers of LDHs. This mechanism of encapsulation of anionic moieties within the positive layers of the inorganic

Computational Pharmaceutics: Application of Molecular Modeling in Drug Delivery, First Edition.
Edited by Defang Ouyang and Sean C. Smith.

LDH, also known as intercalation of organic and biomolecular molecules, makes LDH an excellent delivery carrier for drug and gene therapy applications [19, 20].

Intercalated moieties occupying the interlayer space of LDHs are quite stable in the LDH host but can be displaced through anion exchange reactions with anions with much higher affinity for LDH that are present in surrounding environments. This property of LDHs has been shown to provide sufficient protection [21] and great potential for controlled release properties [22] and is a good parameter especially for site-specific targeting and release into biological systems [23, 24].

LDHs are polycrystalline materials and precise experimental location and structure of interlayer anions are extremely difficult to obtain. Experimental techniques like powder X-ray diffraction (PXRD), TEM, and FTIR [11, 23] have been used to characterize the structure of the hydrotalcite, but disordering within the poorly crystalline hydrotalcite particles makes it difficult to fully elucidate interlayer arrangement of the intercalated anions and water molecules. However, details of the interlamellar properties of hydrotalcite that are difficult to interpret from experimentation can be estimated through the use of computational chemistry tools based on classical force fields and quantum chemical methods of electronic structure calculations and the structure and dynamics of these LDHs evaluated on an atomic scale.

It is significant to note that computational methods allow us not only to understand the molecular nature of dry LDHs but also to obtain an insight into their structure and behavior at different pH levels and in water-rich environments like aqueous solutions, which is difficult to interrogate with experimental methods. Computational methods have been utilized to analyze the structure of the hydrotalcite and to understand the hydration, swelling, bonding, and energetics of the Mg/Al hydroxide layers and the diffusivity of intercalated anions. The calculation of swelling energetics of LDHs resulting in the presence of a distinct minima in the hydration energy indicates the existence of energetically well-defined structural states with specific water content [13, 25] and the LDH system is considered to be stable. However, the absence of the well-defined minima with different water content indicates that the LDH would absorb water continuously in aqueous suspensions, leading to delamination of the double hydroxide layers and the release of anions into solution.

Computational simulations have been used by several groups to determine the stability of the LDH structures intercalated with different types of anions ranging from simple anions such as NO_3^- [26] to DNA [19] or RNA [27] strands.

This chapter will provide an insight into the different computational methodologies used to study the properties and stability of LDH intercalated with different types of anions.

10.2 Basic Structure of LDH

The LDH structure is closely related to brucite – $Mg(OH)_2$. In a brucite layer, each Mg^{2+} ion is octahedrally surrounded by six OH^- ions and the different octahedrons share edges to form a two-dimensional layer, as seen in Figure 10.1. Partial replacements of Mg^{2+} ions by Al^{3+} give the "brucite-like" layers a permanent positive charge,

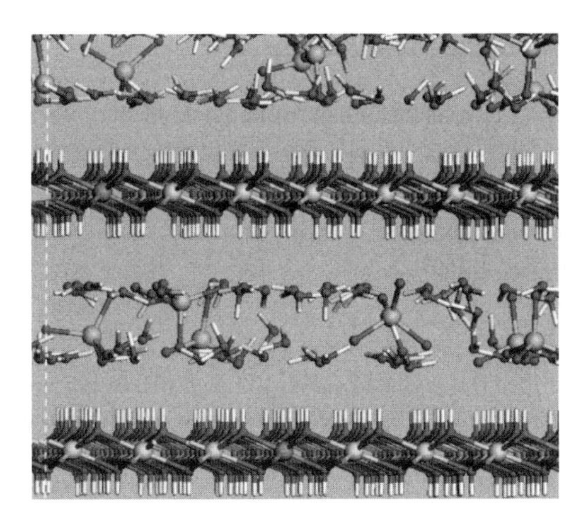

Figure 10.1 *Hydrotalcite 9 × 10 × 2 supercell intercalated with CrO_4^{2-} anions and water molecules. Green balls represent Mg atoms, pink = Al, red = oxygen, gray = hydrogen, and blue = chromium (see color plate section).*

which is balanced by negatively charged anions located in the interlayer region. The general formula for LDHs is $[M^{2+}_{1-x}M^{3+}_{x}(OH)_2][A^{n-}]_{x/x}.yH_2O$, where M^{2+} and M^{3+} are divalent and trivalent metallic cations, respectively, and A is an anion of valence n. The most studied class of LDHs is $Mg_6Al_2(OH)_{16}CO_3$, due to its use as a pharmaceutical antacid, talcid for ulcers and as a product of the alumina industry's alkaline wastewater neutralization process. The interlayer space along with the anions also contains water molecules, hydrogen bonded to the hydroxide layer and/or to the interlayer anions. Through electrostatic interactions and hydrogen bonds, the layers are stabilized in a crystalline form.

10.3 Synthesis of LDH

A common method for the preparation of LDHs is co-precipitation. Co-precipitation is based on the slow addition of a mixed solution of divalent (Mg^{2+}) and trivalent (Al^{3+}) metal salts to an alkaline solution in a reactor, which leads to co-precipitation of the two metallic salts. Formation of the LDH is based on the condensation of hexa–aqua complexes in solution that form the brucite-like layers containing both metallic cations. Interlamellar anions (such as inorganic [28], organic [15] or biomolecular [24]), either arise from the counter-anions of the metallic salts, or anions from the alkaline solution. The precipitated LDH is washed at ambient temperature thoroughly with ultrapure water to remove any residual salts and air dried or oven dried overnight.

The structural characterizations of the dried samples are generally achieved by PXRD, Raman, and IR spectroscopy [29]. However, details of the interlamellar properties of LDH and the interaction between the anions and the LDH layer are difficult to interpret from experiments.

10.4 Molecular Modeling Methodology

The molecular modeling approaches taken to understand the interaction of LDHs with four different types of intercalates will be explained in the following four subdivisions of this section.

1. Intercalation of oxymetal anions
2. Intercalation of organic anions
3. Intercalation of siRNA
4. Intercalation of DNA.

The methodology described for oxymetal anions [25] and siRNA [27] intercalation is original work where the authors are involved, while the work on organic anions [13, 30] and DNA [19] is from other research groups.

10.4.1 Intercalation of Oxymetal Anions into LDH

Here we highlight the general computational methodology that has been used by several research groups [13, 19, 25, 27, 31] to model LDH systems. The methods used by Murthy *et al.* will be discussed specifically to understand the interactions between the LDH layers and the intercalated anions and the water molecules in the gallery space. Properties and predictions that can be made from molecular dynamics (MD) computational simulations will be discussed using carbonate and chromate ions intercalated to LDH.

The model's initial crystal structure was obtained from the previously reported crystal structure of hydrotalcite, $Mg_4Al_2(OH)_{12}CO_3{\cdot}3H_2O$. The unit cell of the host structure is trilayer, the space group is R-3 m with triclinic cell and lattice parameters $a = b = 3.054$ Å, $c = 23.772$ Å; $\alpha = 90°$; $\beta = 90°$; $\gamma = 120°$. After removal of CO_3^{2-} anions and H_2O molecules, a supercell consisting of $9a \times 10b \times 2c$ unit cells was built with lattice parameters $9a = 27.486$ Å, $10b = 30.54$ Å, and $2c = 15.848$ Å. Mg and Al atoms were randomly distributed to obtain the Mg/Al ratio 2 : 1, each hydroxide layer containing 30 Al and 60 Mg, with a charge on each layer of 30 +e. Anions totaling to a charge of 30 –e (15 CO_3^{2-} or 15 CrO_4^{2-} ions) were randomly introduced between two host layers. Each super cell (Figure 10.1) constructed in this way has two metal–hydroxide layers and two interlayer galleries, with oxide anion (CO_3^{2-} or CrO_4^{2-}) charges totaling up to 30 –e and a variable amount of water molecules, $30n$, where $0 \le n \le 15$ (450 water molecules). The typical structural formula of the 2 : 1 Mg/Al hydrotalcite is $Mg_{2n}Al_n(OH)_8Ax^- \cdot nH_2O$.

Molecular dynamic simulations were performed using Forcite in Material Studio (MS) [32] 4.4 and 5.5. COMPASS Force Field [33], which is a general *ab initio* force field) was used for all geometry optimizations and MD simulations. COMPASS is the first high-quality general force field that consolidates parameters for organic and inorganic materials previously found in different force fields. COMPASS force field was assigned to all atoms in the LDH, water, and anions (CO_3^{2-} or CrO_4^{2-}).

Layer charge can have a major influence on the anion packing mode in the interlayer. Hence charges for the LDH and water molecules were modified according to the CLAYFF force field of Cygan *et al.* [34, 35], which is designed to give accurate results for LDH material. Partial charges on all atoms of anions in gaseous and aqueous phase were calculated by DFT methods. IEF-PCM [36], the default solvation method implemented in

Gaussian 03 (G03), was used to calculate the geometry and partial charges of anions in aqueous solution. Geometry optimizations and population analysis of the anions were obtained by using the B3LYP [37, 38] method and the LANL2DZ [39] basis set, suitable for transition metals [40]. The LANL2DZ basis set consists of 6–31G like functions for nontransition metal atoms and a valence double-zeta basis set for 3s, 3p, 3d, and 4s electrons and orbitals along with an effective core potential [41] for the metal. These computations were carried out using the G03 program [42]. Partial charges were obtained from Natural Population Analysis (NPA) [43] using the NBO program in G03.

Geometry optimizations of the hydrotalcite intercalated with anions and water molecules were carried out prior to MD simulations. The electrostatic and van der Waals energies were calculated by the Ewald summation method and minimizations carried out by a Quasi-Newton procedure. Periodic boundary conditions were applied in three dimensions so that the simulation cell was effectively repeated infinitely in each direction. Initially the host layers in the supercell were held as rigid units, allowing the lattice parameter c (gallery height) to vary. The structures of the anions were also held rigid. This enabled the mutual positions of the host layers and positions and orientations of the guests (anions and water molecules) to vary and optimize to a minimum with respect to each other. These optimized structures were then used as the starting configurations for the MD simulations, performed in the *NVT* ensemble (constant volume/constant temperature) where atoms in both the host and guest layers were released. Since all atoms in each system were completely free to move during these simulations, using the constant volume modeling approach with a fixed cell shape did not introduce significant limitations to the resulting interfacial structure, dynamics, and energetics of water.

For all simulations MD simulations were performed in the *NVT* ensemble at 300 K with a time step of 1.0 fs. An initial MD simulation of 30 ps was carried out using an Andersen thermostat for equilibration of thermodynamic parameters, followed by 200 ps simulations with a Nosé thermostat for different hydration states, n, of the system.

Hydration energies (HE) are found to be an effective measure of the affinity of water for the interlayer [13, 25, 44, 45]. Hydration energy of a system is defined as,

$$\Delta U_{\mathrm{H}}\left(N_{\mathrm{w}}\right)=\frac{\left[U_{\mathrm{H}}\left(N_{\mathrm{w}}\right)\right]-\left[U(0)\right]}{N_{\mathrm{w}}} \tag{10.1}$$

where N_{w} is the number of water molecules, and $U(N_{\mathrm{w}})$ and $U(0)$ are the total potential energies of the system with N_{w} and zero water molecules, respectively.

Further, to study the interactions of water molecules and anions in bulk water (i.e., aqueous suspensions) 450 water molecules were placed in each interlayer space and longer *NVE* ensemble MD simulations of 400 ps duration (including an initial 50 ps time for equilibration) were carried out to calculate the mean square displacement (MSD), self diffusion coefficient, and concentration profiles.

10.4.1.1 Modeling Results and Discussion

Optimized structures, where intercalated anions as well as water molecules were allowed to relax, but with fixed LDH layers, are shown in Figure 10.2.

In carbonate-LDH the trigonal planar CO_3^{2-} ions lie horizontally in the interlayer in the absence of water molecules. As the number of water molecules is increased, they orient

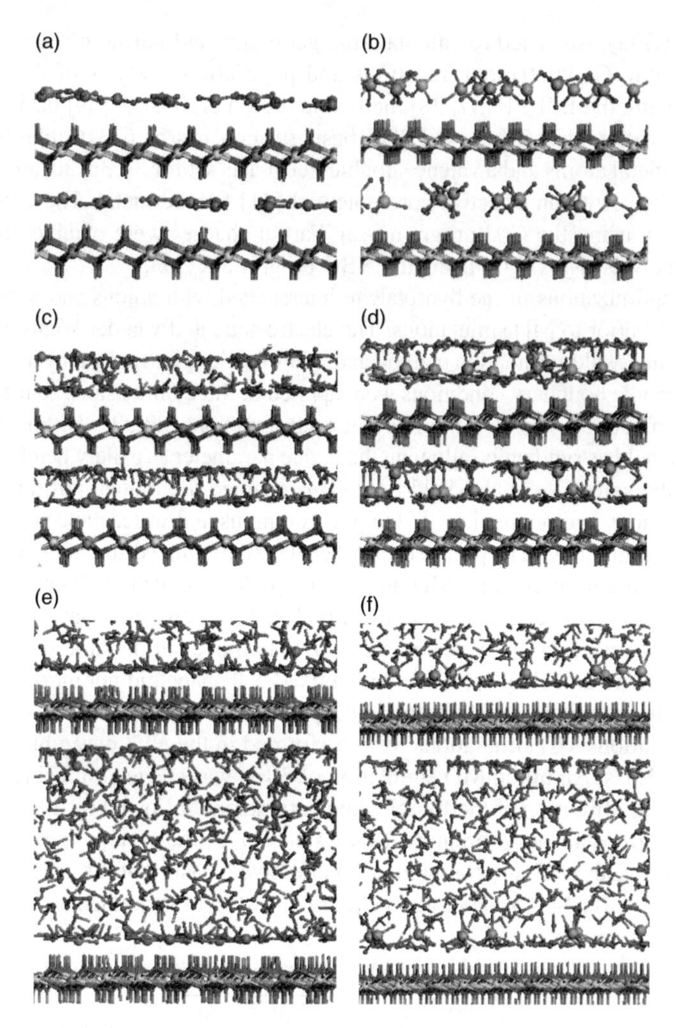

Figure 10.2 *Energy minimized structures of 2 : 1 HT intercalated with carbonates (a,c,e) and chromates (b,d,f) with increasing number of water molecules: (a,b)* n = 0*; (c,d)* n = 3 *(90 water molecules in each interlayer); (e,f)* n = 15 *(450 water molecules in each interlayer). Reprinted with permission from Ref. [25]. Copyright 2011, American Chemical Society.*

themselves perpendicularly to the LDH layer. In the chromate, the tetrahedral CrO_4^{2-} ions have two orientations in the absence of water molecules. One orientation has three oxygen atoms forming hydrogen bonds with one LDH layer and the fourth oxygen with the other LDH layer. The second orientation has two oxygen atoms each hydrogen bonding with OH groups of the LDH layer. As the water content in the interlayer is increased the *d*-spacing increases very slowly as the water molecules fill the empty space between the anions and form hydrogen bonds with the oxygen atoms of the anion and the hydroxide groups in the LDH (Figure 10.2e,f).

As shown in Figure 10.3, once the water content is above $n = 2$ a new water layer is formed and the *d*-spacing takes the pop and fill pattern (Figure 10.4) with increasing water

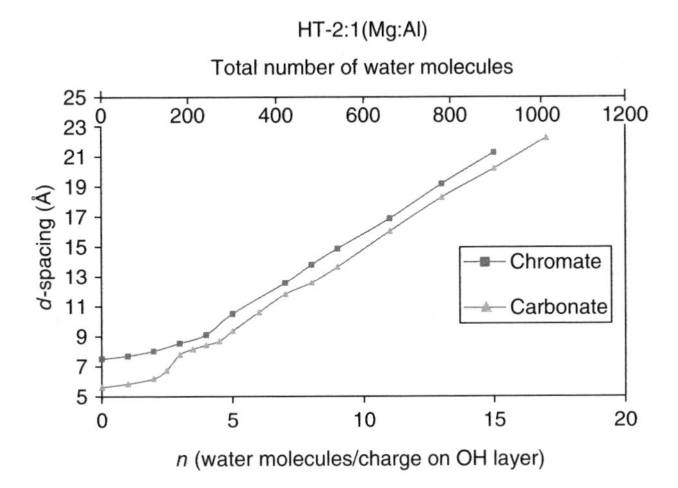

Figure 10.3 *Variation of* d-*spacing (interlayer distance) as a function of number of water molecules (*n*) per charge on the metal hydroxide layer.*

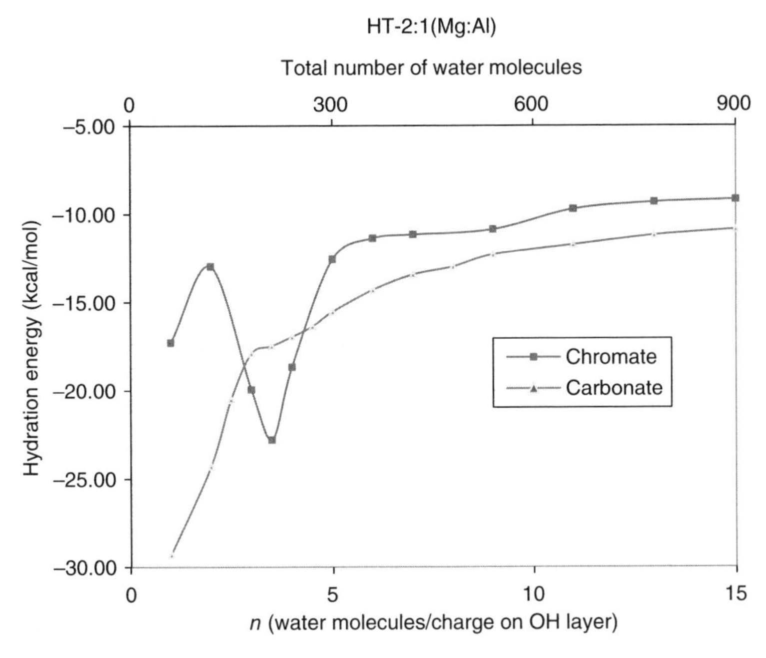

Figure 10.4 *Variation of hydration energy [$\Delta U_H(N_w)$] as a function of number of water molecules (*n*) per charge on the metal hydroxide layer and total number of water molecules in each interlayer.*

content. Carbonate-LDHs show a slightly different trend in the beginning as they switch from lying horizontal, that is, parallel to the metal hydroxide layer until $n=2$ to a perpendicular position afterwards, but continuing with the pop and fill pattern. The water molecules form highly ordered layers with oxygen atoms orienting toward the LDH layer and hydrogen atoms orienting away from the hydroxide layers. This in turn induces further ordered water (OW) layers away from the LDH layer. However, as the distance of the water layers from the LDH layer increases, disorder increases. The average distance between the H atoms in the LDH layer and the oxygen atoms of the first OW layer, water is around 1.80–1.90 Å.

HEs computed are plotted against the number of water molecules (n) per charge on the LDH layer in Figure 10.4. Similar to most of the previous studies on organic anions [13], hydration energy shows the largest negative values at low water content and increases as the water content increases. The hydration energy (Equation 10.1) has several components, including: the potential energy of interlayer water; the effects on the potential energy of changing interactions of the hydroxide layers with themselves and anions, and the anions with themselves. Hydration energy is therefore the sum of energy required to expand an equilibrium structure to create additional interlayer space for more water molecules and the energy gained due to the presence of additional water molecules, rearrangement of interlayer species, reformation of the hydrogen bonding network, and equilibration of the system in a new hydration state. A minimum if observed in the hydration energy on addition of water molecules indicates a preferred hydration state of the system. This can be interpreted as arising from the well-developed hydrogen-bonding network as more water molecules are added.

The $\Delta U_H(N_w)$ values increase rapidly with increasing water content. Chromate-HTs exhibit a decrease in hydration energy after $n=2$, with a minimum at $n=3.5$, while carbonate-HTs do not exhibit a minimum, but only a plateau between $n=3$ and $n=4$. Comparing this with results in Figure 10.3, it is seen that the minima coincide with plateauing of the d-spacing or filling up of the first OW layer. After $n=5$, as the water content further increases, the $\Delta U_H(N_w)$ values increase at a very slow rate and plateau almost parallel to each other. The HE of both Huntingtons (HTs) approach the bulk water potential energy value (~−10 kcal/mol [13, 44]). Carbonate-HT has a lower $\Delta U_H(N_w)$ value compared to chromate, which indicates its stronger affinity to water at all phases of water content.

To model the relative positions of the anions and water molecules in water-rich environments such as high humidity conditions or in aqueous suspensions, longer simulations are carried out with 450 water molecules in each interlayer. The atomic density profiles along the z-direction, obtained from 500 ps of fully relaxed (*NVE* ensemble) MD simulations of the LDH with 450 water molecules in each interlayer is shown in Figure 10.5.

Water molecules form two distinct OW layers in all LDHs. Results show that most of the anions (represented by the blue line) prefer to position themselves close to the first OW layer. The chromate anions are found to be positioned between the first and the second OW layers. One or three oxygen atoms are in plane with the first OW layer and form hydrogen bonds with both the water molecules and the LDH layer, while the remaining oxygen atoms (three or one) are situated in the second OW layer. The majority of the anions prefer the orientation where the three oxygen atoms are in the first OW layer. As a result, the metal atom of the anion is located just above the first OW layer, as seen in both Figure 10.2 and the atomic density profiles in Figure 10.5.

The carbon and oxygen atoms in the planar carbonate ions orient themselves parallel to the LDH layer and along the first OW layer in the water-rich LDHs, but in the preferred or optimal water content situations (Figure 10.2c) they are found to orient predominantly perpendicularly to the LDH layer. Simulations indicate that in general

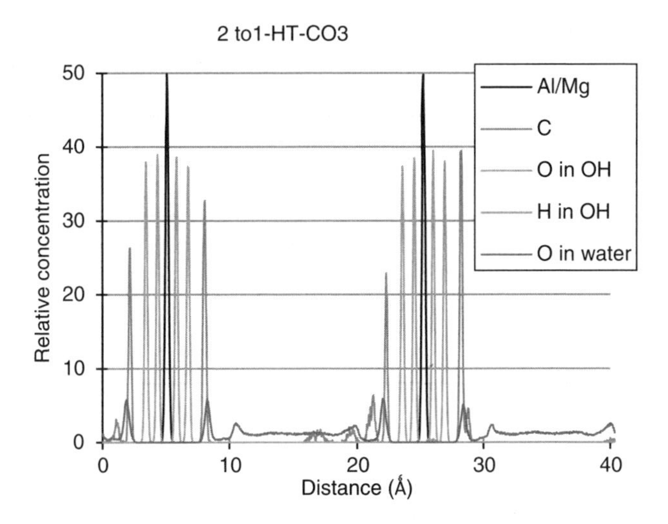

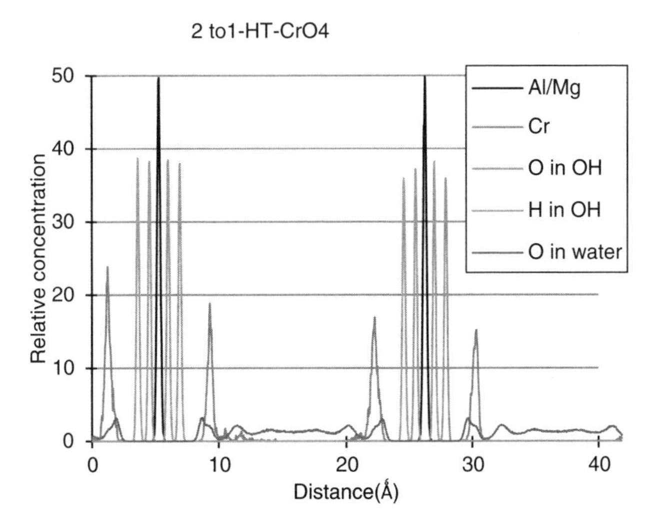

Figure 10.5 *Atomic density profiles along the z-direction of supercells obtained from 500 ps of fully relaxed MD simulations of the LDH with 450 water molecules in each interlayer. Reprinted with permission from Ref. [25]. Copyright 2011, American Chemical Society.*

the chromate ions are found only in the first OW layer. In comparison, the majority of the carbonate ions are also found in the first OW layer, but some are found in the second OW layer.

To study the relative mobility of the anions, self diffusion coefficients (D) were calculated from the MSDs of anions and water molecules in different OW layers; and in the middle of the interlayer space using fully relaxed NVE models, with MD simulation of 300 ps, at 300 K. Calculated D from the simulations are presented in Table 10.1 and plotted

Table 10.1 *Self diffusion coefficient (D) of anions and water molecules in bulk water (450 water molecules in each interlayer) calculated from fully relaxed MD simulation for 300 ps, at 300 K.*

Self diffusion coefficient (cm^2/s)		
2- to 1-LDH	CO_3^{-2}	CrO_4^{-2}
Ions in layer 1	2.02×10^{-7}	7.84×10^{-7}
Ions in midgallery	1.11×10^{-6}	—
Water in layer 1	7.34×10^{-6}	8.90×10^{-6}
Water in layer 2	1.72×10^{-5}	2.17×10^{-5}
Water in midgallery	1.89×10^{-5}	2.34×10^{-5}

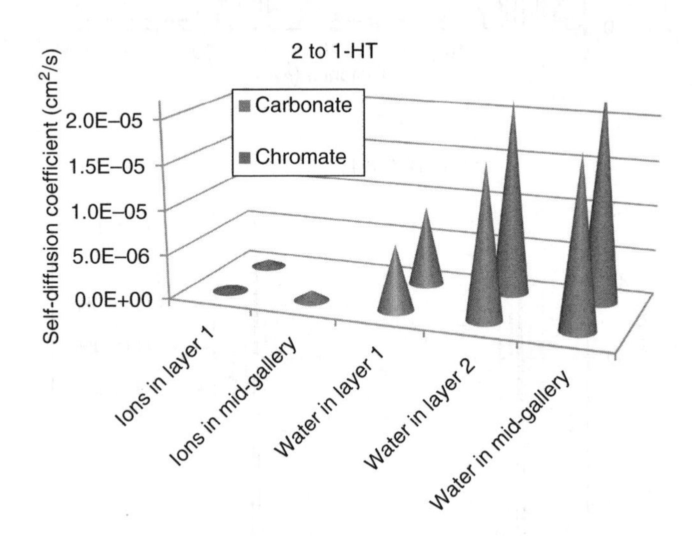

Figure 10.6 *Comparative self diffusion coefficient (D) in cm^2/s of anions and water molecules calculated from fully relaxed MD simulation for 500 ps, at 300 K for LDHs.*

Table 10.2 *Self diffusion coefficient (D) of anions and water molecules in preferred water state LDHs calculated from fully relaxed MD simulation for 300 ps, at 300 K.*

Self diffusion coefficient (cm^2/s)				
	CO_3^{-2}	Water	CrO_3^{-2}	Water
2:1 LDH	6.67×10^{-9}	1.50×10^{-7}	3.00×10^{-8}	4.02×10^{-7}

in Figure 10.6 for comparison. MSDs were also calculated for the optimal or preferred water states for all LDHs (Table 10.2).

The results indicate that the self diffusion coefficient (*D*) of anions increases as the distance of the anion from the LDH layer increases. The diffusion of water molecules in various ordered and disordered layers follows the same trend. The bulk water self diffusion coefficient

Table 10.3 *Comparison of water content (n) in preferred hydration states along with total number of water molecules and anions in each interlayer.*

	2:1 HT				
Anions	*n*	H_2O molecules	Anions	O atoms in anions	Total O atoms (H_2O+anion)
CrO_4^{2-}	3.5	105	15	60	165
CO_3^{2-}	3.5	105	15	45	150

calculated under the same simulation conditions [as stated in the theoretical methods section (at 300 K)] is determined to be 3.2×10^{-5} cm²/s. This value of D is slightly higher than the experimental self diffusion coefficient of water between 298 and 305 K, reported as 2.3×10^{-5} to 2.9×10^{-5} cm²/s [46, 47]. A D value of 3.2×10^{-5} cm²/s using COMPASS FF has also been observed elsewhere [48]. Simulations were also run for salt solution ($MgCl_2$) of the same ionic strength as the LDH systems studied here. The calculated D for the salt solution was found to be 2.34×10^{-5} cm²/s. D values were calculated from MSDs for the optimal or preferred water states (see Figure 10.4) for all LDHs.

The preferred hydration states are tabulated in Table 10.3. The water molecules in these states are found to form two water layers (as shown in Figure 10.2c,d), forming a network of hydrogen bonds with the H atoms on the metal hydroxide layer. The anions are found to position themselves between these two water layers with the oxygen atoms embedded in the water layers.

The relative mobility of the anions and water molecules in the interlayer of hydrotalcites in water-rich environments, such as aqueous suspensions, can be studied by comparing the self diffusion coefficients (D). Chromate ions have a smaller D value (Figure 10.6) compared to carbonate ions. Water molecules in the first OW of chromate-HT are also less mobile than waters in the first OW layer of carbonate-HT. The ions in the first layer are stabilized by forming stronger hydrogen bonds with the metal hydroxide layer. A few carbonate ions are found to migrate to the middle of the gallery, forming hydrogen bonds with the water molecules.

Hydration energy calculations performed with different water contents show a minimum for chromate-LDH and it is found to be lower than the corresponding carbonate system in the same water regime, indicating that there is a greater degree of stabilization of chromate-LDH in the preferred hydration state. This well-defined minimum in the hydration energy of chromate-LDH suggests that it has an energetically well-defined structural state at specific water content and that chromate-LDH is stabilized by the formation of hydrogen bonds between the metal hydroxide and chromate ions with the water molecules.

10.4.2 Intercalation of Organic Anions into LDH

Kalinichev *et al.* [30] and Kumar *et al.* [13] performed MD computer simulations to investigate the interlayer expansion, HEs and effects of hydrogen bonding on the properties of Mg_xAl_y-LDHs intercalated with mono- and polycarboxylate anions such as formate, acetate, propanoate, lactate, citrate, and glutamate anions over a wide range of water contents.

For modeling the interactions of the LDH layers with the intercalated organic acids they used CLAYFF force field to describe the interatomic interactions of the LDH layers, the SPC model for water, but used the CVFF force field [48] to describe the organic acid anions, which has proven to provide reliable molecular description of various organo-inorganic systems [37, 38, 49, 50]. They used the OFF module of the Cerius2 molecular modeling package in all simulations. The MD results were analyzed to obtain the basal spacing (*d*-spacing) and compute the HEs similar to the above section on chromate-LDHs. Figure 10.7 is obtained from the Kalinichev *et al.* [30] result section.

Unlike the chromate ions the HEs of the organic acid anions, as shown in Figure 10.7, do not exhibit distinct minima and indicate the absence of any stable well-defined structural states in these materials at any specific hydration level. It also suggests that LDHs intercalated with such organic anions, unlike most inorganic-intercalated LDHs, absorb water in a continuous fashion in water-rich environments and are found to be consistent with experimental observations of the expansion, exfoliation, and delamination of various organic-intercalated LDHs in aqueous solutions [17, 20–22, 26, 51].

Kalinichev *et al.* suggest that the swelling behavior of all these phases is due largely to the affinity of the $-COO^-$ groups for H-bonds donated by water molecules, which can better solvate them in well-integrated H-bond network than the fixed –OH sites of the hydroxide layers. Thus, the H-bond network is much better interconnected and less strained when water is present than in the dry phase.

The higher-charge anions like citrate (3^-) or glutamate (2^-) exhibit superior swelling behavior, that is, more negative HE compared to the LDHs intercalated with monocarboxylic anions, resulting slower basal expansion under hydration. Hence they suggest that the swelling behavior of LDHs in water can be improved by starting with larger interlayer anions containing multiple carboxylic groups.

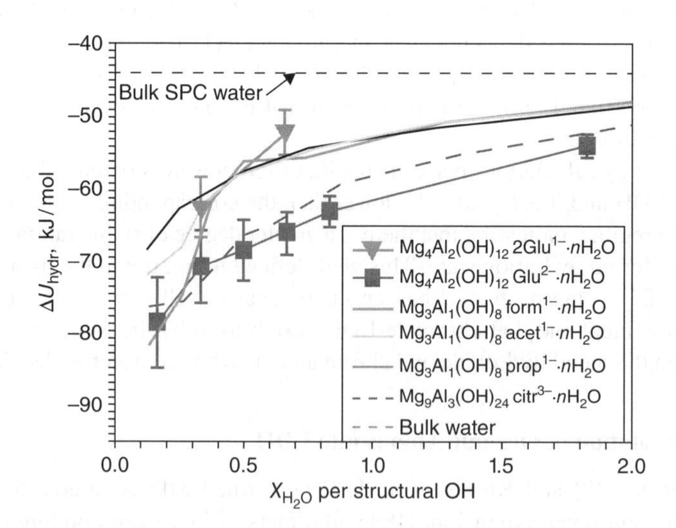

Figure 10.7 *Hydration energy of formate, acetate, propanoate, citrate, and glutamate-intercalated Mg, Al-LDH as a function of water content in the interlayer [30]. Reprinted from Ref. [30] with the permission of Taylor and Francis Ltd, www.tandfonline.com.*

10.4.3 Intercalation of siRNA into LDH

The intercalation of small interfering RNA (siRNA) into LDH for gene therapy applications has shown great potential in recent years. The introduction of small interfering RNA into neurons potentially allows the treatment of HT disease through RNA interference to silence the HT gene [20]. The LDH system has been identified as a promising candidate to create a delivery system that will be able to carry the interfering RNA across the blood–brain barrier and into affected neurons. Interestingly, the RNA World hypothesis postulates that, since RNA itself can act as a catalyst (in addition to carrying genetic information), the earliest forms of life were built upon RNA [33]. Deep ocean hydrothermal vents have been suggested as possible sources for precursors of prebiological molecules, and clay-like particles present may have acted as structures to support and protect the nucleic acids formed from the elevated temperatures and pressures around these vents.

Hence, understanding on the atomistic level both the structure and dynamics of RNA molecules intercalated within LDH layers can give a clearer picture of their role and function as therapeutic agents in gene therapy applications, as well as their possible role in prebiotic synthesis.

Xu *et al.* addressed this by a combination of experimental approaches and MD simulations, first by studying a simpler LDH-sulfonate system [15, 31] and then extending to the preparation and characterization of LDH-siRNA nanoparticles [20, 27]. Their experiments showed that the average sizes of LDH nanoparticles were about 100 nm and the nanoparticles were MgAl-LDH-type materials. Based on the zeta potential values of the nanomaterials, it was estimated that over 85% of the total absorbed RNAs were intercalated into the interlays of LDH due to the electrostatic interaction between the nanomaterials and nucleic acids [20]. LDHs are polycrystalline materials and precise experimental location of interlayer RNA anions is extremely difficult to obtain. Computer simulations were conducted to provide an insight into the structure and stability of siRNA while intercalated in layered materials to complement the limited information that could be obtained from experiments.

Using MD simulations, both the structural and the dynamical details of the LDH-siRNA hybrid system were investigated. A combination of CLAYFF and COMPASS force fields were used to obtain the simulations [27].

The siRNA molecule was optimized and introduced into the central LDH interlayer. The sequence of 21 bp siRNA was taken from an earlier study [49] and is as follows:

- Sense 5′-GCAACAGUUACUGCGACGUUU-3′
- Antisense 3′-UUCGUUGUCAAUGACGCUGCA-5′

Both 3′-terminal UU of the RNA duplex were cut to form a canonical RNA duplex and then A-RNA and A′-RNA was generated in the NUCGEN module of AMBER. The phosphate groups of all RNA strands were unprotonated with a unit negative charge. The Discover module in MS 4.4 was employed to perform MD simulations. The MSD, self diffusion coefficient and concentration profiles, and so on, were calculated using the analysis part of the Discover module in MS 4.4 using the trajectory files generated from 500 ps fully relaxed MD simulations. Figure 10.8 shows the minimized structure for partially constrained hybrid systems.

Inspection of the structures for the hybrid system, as shown in Figure 10.8b, shows that the RNA double helix orients parallel to the hydroxide layers. Compared with the structures

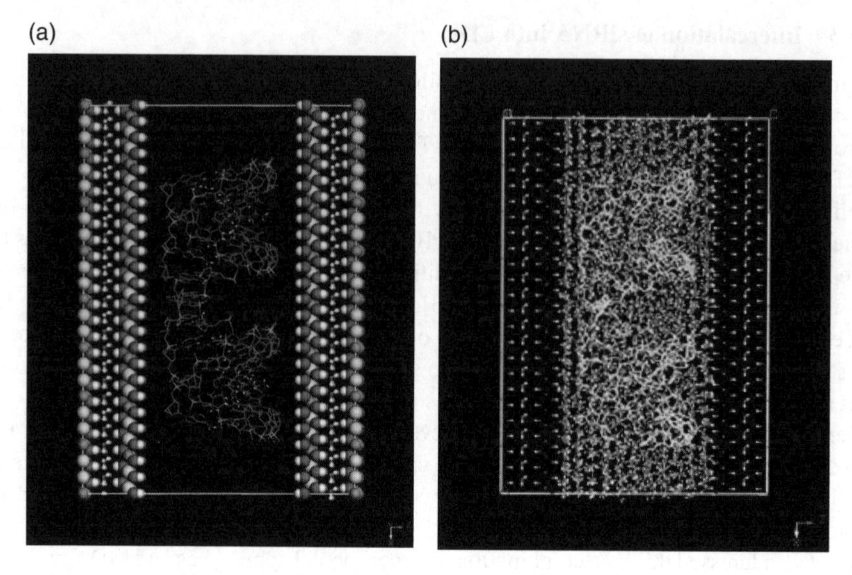

Figure 10.8 A-RNA plus LDH system prior to relaxation [27]: (a) water molecules omitted for clarity; (b) with water molecules. (Reprinted from Ref. [27] as per MDPI open access policy.)

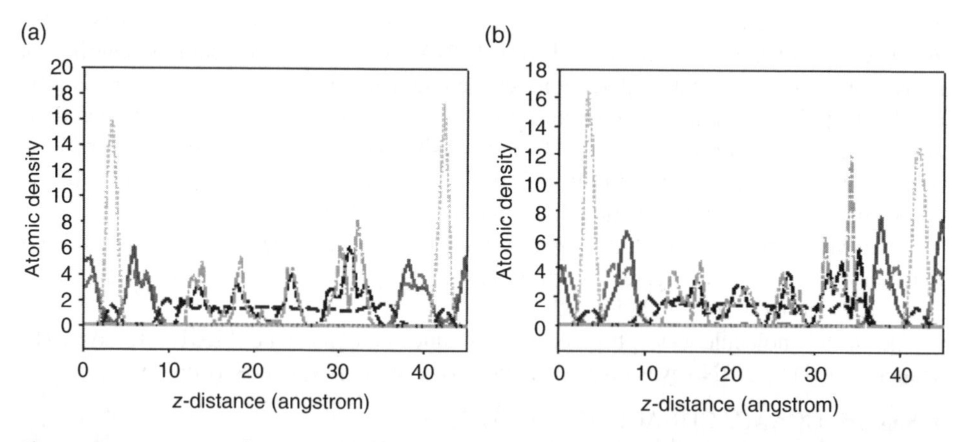

Figure 10.9 Atomic density profiles for: (a) LDH+A-RNA and (b) LDH+A′-RNA. The solid line represents Mg atoms in the LDH layer, the long dashes line represents the oxygen atoms in the LDH layers, the short dashes line represents oxygen atoms of water, the medium dashes line represents oxygen atoms of phosphate groups in siRNA, the dotted line represents Cl anions, and the dash dot dash line represents P atoms in siRNA. (Reprinted from Ref. [27] as per MDPI open access policy.)

"as built" or the gas phase optimized structures, there are additionally strong duplex-RNA deforming characteristics in the presence of LDH.

Figure 10.9 shows the atomic density profiles orthogonal to LDH layers for selected atoms in the hybrid systems. These analyses are based on trajectories from 500 ps of fully relaxed MD simulations for the LDH-siRNA hybrid system at 300 K. From this figure we can see the hybrid system still has a well-defined structure, with oxygen atoms of OH in the LDH layer closest to the Mg/Al sheet (the z distance is zero for the first Mg/Al sheet in LDH layer), followed by oxygen atoms of water as well as oxygen atoms in the phosphate groups in siRNA, and finally the P atoms of phosphate groups in siRNA. The z distance of some oxygen atoms in water and in the phosphate groups of siRNA are actually very close, and can be regarded as the same layer in general. Our results indicate that Cl^- anions in the two edge LDH layers are well structured due to their sharp peaks (together with water molecules to form one single layer), followed by Mg atoms in the LDH layers, and then by P atoms in the phosphate groups of siRNA. Finally the water molecules in the central layer are less structured, albeit demonstrating some peaks in their distribution.

Analysis of dynamic properties, like self diffusion coefficients (D), for the intercalated siRNA and water can provide some useful information for the stability of the hybrid system, which is in turn very important for the efficient delivery of siRNA using LDH nanoparticles as carriers.

The A-form and the A′-form RNA self diffusion coefficient D was found to be $7.24 \times 10^{-8}\,cm^2/s$ and $4.24 \times 10^{-8}\,cm^2/s$, respectively at 300 K. However, the variability in the D values of A-RNA and A′-RNA seen in LDH was not observed in water. The diffusion coefficients of siRNA are generally one order smaller than the constrained water molecules in the LDH layers, therefore they are much more stable when complexed with LDH layers. This stability is required for siRNA to be protected in the delivery processes, using nanoparticles as carriers, but needs to be released once siRNA reaches the target. Strong interaction between carrier and nucleic acid will hinder the release of the gene from the complex in the cytosol, thereby adversely affecting transfection efficiency [26]. However, for MgAl-LDH-siRNA systems, the release of siRNA can be achieved by an acidic environment such as in the later endosomal or lysosomal compartment (i.e., a change in pH to 3.0–5.0), which promotes hybrid disruption and the release of DNA from the complexes. In the acidic environment, ions such as Mg^{2+}, Al^{3+}, Cl^-, and siRNA are released and small ions, like Na^+, K^+, and Cl^-, may leave the cell through the so-called ion tunnels. MgAl-LDHs therefore have the right balance between chemical stability and biodegradability, and thus are very promising for cellular delivery.

Parallel simulations were performed for both A-RNA and A′-RNA forms as a probe of the ability of the LDH-intercalated environment to support structural diversity in RNA. These two crystalline forms differ in the degree of helical twisting. The diffusion coefficients are distinctly different and the structures, as parameterized by the major groove width averaged along the length of the strand, are clearly distinguishable by simulations.

Molecular modeling revealed the arrangement of the guest RNA, layer stacking and spatial distribution of water molecules in the interlayer gallery of the host structure. The simulations showed that RNA double helices are oriented parallel to the hydroxide layers. Due to strong electrostatic forces acting between the LDH sheets with permanent positive charge and the intercalated RNA, the siRNA molecules have apparent deformations compared to siRNA in bulk water. The LDH layers also adjust their structure in order to host the siRNA molecules, but the hybrid system still has a well-defined layered structure despite the distortions observed. This work supports the concept that clay-like particles may have acted as hosts which support and protect diverse RNA structures once formed. These findings also support the proposal that the MgAl-LDH host is potentially a good candidate for the delivery of functional oligonucleotides such as RNA for gene therapy applications.

10.4.4 Intercalation of DNA into Layered Double Hydroxides

DNA also exists in closed double stranded (ds) loops, called plasmids, which are commonly used in gene therapy. Plasmids can exist as supercoiled structures, which often form when the DNA strand is under a twisting strain. The experimental synthesis technique employed when making hybrid LDH-DNA systems has been shown to dictate the size of intercalated DNA [50]. LDHs have been found to intercalate larger linear and plasmid ds DNA between 100 and 8000 bp long using the co-precipitation method [50]. This method directly forms LDH around DNA, as opposed to anion exchange of DNA into pre-existing LDHs [51].

To provide an insight into the structure and stability of DNA while intercalated in layered materials is very difficult to obtain from experiment. Thyveetil *et al.* [19] performed large-scale molecular dynamic simulations to extract detailed information on the structure and dynamics of DNA intercalated within Mg_2Al-LDH, as well as to see how the material properties of Mg_2Al-LDH are modified when intercalated with DNA. They used four different DNA molecules starting from 12 to 480 bp to simulate intercalation into LDH. The models of LDH they used were created by replication of the primitive crystallographic cells. The aim was to measure the in-plane elastic coefficients, so the simulation cell was replicated laterally along the basal plane, rather than increasing the thickness of the LDH. Figure 10.10 shows the initial structure of the DNA molecule intercalated LDH system.

The Amber ff99 force field was used to obtain parameters for the partial charges and bonded interactions within DNA molecules, while the ClayFF force field furnished parameters for atoms within LDH. Water molecules were described using the flexible single-point charge (SPC) model. In addition to ambient conditions of 300 K and 1 atm, simulations were run at higher temperatures and pressures aiming to reproduce the high temperatures and pressures around deep ocean hydrothermal vents. Although detailed structural information may be less reliable at higher temperatures and pressures, the simulations are expected to provide valuable insight into the behavior of LDH-DNA systems.

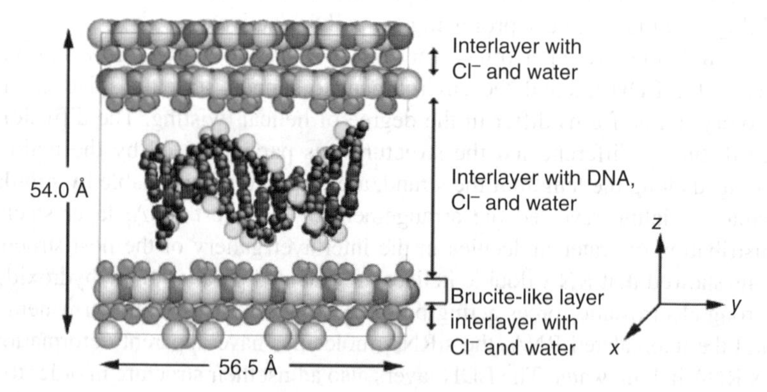

Figure 10.10 *Initial structure of LDH intercalated with DNA molecule with 12 bp [19]. Magnesium, aluminum, chlorine, phosphorus, carbon, and nitrogen atoms are represented as light-gray, pink, green, yellow, dark-gray, and blue spheres, respectively. Oxygen and hydrogen atoms have been removed in order to aid viewing. Reprinted with permission from Ref. [19]. Copyright 2008 American Chemical Society (see color plate section).*

Simulations of 1 ns were performed following energy minimization and thermalization. Simulations run at 300 and 350 K were maintained at 1 atm pressure, the 400 K simulation was kept at 50 atm pressure, and the 450 and 500 K simulations were held at 100 atm. A control simulation of DNA in bulk water was also studied under ambient conditions and at higher temperatures and pressures. The systems were judged to have reached equilibration before 500 ps by monitoring the potential energy and the cell parameters.

Root mean square deviation (rmsd) and the radius of gyration were obtained from simulation results. While rmsd quantifies the variation of the DNA from its initial structure and hence can be used to measure the structural stability, the radius of gyration quantifies how much stretching or compression the DNA molecules undergo.

The rmsd is calculated using:

$$r_{\mathrm{rmsd}}(\mathbf{r},\mathbf{r}0)=\sqrt{\frac{1}{N}\sum_{i=1}^{N}\left(r_i-r_{i,0}\right)}$$

where **r** is the current position of atom *i* and **r** *i*,0 is its initial position. The summation is carried out over all *N* DNA atoms.

The radius of gyration is calculated using:

$$r_{\mathrm{G}}=\sqrt{\frac{1}{N}\sum_{i=1}^{N}\left(r_i-r\right)^2}$$

where *r* is the mean position of all *N* DNA atoms.

The elastic properties of clay sheets can be determined theoretically through uniaxial expansion and contraction of the simulation cell. The elastic modulus tensor, S_{ijkl} is defined by the three-dimensional version of Hooke's Law:

$$\sigma_{ij}=\sum_{k=1}^{3}\sum_{l=1}^{3}S_{ijkl}\,\varepsilon_{kl}$$

where σ_{ij} and ε_{kl} are elements of the stress and strain tensors, respectively. The atomic coordinates are rescaled in order to fit the new geometry of the simulation box.

Simulations and analysis conducted by Thyveetil *et al.* showed that strong electrostatic forces act between LDH and DNA. This makes the intercalated DNA molecules to be significantly restricted in movement compared to DNA in bulk water. Structural analysis of intercalated DNA molecules averaged over 1 ns indicates that the motion of phosphate groups in the DNA backbone is extremely restricted, as compared to simulations of DNA in bulk water. The hydration energy values demonstrate that hydration plays a crucial part in the stability of the molecule. When no water molecules are present, all Watson–Crick hydrogen bonds are disrupted. The basal spacing is dependent on the extent of hydration of the interlayer, which can be seen by steps in the swelling curve, corresponding to the formation of water layers.

Simulations performed by increasing the temperature and pressure suggest that the structure of DNA is stabilized by intercalation. Simulations at higher temperatures demonstrate an increase in average rmsd relative to their initial structure. There is some difference in the behavior of larger and smaller systems so far as the effect of temperature on the number of Watson–Crick base pairs is concerned. When DNA containing 108 bp is subjected to higher temperatures and pressures, an increase in the LDH basal spacing is observed, together with an increase in the number of Watson–Crick hydrogen-bonded base pairs, whereas the number of Watson–Crick hydrogen-bonded base pairs decreases when present in bulk water. This indicates that DNA intercalated into LDHs has enhanced structural stability. Moreover, the largest LDH-DNA model, containing a DNA plasmid with 480 bps, exhibits larger amplitude height fluctuations in the LDH sheets than are seen in smaller models. The Young's moduli of these LDH-DNA systems decreases in all three dimensions, showing that the system becomes more flexible.

Hence it is evident that simulations provide some support for the origins of life theory that LDHs could have acted as a protective environment for the first nucleic acids in extreme environmental conditions, such as those found around deep ocean hydrothermal vents. The simulations also indicate that plasmids are structurally supported when intercalated under ambient conditions, explaining the high efficacy rate of LDHs in gene transport observed experimentally.

10.5 Conclusions

In this chapter we have summarized the molecular modeling approaches taken to understand the interaction of LDHs with four different types of intercalates. All four modeling studies show that the combination of experimental work and MD modeling approaches are highly effective in examining the structure and dynamics of ionic and molecular species in the interlayers of the LDH material.

The *CLAYFF* force field [34] used in all four simulations (in combination with other force fields for intercalates) produces results that are in good agreement with experimentally measured PXRD and vibrational spectra for LDH. Results show that molecular modeling aids significantly to probe the intermolecular and H-bonding interactions that dominate the LDH materials and help understand the atomic/molecular motions present in these complex situations. Molecular modeling simulation provides further evidence that LDH material can be used as an effective drug delivery material for small molecules up to large biomolecules.

References

[1] Cavani, F., F. Trifirò, and A. Vaccari, Hydrotalcite-type anionic clays: preparation, properties and applications. *Catalysis Today*, 1991. **11**(2): 173–301.

[2] Centi, G. and S. Perathoner, Catalysis by layered materials: a review. *Microporous and Mesoporous Materials*, 2008. **107**(1/2): 3–15.

[3] Mandal, S., Tichit D, Lerner DA *et al.*, Azoic dye hosted in layered double hydroxide: physicochemical characterization of the intercalated materials. *Langmuir*, 2009. **25**(18): 10980–10986.

[4] Wen, R., Z. Yang, H. Chen *et al.*, Zn-Al-La hydrotalcite-like compounds as heating stabilizer in PVC resin. *Journal of Rare Earths*, 2012. **30**(9): 895–902.

[5] Sasai, R., W. Norimatsu, and Y. Matsumoto, Nitrate-ion-selective exchange ability of layered double hydroxide consisting of MgII and FeIII. *Journal of Hazardous Materials*, 2012. **215–216** p. 311–314.

[6] Wang, Z., F. Liu, and C. Lu, Chemiluminescence flow biosensor for glucose using Mg-Al carbonate layered double hydroxides as catalysts and buffer solutions. *Biosensors and Bioelectronics*, 2012. **38**(1): 284–288.

[7] Parello, M.L., R. Rojas, and C.E. Giacomelli, Dissolution kinetics and mechanism of Mg–Al layered double hydroxides: a simple approach to describe drug release in acid media. *Journal of Colloid and Interface Science*, 2010. **351**(1): 134–139.
[8] Kong, X., L. Jin, M. Wei *et al.*, Antioxidant drugs intercalated into layered double hydroxide: structure and in vitro release. *Applied Clay Science*, 2010. **49**(3): 324–329.
[9] Palmer, S.J. and R.L. Frost, Use of hydrotalcites for the removal of toxic anions from aqueous solutions. *Industrial and Engineering Chemistry Research*, 2010. **49**(19): 8969–8976.
[10] Parker, L.M., N.B. Milestone, and R.H. Newman, The use of hydrotalcite as an anion absorbent. *Industrial and Engineering Chemistry Research*, 1995. **34**(4): 1196–1202.
[11] Palmer, S., R. Frost, and T. Nguyen, Thermal decomposition of hydrotalcite with molybdate and vanadate anions in the interlayer. *Journal of Thermal Analysis and Calorimetry*, 2008. **92**(3): 879–886.
[12] Goh, K.-H., T.-T. Lim, and Z. Dong, Application of layered double hydroxides for removal of oxyanions: a review. *Water Research*, 2008. **42**(6/7): 1343–1368.
[13] Kumar, P.P., A.G. Kalinichev, and R.J. Kirkpatrick, Molecular dynamics simulation of the energetics and structure of layered double hydroxides intercalated with carboxylic acids. *Journal of Physical Chemistry C*, 2007. **111**(36): 13517–13523.
[14] Kovář, P., Pospísil M, Nocchetti M *et al.*, Molecular modeling of layered double hydroxide intercalated with benzoate, modeling and experiment. *Journal of Molecular Modeling*, 2007. **13**(8): 937–942.
[15] Xu, Z.P. and P.S. Braterman, Competitive intercalation of sulfonates into layered double hydroxides (LDHs): the key role of hydrophobic interactions. *Journal of Physical Chemistry C*, 2007. **111**(10): 4021–4026.
[16] Aisawa, S., S. Sasaki, S. Takahashi *et al.*, Intercalation of amino acids and oligopeptides into Zn-Al layered double hydroxide by coprecipitation reaction. *Journal of Physics and Chemistry of Solids*, 2006. **67**(5/6): 920–925.
[17] Nakayama, H., N. Wada, and M. Tsuhako, Intercalation of amino acids and peptides into Mg–Al layered double hydroxide by reconstruction method. *International Journal of Pharmaceutics*, 2004. **269**(2): 469–478.
[18] Desigaux, L., Belkacem MB, Richard P *et al.*, Self-assembly and characterization of layered double hydroxide/DNA hybrids. *Nano Letters*, 2005. **6**(2): 199–204.
[19] Thyveetil, M.-A., Coveney PV, Greenwell HC *et al.*, Computer simulation study of the structural stability and materials properties of DNA-intercalated layered double hydroxides. *Journal of the American Chemical Society*, 2008. **130**(14): 4742–4756.
[20] Xu, Z.P., Q. H. Zeng, G. Q. Lu *et al.*, Inorganic nanoparticles as carriers for efficient cellular delivery. *Chemical Engineering Science*, 2006. **61**(3): 1027–1040.
[21] Saunders, N.R., Ek CJ, Habgood MD *et al.*, Barriers in the brain: a renaissance? *Trends in Neurosciences*, 2008. **31**(6): 279–286.
[22] Rives, V., M. del Arco, and C. Martín, Layered double hydroxides as drug carriers and for controlled release of non-steroidal antiinflammatory drugs (NSAIDs): a review. *Journal of Controlled Release*, 2013. **169**(1/2): 28–39.
[23] Ladewig, K., Niebert M, Xu ZP *et al.*, Efficient siRNA delivery to mammalian cells using layered double hydroxide nanoparticles. *Biomaterials*, 2010. **31**(7): 1821–1829.
[24] Wong, Y., Markham K, Xu ZP *et al.*, Efficient delivery of siRNA to cortical neurons using layered double hydroxide nanoparticles. *Biomaterials*, 2010. **31**(33): 8770–8779.
[25] Murthy, V., Smith HD, Zhang H *et al.*, Molecular modeling of hydrotalcite structure intercalated with transition metal oxide anions: CrO_4^{2-} and VO_4^{3-}. *Journal of Physical Chemistry A*, 2011. **115**(46): 13673–13683.
[26] Tran, P., S. Smith, H. Zhang *et al.*, Molecular dynamic simulations of interactions between LDH and NO_3^- intercalates in aqueous solution. *Journal of Physics and Chemistry of Solids*, 2008. **69**(5/6): 1044–1047.
[27] Zhang, H., Ouyang D, Murthy V *et al.*, Hydrotalcite intercalated siRNA: computational characterization of the interlayer environment. *Pharmaceutics*, 2012. **4**(2): 296–313.
[28] Palmer, S.J. and R.L. Frost, Effect of pH on the uptake of arsenate and vanadate by hydrotalcites in alkaline solutions: a Raman spectroscopic study. *Journal of Raman Spectroscopy*, 2010: 243–248.

[29] Smith, H.D., G.M. Parkinson, and R.D. Hart, In situ absorption of molybdate and vanadate during precipitation of hydrotalcite from sodium aluminate solutions. *Journal of Crystal Growth*, 2005. **275**(1/2): e1665–e1671.

[30] Kalinichev, A.G., P. Padma Kumar, and R. James Kirkpatrick, Molecular dynamics computer simulations of the effects of hydrogen bonding on the properties of layered double hydroxides intercalated with organic acids. *Philosophical Magazine*, 2010. **90**(17/18): 2475–2488.

[31] Zhang, H., Z. P. Xu, G. Q. Lu *et al.*, Intercalation of sulfonate into layered double hydroxide: comparison of simulation with experiment. *Journal of Physical Chemistry C*, 2008. **113**(2): 559–566.

[32] Accelrys Software Inc., Material Studio, 2010.

[33] Sun, H., COMPASS: an ab initio force-field optimized for condensed-phase applications overview with details on alkane and benzene compounds. *Journal of Physical Chemistry B*, 1998. **102**(38): 7338–7364.

[34] Cygan, R.T., J.-J. Liang, and A.G. Kalinichev, Molecular models of hydroxide, oxyhydroxide, and clay phases and the development of a general force field. *Journal of Physical Chemistry B*, 2004. **108**(4): 1255–1266.

[35] Cygan, R.T., Greathouse, J.A., Heinz, H. *et al.*, Molecular models and simulations of layered materials. *Journal of Materials Chemistry*, 2009. **19**(17): 2470–2481.

[36] Tomasi, J., B. Mennucci, and E. Cancès, The IEF version of the PCM solvation method: an overview of a new method addressed to study molecular solutes at the QM ab initio level. *Journal of Molecular Structure: THEOCHEM*, 1999. **464**(1/3): 211–226.

[37] Becke, A.D., Correlation energy of an inhomogeneous electron gas: a coordinate-space model. *Journal of Chemical Physics*, 1988. **88**(2): 1053–1062.

[38] Becke, A.D., Density-functional thermochemistry. III. The role of exact exchange. *Journal of Chemical Physics*, 1993. **98**(7): 5648–5652.

[39] Wadt, W.R. and P.J. Hay, Ab initio effective core potentials for molecular calculations. Potentials for main group elements Na to Bi.*Journal of Chemical Physics*, 1985. **82**(1): 284–298.

[40] Hedegård, E.D., J. Bendix, and S.P.A. Sauer, Partial charges as reactivity descriptors for nitrido complexes. *Journal of Molecular Structure: THEOCHEM*, 2009. **913**(1/3): 1–7.

[41] Hay, P.J. and W.R. Wadt, Ab initio effective core potentials for molecular calculations. Potentials for K to Au including the outermost core orbitale. *Journal of Chemical Physics*, 1985. **82**(1): 299–310.

[42] Frisch, M.J. and G.W. Trucks, *Gaussian 03*. 2003. Gaussian, Inc., Pittsburgh, PA.

[43] Reed, A.E., L.A. Curtiss, and F. Weinhold, Intermolecular interactions from a natural bond orbital, donor–acceptor viewpoint. *Chemical Reviews*, 1988. **88**(6): 899–926.

[44] Smith, D.E., Molecular computer simulations of the swelling properties and interlayer structure of cesium montmorillonite. *Langmuir*, 1998. **14**(20): 5959–5967.

[45] Wang, J., Kalinichev AG, Kirkpatrick RJ, *et al.*, Structure, energetics, and dynamics of water adsorbed on the muscovite (001) surface: a molecular dynamics simulation. *Journal of Physical Chemistry B*, 2005. **109**(33): 15893–15905.

[46] Mills, R., Self-diffusion in normal and heavy water in the range 1–45.deg. *Journal of Physical Chemistry*, 1973. **77**(5): 685–688.

[47] Zhao, L. and P. Choi, Molecular dynamics simulation of the coalescence of nanometer-sized water droplets in n-heptane. *Journal of Chemical Physics*, 2004. **120**(4): 1935–1942.

[48] Li, C., Z. Li, and P. Choi, Stability of water/toluene interfaces saturated with adsorbed naphthenic acids–A molecular dynamics study. *Chemical Engineering Science*, 2007. **62**(23): 6709–6715.

[49] Gu, Z., H. Zuo, A. Wu *et al.*, Stabilisation of Layered double hydroxide nanoparticles by bovine serum albumin pre-coating for drug/gene delivery, Australian Institute for Bioengineering and Nanotechnology, The University of Queensland, Brisbane, 2014.

[50] V. Ramakrishna, B.J. Duke, and M.L. McKee, Gallium derivatives of tetraborane(10): can bis(digallanyl) isomers exist? *Molecular Physics*, 2000. **98**(11): 745–750.

[51] Chi, L., Murthy, V., and Smith, H., Long-term environmental stability of hydrotalcite precipitated in industrial systems – a molecular modeling approach. Proceedings of the 9th International Alumina Quality Workshop, Perth, Australia, 2012.

11

Molecular Modeling as a Tool to Understand the Role of Poly(Ethylene) Glycol in Drug Delivery

Alex Bunker

Centre for Drug Research, Faculty of Pharmacy, University of Helsinki, Finland

11.1 PEGylation in Drug Delivery

Fundamentally, pharmaceutical research concerns the development of new drug molecules. Molecules are found which can be synthesized, perform a specific function at a specific place in the body, have a solubility profile that allows them to be carried through the bloodstream to this location, and at their efficacious dose have a level of side effects and toxicity that is tolerable. Drug molecules would be made based on this compromise, with the dosage set according to the maxim of Paracelsus: "*Dosis facit venenum*" – the dose makes the poison. Within this paradigm drug design is essentially a balancing act. The drug molecule enters the body and diffuses, hopefully to reach the desired location intact, in significant quantities to have the desired effect. At the same time it is hoped that the drug does not build up in sufficient quantities in areas of the body where any undesired effect could occur, resulting in intolerable side effects. In many cases drugs that are extremely effective (once they reach their target) cannot be administered in such a fashion since this balancing act is not achieveable. This can arise either due to drug toxicity, or as a result of the drug target area being extremely difficult to reach.

Recent developments in pharmaceutical research have allowed us to move beyond this. Specific mechanisms can be developed to: (i) target the drug molecule to the specific area

Computational Pharmaceutics: Application of Molecular Modeling in Drug Delivery, First Edition.
Edited by Defang Ouyang and Sean C. Smith.

of the body that one desires to reach and (ii) prolong the bloodstream circulation of the drug by interfering with its uptake by the reticuloendothelial system (RES): that is, targeting and protecting. This can be achieved either through the creation of drug nanocarriers, or through the conjugation of the drug molecule with another molecule that protects and/or targets the drug.

Drug nanocarriers, also referred to as "nanovectors" [1], are nanoscale (100nm diameter) devices that encapsulate, protect, and in some cases target drug molecules. Drug nanocarriers have been used successfully for many different therapeutic applications, including combating infectious agents [2], cancer therapy [3, 4], medical imaging [5, 6], gene therapy [7], and delivery of protein and peptide-based drugs [8]. While there are several forms of nanocarriers [3, 4], for example, liposomes, polymeric micelles, dendrimers, and nanoparticles, they all have a common basic structure. They are composed of: (i) a core compartment where the drug to be delivered is stored, (ii) a protective "stealth sheath" composed of polymers tethered to the nanocarrier core, and (iii) possibly targeting moieties conjugated to the ends of some of the protective polymers, to enable targeting to specific cell types. The exact same effect can be achieved when the drug is conjugated to a single polymer: the polymer can wrap around the drug, creating the same effect as the stealth sheath of the nanocarrier. In addition the polymer can in turn be functionalized, with one end of the polymer bound to the drug molecule and the other bound to a targeting moiety, thus achieving targeting as well as protection.

The current gold standard regarding the polymer used to form the protective "stealth sheath" is poly(ethylene) glycol (PEG). The monomer of PEG is comprised of a nonpolar ethylene group (C_2H_4) and a polar oxygen atom. As a result, PEG, in addition to being highly water soluble, is also soluble in a wide variety of both polar and nonpolar solvents [9]. The PEG polymer shows little toxicity, is rapidly cleared from the body without structural change [10], has good excretion kinetics, and lacks immunogenicity. As a result PEG has been approved by the United States Food and Drug Administration (FDA) for internal use. When PEG is either covalently bound to a drug molecule, therapeutic protein, or surface of a nanocarrier, the alteration is known as "PEGylation" and the molecule or nanocarrier to which the PEG is conjugated is "PEGylated." PEGylation increases the bloodstream circulation time mainly as a result of steric shielding [11–16]. In addition to the "stealth" effect of decreased clearance and immunogenicity resulting in increased bloodstream lifetime, PEGylation has also been shown to result in other beneficial effects. These include an altered biodistribution that results in a higher drug concentration at the site of action [17, 18], an increase in the enhanced permeability and retention effect [10], an altered membrane permeability [19, 20], and an increase in the drug concentration at the site of action [18].

The role of PEGylation in drug delivery is covered in many review articles [10, 12, 21–32]. The first successful application of PEGylation was achieved by Davies and Abuchowski in 1977 [33, 34]. This involved the PEGylation of two bovine proteins: bovine serum albumin and bovine liver catalase. Davies and Abuchowski were able to show that PEGylation resulted in both a decrease in immunogenicity and an increase in blood circulation time. Currently there are ten PEGylated drugs in clinical practice [31]. In addition to proteins [35] and other drug molecules, [36] PEGylation has been used successful in drug delivery liposomes (DDLs) [37, 38] and nanoparticles [39].

Liposome-based systems have been used in drug delivery since the 1970s [40]. A liposome consists of a phospholipid bilayer (membrane) formed into an enclosed sack. As such they are extremely versatile drug delivery vectors, as they are capable of transporting both hydrophobic drugs within the liposome membrane and hydrophilic drugs within the internal liposome cavity. Arguably the most successful application of PEGylation is the drug Doxil [41], the drug doxorubicin encapsulated in a PEGylated liposome. PEGylation of a DDL is achieved through attaching the PEG to the headgroups of a subset of the phospholipids from which the liposome is composed. For the case of Doxil, 5 mol% of the phospholipids have their phosphatidyl choline (PC) headgroup replaced by a phosphatidyl ethanolamine–polyethylene glycol (PE-PEG) group with a PEG polymer of molecular weight 2 kDa, ~45 monomer units long. This length of PEG is referred to as PEG2000. Liposomes with 5000 kDa PEG are also used as DDLs, and these are referred to as PEG5000. DDLs often also have cholesterol incorporated into the liposome formulation. Cholesterol plays a role in lipid stability and packing and is used to engineer the permeability, and thus the drug release rate, of DDLs [42].

As a protective coating PEGylation has seen considerable success. For example, PEGylation has increased the time DDLs circulate in the bloodstream from 1 h to the range of 1–2 days [38]. There remains, however, considerable room for improvement; red blood cells, blood platelets, and some antibodies circulate in the bloodstream for 1–2 months. As a result the search for possible alternative protective polymer coatings to PEG is an active field of research, as outlined in the comprehensive review paper of Knop *et al.* [43]. Additionally interactions between the protective PEG corona and the targeting moieties can interfere with drug targeting.

In order to follow a rational design approach to developing possible alternatives to PEG we first must understand the mechanisms through which PEGylation performs its protective role in drug delivery. While some progress has been made on this front through experimental investigation, the results remain conflicting, and the extent to which this can be investigated experimentally remains limited. For example, three different studies found three different results regarding the effect of PEGylation on complement activation, the first step of uptake by the RES. Some studies have shown that PEGylation inhibited protein adhesion [44, 45], and thus complement activation, while a separate study by Szebeni *et al.* [46] found that PEGylation actually accelerated it, and a third study by Price *et al.* [47] found that PEGylation had no effect. Other mechanisms have been proposed, including the inhibition of liposome fusion and the PEG corona forming a steric barrier against macrophages [47]. Calcium ions are known to accelerate liposome fusion by crosslinking phospholipid headgroups that are known to strongly bind Ca^{2+} ions [48–50] and experimental evidence exists that PEGylation inhibits this [51].

Through molecular dynamics simulation it has been possible to significantly extend our understanding of the structure and protective mechanisms involved in PEGylation beyond that which experiment alone has provided. Since PEG is an important polymer with many applications there is a long history of modeling PEG, and molecular modeling has been successfully used in relation to all aspects of PEGylation, from PEGylated proteins and drugs to PEGylated liposomes. As such this can be seen as a case study in how molecular dynamics simulation can be used as a tool to provide the mechanistic understanding needed for the rational design of drug delivery agents. The rest of the chapter consists of a brief overview of the history of the molecular modeling of PEG, followed by a description of several instances where molecular modeling has gained an important insight into the structure and function of PEGylated drugs and nanocarriers.

11.2 A Brief History of the Computational Modeling of PEG

Several force fields have been constructed to model PEG. Force field construction involves first performing quantum mechanics (QM) based calculations to generate either all-atom (AA) or united atom (UA) force fields, where the CH_2 group is represented by a single bead. These force fields have then been used to develop coarse grained (CG) force fields. The details of the force field parameter sets will not be discussed here, but can all be found in the references. Here we will provide a brief overview of the models developed and the systems that have been studied computationally that involve PEG.

The earliest computational models of PEG were constructed in the early 1990s. A force field for 1,2-dimethoxyethylene was developed by Smith *et al.* [52] and this force field was used to build a model of PEG in aqueous solution [53, 54]. Other early work involved the construction of UA models of PEG to demonstrate its property as a polymer electrolyte [55, 56]. Since the PEG monomer is composed of a nonpolar $(CH_2)_2$ group and an electronegative oxygen atom, it will bind to cations found in the solution. This early work with UA models showed that PEG wraps around cations, with four to five PEG oxygens interacting with each ion. These early simulations particularly focused on the interaction of Li^+ ions with PEG in a solution with dissolved LiI. The salt LiI was chosen as the interest was in the possible use of PEG as an electrolyte in lithium ion batteries. Other early simulations using the initial models studied polymer melts of amorphous PEG [57, 58], crystalline PEG [59], PEG surfaces [60], the elastic properties of a single PEG chain [61], the interaction between PEG and water in dilute solution [62] and PEG dissolved in the nonpolar solvent benzene [63].

A particularly noteworthy early study was carried out by Rex *et al.* [64]. They combined an experimental study of the interaction of PEGylated liposomes with fluorescent acylated PEG with a Monte Carlo (MC) simulation of a PEGylated membrane. They saw the behavior of PEG as being in line with the behavior of a surface decorated with tethered polymers as described by the Alexander–de Gennes theory [65–68]. According to this analytical model a surface with polymers grafted on it is in one of two regimes. At low grafting density and polymer length, the surface is in what is known as a "mushroom" regime: the interaction between the different polymers can be ignored; each polymer forms an approximately dome-shaped self avoiding random walk, and the polymer layer thickness scales as the square root of polymer length. As the grafting density and polymer length increases the surface undergoes a transition from the "mushroom" to the "brush" regime where neighboring grafted polymers interact with each other and the polymer layer thickness scales linearly with polymer length.

Several other groups have since developed their own potential sets for PEG [69–72]. Recently PEG force fields have been developed that are compatible with the CHARMM [73] and OPLS [74, 75] potential sets. A polarizable forcefield for PEG has also been developed [57, 76–78]. In addition to the AA potentials, CG potentials for PEG have also been developed. In a CG potential the atoms are replaced with interacting particles that represent atom groups. This approach allows for larger time and length scales to be accessed but at the cost of losing many details specific to the atomistic level interactions [79, 80], for example, H bonds.

The first CG models of PEG were implicit solvent models; the effect of the solvent is modeled through adjustments to the solute–solute particle interactions. These gave results for properties including aggregation number, chain dimensions, and critical micelle concentration in agreement with experiment [81–83]. Due to the extreme approximation that has been made the applicability of implicit solvent models is limited, though more recently efforts have been made to combine implicit with explicit solvent models of PEG [84]. A CG model of PEG with explicit solvent was then developed, and the model was used to simulate the formation of diblock copolymers and their interaction with lipid bilayers [85–87]. A generalized scheme to develop CG interaction potentials for molecules that has become extremely popular is the "MARTINI" CG force field [88, 89]. Four independent MARTINI force fields have been developed for PEG [90–93], and these have been used to model the formation of micelles, bicelles, and liposomes in mixed systems composed of PEGylated and regular phospholipids. Recently a CG force field was created that maps two PEG monomers into a single particle [94].

Molecular modeling, both AA and CG, has been used to study the properties of PEG in many applications and environments, outside of its use in drug delivery. Examples include the adsorption of PEG onto a free water surface [95, 96], polymer blends where PEG is a component [58, 69, 97–106], PEG copolymers [91, 107–111], the properties of single PEG molecules [61, 112, 113], PEG in solution [53, 54, 62, 70, 76, 82, 102, 114–119], PEG and carbon nanotubes [120, 121], and PEG combined with crystalline materials to make nanocomposites [60, 122–126]. A considerable number of computational studies have been made of the PEG interaction with lithium ions in the context of its role as a polymer electrolyte in batteries [57, 127–142], including a study of ion transport in a PEG matrix [132, 133, 137, 139–141, 143] and polymer electrolytes composed of PEG and ionic liquids [135, 136].

11.3 Molecular Modeling Applied to the Role PEG Plays in Drug Delivery

PEGylation has been applied to a broad range of drugs and drug delivery devices, and molecular modeling has been used to develop a mechanistic understanding of most of these devices. Previous reviews have been written which cover subsets of this work, for example, PEGylated small molecular drugs and liposomes [144], polymeric micelles [145], and PEGylated peptides, dendrimers, and carbon nanotubes [146]. We will attempt here to briefly outline a broad range of the work that has been undertaken to computationally model the effect of PEGylation on drugs and drug carriers. A large amount of the work performed in this area has made use of CG models, mainly the MARTINI model. This includes the study of PEGylated dendrimers [147–149], phospholipid membranes [92, 150, 151], polymeric micelles and polymersomes [110], and nanoparticles [152]. In many cases computational modeling has been performed on equivalent systems using AA models, and contrasting the two sets of results demonstrates both the limitations and advantages of CG modeling. Many interaction details are lost in the MARTINI model [153], and this can lead to erroneous results; and we will see several examples of this here.

Dendrimers are symmetric hyperbranched molecules that are extremely good drug delivery vehicles due to the ability to fine tune their size, surface valency, and functionality [154]. Dendrimers have been studied computationally in their capacity as both drug delivery [155] and gene therapy agents [156]. A dendrimer is PEGylated by binding PEG polymers to the outer ends of the branches of the dendrimer. A series of CG simulations performed by Lee *et al.* [147–149] studied the effect of PEGylation on dendrimer-induced pore formation in lipid membranes and the resulting structure with varying dendrimer generation, PEG polymer length, and grafting density. Lee *et al.* studied PEG of length up to 5 kDa (PEG5000) and found that at this length of PEG the vacant interior of the dendrimer, used for drug storage, started to become occupied by the PEG. This result is particularly noteworthy as, while 2 kDa PEG (PEG2000) has been simulated successfully with AA models, simulating PEG5000 with AA resolution would not be practical, as at this length the relevant dynamics slow significantly due to entanglement, thus the timescale needed to study the system exceeds that which is accessible with AA models.

Simulations of PEGylated dendrimers with AA resolution have, however, been able to investigate several properties not accessible with CG models. In AA simulations of a dendrimer covalently bound to a single PEG chain, Tanis and Karatasos [157] were able to study the effect of pH on the structure and H-bonding behavior of PEGylated dendrimers of different generations. AA simulations of PAMAM dendrimers with PEG spacers inserted into the dendrimer structure, rather than being grafted onto the dendrimer termini, have been combined with fluorescence sensing to study the effect of dendrimer architecture on the ability to carry fluorescent dye molecules for application as a fluorescent sensor [158]. Karatasos [159] made an AA simulation of PEGylated dendrimers with doxorubicin molecules to study the effect of PEGylation on the drug loading of dendrimer carriers. Simulation of PEGylated dendrimers was also carried out by Pavan *et al.* [160] where the metadynamics simulation method [161] was applied to an AA model to obtain very good statistics for dendrimer structure.

Clearly, both CG and AA simulations have their different strengths and weaknesses. Important interaction details can be missed by CG simulations while AA simulations may not be able to access the necessary length and time scales. Combining the two methods in a single study, a technique often referred to as "multiscale modeling" [79] can thus prove fruitful. The computational study of micelles and polymeric micelles and polymersomes composed of poly(β-amino ester) (PAE)–PEG copolymer carried out by Luo and Jiang [110] is a good example of this. The Flory-Huggins parameters and miscibility of PAE and PEG were estimated through AA simulations. The results of these AA simulations were used to construct a CG model of PAE-PEG polymers in solution to investigate the formation of micelles and polymersomes. This CG model was then used to study the formation of micelles and polymersomes and loading of the drug camptothecin (CPT), all phenomena that occur on a timescale too long to be accessed by AA resolution simulation. Cheng and Cao [152] combined AA and CG simulation to study PEGylated gold nanoparticles. They used a hierarchical modeling method where an AA model of the PEGylated nanoparticle was first built and then this model was used to construct a CG potential. The CG model was then used to study the effect of grafting density and PEG length on the concentration at which the nanoparticles aggregated. Other groups have studied PEGylated gold nanoparticles using AA models [162, 163], and the multiscale modeling technique has been applied to a similar system, PEGylated fullerenes [164–166].

The difference in the domains that can be explored with CG and AA models becomes very clear when we consider simulations performed on PEGylated lipid membranes, and by extension the structures comprised of these membranes, like liposomes and bicelles. As described in the previous section, the very first attempt at computationally modeling a PEGylated membrane, carried out by Rex *et al.* [64], concluded that this structure is effectively described by the "mushroom" versus "brush" paradigm of the Alexander–de Gennes theory [65–68]. More recently the MARTINI model has been used by two separate groups

[92, 151] to simulate the PEGylated membrane. These simulations were successfully able to study the relationship between PEG length, grafting density, and phase behavior, that is, whether or not the system formed bicelles, micelles, or liposomes and how the system reacted to changes in pressure. Such large-scale phenomena are not accessible within the time and length scale that can be simulated with AA resolution. These simulations, however, miss some very important details, particularly regarding the effect PEGylation has on the DDL surface in the bloodstream.

As already discussed in the first section, PEG is not a generic hydrophilic polymer, it has very specific properties: PEG is soluble in both polar and nonpolar solvents [9] and PEG is a polymer electrolyte that binds cations [55, 56]. This behavior is dependent on the fact that the PEG monomer comprises both a polar oxygen atom and a nonpolar $(CH_2)_2$ group. Clearly when the two are combined into a single particle, as it is in the MARTINI model, then this property is lost. We have carried out a series of simulations [75, 144, 167–169] where we have modeled PEGylated membranes with AA resolution and have demonstrated that these properties of PEG play an important role in the behavior of PEGylated membranes.

In our first piece of work on this topic [75], we simulated PEGylated membranes in both gel and liquid crystalline states. The gel membrane was distearyl phosphatidylcholine (DSPC), and the liquid crystalline membrane was dilinoleyl phosphatidylcholine (DLPC). By combining the results with results that we published in a previous work [170], where we modeled the same systems without PEGylation, we were able to determine the changes PEGylation makes to the liposome surface at these two formulation extremes. We observed, as expected, the PEG to strongly bind to the Na^+ ions, with the PEG coiling around the ions. We found some very interesting differences between the behavior of the PEG in the gel and liquid crystalline membranes. At 10 molar% PE-PEG density the PEG layer of the gel membrane was very dense. As a result the Cl^- ions, with their tightly bound water shells, were expelled from the PEG layer. For the liquid crystalline membrane at the same molar concentration of PEGylated lipids, the area per lipid headgroup was, as expected, considerably larger and we observed about 10% of the PEG polymers to enter into the membrane core. Since PEG is soluble in both polar and nonpolar solvents, this should not be an unexpected result. As a result, the area per lipid was seen to increase and the PEG layer was considerably less dense, allowing the Cl^- ions with their water shells to sit within the PEG layer. Vuković *et al.* [171] used a UA model to study a micelle of PEGylated lipids, and they obtained similar results for ion binding.

In subsequent work [167], we replaced the NaCl with $CaCl_2$ and KCl and reduced the PEGylated lipid density. Reducing the PEGylated lipid density also resulted in allowing the Cl^- ions to enter the PEG layer. This has important implications for the protective properties of the PEG layer. Since surface charge plays a role in complement activation [172], we see that there is an optimum density of PEG where the surface charge is minimized. This could possibly explain why PEG has an optimum density of 5% in DDLs. Another striking result that we saw was that the PEG polymer did not interact with the Ca^{2+} ions, which instead showed a preference for binding to the lipid headgroups. This may provide a mechanism that explains the experimentally observed result [51] that PEGylation inhibits calcium-induced membrane fusion; the PEG provides a steric barrier against this.

The PEGylated membrane was then simulated with targeting moieties [168]. We found that a new targeting moiety, the AETP moiety (successful in phage display), did not provide any benefit when a PEGylated liposome was functionalized with it. A possible problem with this targeting moiety is its hydrophobicity, however there was no experimental method capable of determining exactly what it was doing on the liposome surface. We simulated a PEGylated membrane with this moiety and also with the RGD peptide that was previously found to work. The result was unexpected: the more hydrophobic AETP moiety did not enter into the membrane core, but rather was obscured by the PEG polymer. Since PEG is not completely hydrophilic the AETP moiety was attracted to it, while the free energy barrier against entry into the membrane core was too great. This indicates a possible route to increasing the effectiveness of more hydrophobic targeting moieties: replace the PEG layer with a different, more hydrophilic protective polymer, like, for example, polymethyl oxazoline [173].

Finally the effect of an inclusion of cholesterol in the PEGylated membrane was investigated [169]. Cholesterol is present in all currently approved liposome-based therapies. We simulated varying the level of PEGylated lipid and cholesterol density and found, once again, that the presence of PEG has unexpected consequences. Cholesterol normally plays a role in structuring and compacting the membrane, thus lowering membrane permeability. We found that, like the liquid crystalline DLPC membrane, in the membrane with cholesterol included the PEG entered the membrane. Unlike the case for the pure DLPC membrane, the cholesterol entered into the membrane core in a specific fashion: it wound along the β surface of the cholesterol molecule. As a result, while increasing cholesterol normally decreases the area per lipid of the membrane, successfully compacting the membrane, when the membrane is PEGylated the area per lipid instead increases. The PEG is thus disrupting the structure of the membrane; the β surface plays an important role in the packing of acyl chains and the strength of the ordering effect of cholesterol [42, 174]. What is particularly noteworthy is that the PEGylated membrane with cholesterol has previously been simulated with a CG MARTINI model [175], and this effect was completely missed in this study.

The simplest way to perform PEGylation is to simply covalently bind a single PEG molecule to the drug molecule to be delivered. This has been performed with proteins and small peptides [10, 21, 24], and small molecule-based drugs. Examples of such systems that have been studied computationally include cecropin P1 linked to a silica surface via a PEO chain [176], PEGylated tachyplesin I and magainin II interacting with lipid layers [177], PEGylated insulin [178], and a PEGylated coiled-coil peptide [179]. We have studied PEGylated hematoporphyrin, first comparing its interaction with PEG to that of two other drugs where PEG is used as a dissolution aid, paclitaxel, and piroxicam [180], then studying the effect of PEGylation on the interaction between hematoporphyrins and biomembranes [181]. We found that, while there is no specific interaction between PEG and paclitaxel and piroxicam, there is a strong attractive interaction between the nonpolar CH_2 groups of PEG and the hydrophobic center of the porphin ring of the hematoporphyrin. We also found that the cations bound by the PEG results in an additional electrostatic repulsion between the PEGylated hematoporphyrin and the membrane, in addition to the entropic repulsion present when any polymer is bonded to a molecule.

11.4 Future Directions

While the use of computational methods in drug discovery can now be seen as a mature field, the application of computational methods to drug delivery mechanisms is a far less developed field. The main reason for this is the computational resource required. Our work using AA models to describe PEGylated systems represents several million CPU hours of supercomputing resources. Computational drug design, on the other hand, can usually be carried out using software running on regular laptops and desktops. Since the amount of available computational resources is exponentially increasing this is set to change, and in the foreseeable future the routine use of this toolkit will be possible.

Regarding PEGylation, we now see that a framework has emerged to use both CG and AA simulations to study the behavior of PEGylated drug delivery devices. Computational modeling has shown us that the effect of PEGylation is more complex and specific to the properties of PEG than previously thought, and the use of both AA and CG models is necessary to understand its behavior. These factors must be taken into account in the search for possible alternate protective polymers to PEG, as the alternatives will probably differ in most, if not all, of the observed behaviors. Clearly the next step is to apply this framework of computational analysis to the possible alternative protective polymers. Regarding the development of alternatives to PEG there are two main routes to take, polyoxazolines [173] and carbohydrates, and the same analysis described above can easily be applied to lipid membranes functionalized with the new polymers.

References

[1] Riehemann, K., Schneider, S.W., Luger, T.A. *et al.* (2009) Nanomedicine – challenge and perspectives. *Angewandte Chemie International Edition*, **48**, 872–897.

[2] Bakker-Woudenberg, I.A.J.M. (2002) Long-circulating sterically stabilized liposomes as carriers of agents for treatment of infection or for imaging infectious foci. *International Journal of Antimicrobial Agents*, **19** (2), 299 – 311.

[3] Misra, R., Acharya, S., and Sahoo, S.K. (2010) Cancer nanotechnology: application of nanotechnology in cancer therapy. *Drug Discovery Today*, **15**, 842 – 850.

[4] Loomis, K., McNeeley, K., and Bellamkonda, R.V. (2011) Nanoparticles with targeting, triggered release, aid imaging functionality for cancer applications. *Soft Matter*, **7**, 839–856.

[5] Torchilin, V.P. (1996) Liposomes as delivery agents for medical imaging. *Molecular Medicine Today*, **2** (6), 242–249.

[6] Lammers, T., Aime, S., Henninck, W.E., Storm, G., and Kiessling, F. (2011) Theranostic nanomedicine. *Accounts of Chemical Research*, **44** (10), 1029–1038.

[7] Felgner, P.L. and Ringold, G.M. (1989) Cationic liposome-mediated transfection. *Nature*, **337**, 387–388.

[8] Tan, M.L., Choong, P.F.M., and Dass, C.R. (2010) Recent developments in liposomes, microparticles and nanoparticles for protein and peptide drug delivery. *Peptides*, **31**, 184–193.

[9] Dinç, C.Ö, Kibarer, G., and Güner, A. (2010) Solubility profiles of poly(ethylene glycol)/ solvent systems. II. Comparison of thermodynamic parameters from viscosity measurements. *Journal of Applied Polymer Science*, **117**, 1100–1119.

[10] Parveen, S. and Sahoo, S.K. (2006) Clinical applications of polyethylene glycol conjugated proteins and drugs. *Clinical Pharmacokinetics*, **45** (10), 965–988.

[11] Harris, J.M., Martin, N.E., and Modi, M. (2001) PEGylation: a novel process for modifying pharmacokinetics. *Clinical Pharmacokinetics*, **40** (7), 539–551.

[12] Harris, J.M. and Chess, R.B. (2003) Effect of PEGylation on pharmaceuticals. *Nature Reviews Drug Discovery*, **2** (3), 214–221.

[13] Allen, T.M., Hansen, C., Martin, F. *et al.* (1991) Liposomes containing synthetic lipid derivatives of pole(ethylene glycol) show prolonged circulation half lives in vivo. *Biochimica et Biophysica Acta*, **1066** (1), 29–36.

[14] Allen, T.M. and Hansen, C. (1991) Pharmacokinetics of stealth versus conventional liposomes: effect of dose. *Biochimica et Biophysica Acta*, **1068** (2), 133–141.

[15] Papahadjopoulos, D., Allen, T.M., Gabizon, A. *et al.* (1991) Sterically stabilized liposomes: improvements in pharmacokinetics and antitumor therapeutic efficacy. *Proceedings of the National Academy of Science of the United States of America*, **88** (24), 11460–11464

[16] Klibanov, A.L., Maruyama, K., Torchilin, V.P. *et al.* (1990) Amphipathic polyethelyneglycols effectively prolong the circulation time of liposomes. *FEBS Letters*, **268** (1), 235–237.

[17] Kanjickal, D.G. and Lopina, S.T. (2004) Modeling of drug release from polymeric delivery systems: a review. *Critical Reviews in Therapeutic Drug Carrier Systems*, **21**, 345–386.

[18] Kumar, M.N. and Kumar, N. (2001) Polymeric controlled drug-delivery systems: perspective issues and opportunities. *Drug Development and Industrial Pharmacy*, **27**, 1–30.

[19] Nicholas, A.R., Scott, M.J., Kennedy, N.I. *et al.* (2000) Effect of grafted polyethylene glycol (PEG) on the size, encapsulation efficiency and permeability of vesicles. *Biochimica et Biophysica Acta*, **1463**, 167–178.

[20] Nikolova, A.N. and Jones, M.N. (1996) Effect of grafted PEG-2000 on the size and permeability of vesicles. *Biochimica et Biophysica Acta*, **1304**, 120–128.

[21] Veronese, F.M. (2001) Peptide and protein PEGylation: a review of problems and solutions. *Biomaterials*, **22**, 405–417.

[22] Molineaux, G. (2002) PEGylation: engineering improved pharmaceuticals for enchanced therapy. *Cancer Treatment Reviews*, **28** (A), 13–16.

[23] Chapman, A.P. (2002) PEGylated antibodies and antibody fragments for improved therapy: a review. *Advanced Drug Delivery Reviews*, **54**, 531–545.

[24] Roberts, M.J., Bentley, M.D., and Harris, J.M. (2002) Chemistry for peptide and protein PEGylation. *Advanced Drug Delivery Reviews*, **54**, 459–476.

[25] Otsuka, H., Nagasaki, Y., and Kataoka, K. (2003) PEGylated nanoparticles for biological and pharmaceutical applications. *Advanced Drug Delivery Reviews*, **55**, 403–419.

[26] Veronese, F.M. and Pasut, G. (2005) PEGylation, successful approach to drug delivery. *Drug Discovery Today*, **10** (21), 1451–1458.

[27] Jain, A. and Jain, S.K. (2008) PEGylation: an approach for drug delivery. A review. *Critical Reviews in Therapeutic Drug Carrier Systems*, **25** (5), 403–447.

[28] Ryan, S.M., Mantovani, G., Wang, X., Haddleton, D.M., and Brayden, D.J. (2008) Advances in pegylation of important biotech molecules: delivery aspects. *Expert Opinion on Drug Delivery*, **5**, 371–383.

[29] Kang, J.S., DeLuca, P.P., and Lee, K.C. (2009) Emerging pegylated drugs. *Expert Opinion on Emerging Drugs*, **14** (2), 363–380.

[30] Baillon, P. and Won, C.Y. (2009) Peg-modified biopharmaceuticals. *Expert Opinion on Drug Delivery*, **6** (1), 1–16.

[31] Pasut, G. and Veronese, F.M. (2012) State of the art in PEGylation: the great versatility achieved after forty years of research. *Journal of Controlled Release*, **161**, 461–472.

[32] Milla, P., Dosio, F., and Cattel, L. (2012) Pegylation of proteins and liposomes: a powerful and flexible strategy to improve the drug delivery. *Current Drug Metabolism*, **13**, 105–119.

[33] Abuchowski, A., van Es, T., Palczuk, N.C., and Davis, F.F. (1977) Alteration of immunological properties of bovine serum albumin by covalent attachment of polyethylene glycol. *Journal of Biological Chemistry*, **252**, 3578–3581.

[34] Abuchowski, A., McCoy, J.R., Palczuk, N.C. *et al.* (1977) Effect of covalent attachment of polyethylene glycol on immunogenicity and circulating life of bovine liver catalase. *Journal of Biological Chemistry*, **252**, 3582–3586.

[35] Veronese, F.M. and Pasut, G. (2009) PEGylation for improving the effectiveness of therapeutic biomolecules. *Drugs Today*, **45** (9), 687–695.

[36] Nawalany, K., Kozik, B., Kepczynski, M. *et al.* (2008) Properties of polyethylene glycol supported tetraaryporphyrin in aqueous solution and its interaction with liposomal membranes. *Journal of Physical Chemistry B*, **112**, 12231–12239.

[37] Torchilin, V.P. (2005) Recent advances with liposomes as pharmaceutical carriers. *Nature Reviews Drug Discovery*, **4** (2), 145–160.

[38] Moghimi, S.M. and Szebeni, J. (2003) Stealth liposomes and long circulating nanoparticles: critical issues in pharmacokinetics, opsonization and protein-binding properties. *Progress in Lipid Research*, **42** (6), 463–478.

[39] Olivier, J.C. (2005) Drug transport to brain with targeted nanoparticles. *NeuroRx*, **2** (1), 108–119.
[40] Lasic, D.D. (1998) Novel applications of liposomes. *Trends in Biotechnology*, **16** (7), 307–321.
[41] Gabizon, A., Catane, R., Uziely, B. *et al.* (1994) Prolonged circulation time and enhanced accumulation in malignant exudates of doxorubicin encapsulated in polyethylene-glycol coated liposomes. *Cancer Research*, **54** (4), 987–992.
[42] Martinez-Seara, H., Róg, T., Karttunen, M. *et al.* (2010) Cholesterol induces specific spatial and orientational order in cholesterol/phospholipid membranes. *PLoS ONE*, **5** (6), e11162.
[43] Knop, K., Hoogenboom, R., Fischer, D., and Schubert, U. (2010) Poly(ethylene glycol) in drug delivery: pros and cons as well as potential alternatives. *Angewandte Chemie International Edition*, **49**, 6288–6308.
[44] Bradley, A.J., Devine, D.V., Ansell, S.M. *et al.* (1998) Inhibition of liposome-induced complement activation by incorporated poly(ethylene glycol)-lipids. *Archives of Biochemistry and Biophysics*, **357** (2), 185–194.
[45] Du, H., Chandaroy, P., and Hui, S.W. (1997) Grafted poly-(ethylene glycol) on lipid surfaces inhibits protein adsorption and cell adhesion. *Biochimica et Biophysica Acta*, **1326**, 236–248.
[46] Szebeni, J., Baranyi, L., Savay, S. *et al.* (2002) Role of complement activation in hypersensitivity reactions to doxil and hynic PEG liposomes: experimental and clinical studies. *Journal of Liposome Research*, **12** (1/2), 165–172.
[47] Price, M.E., Cornelius, R.M., and Brash, J.L. (2001) Protein adsorption to polyethylene glycol modified liposomes from fibrinogen solution and from plasma. *Biochimica et Biophysica Acta*, **1512**, 191–205.
[48] Pöyry, S., Róg, T., Karttunen, M., and Vattulainen, I. (2009) Mitochondrial membranes with mono- and divalent salt: changes induced by salt ions on structure and dynamics. *Journal of Physical Chemistry B*, **113**, 15513–15521.
[49] Issa, Z.K., Manke, C.W., Jena, B.P., and Potoff, J.J. (2010) Ca^{2+} bridging of apposed phospholipid bilayers. *Journal of Physical Chemistry B*, **114**, 13249–13254.
[50] Iraolagoitia, X.L.R. and Martini, M.F. (2010) Ca^{2+} adsorption to lipid membranes and the effect of cholesterol in their composition. *Colloids and Surfaces, B: Biointerfaces*, **76**, 215–220.
[51] Holland, J.W., Hui, C., Cullis, P.R., and Madden, T.D. (1996) Poly(ethylene glycol)lipid conjugates regulate the calcium-induced fusion of liposomes composed of phosphatidylethanolamine and phosphatidylserine. *Biochemistry*, **35**, 2618–2624.
[52] Smith, G.D., Jaffe, R.L., and Yoon, D.Y. (1993) A force-field for simulations of 1,2-dimethoxyethane and poly(oxyethylene) based upon ab-initio electronic-structure calculations on model molecules. *Journal of Physical Chemistry*, **97**, 12752–12759.
[53] Bedrov, D. and Smith, G.D. (1998) Anomalous conformational behavior of poly(ethylene oxide) oligomers in aqueous solutions. A molecular dynamics study. *Journal of Chemical Physics*, **109** (18), 8118–8123.
[54] Smith, G.D., Bedrov, D., and Borodin, O. (2000) Conformations and chain dimensions of poly(ethylene oxide) in aqueous solution: a molecular dynamics simulation study. *Journal of the American Chemical Society*, **122**, 9548–9549.
[55] Müller-Plathe, F. and van Gunsteren, W.F. (1995) Computer simulation of a polymer electrolyte: lithium iodide in amorphous poly(ethylene oxide). *Journal of Chemical Physics*, **103** (11), 4745–4756.
[56] Laasonen, K. and Klein, M.L. (1995) Molecular dynamics simulations of the structure and ion diffusion in poly(ethylene oxide). *Journal of the Chemical Society, Faraday Transactions*, **91** (16), 2633–2638.
[57] Borodin, O. and Smith, G.D. (1998) Molecular dynamics simulations of poly(ethylene oxide)/LiI melts 1. Structural and conformational properties. *Macromolecules*, **31**, 8396–8406.
[58] Neyertz, S. and Brown, D. (1995) A computer simulation study of the chain configurations in poly(ethylene oxide)-homolog melts. *Journal of Chemical Physics*, **102** (24), 9725–9735.
[59] Neyertz, S., Brown, D., and Thomas, J.O. (1994) Molecular dynamics simulation of crystalline poly(ethylene oxide). *Journal of Chemical Physics*, **101** (11), 10064–10073.
[60] Aabloo, A. and Thomas, J. (1997) Molecular dynamics simulations of a poly(ethylene oxide) surface. *Computational and Theoretical Polymer Science*, **7** (1), 47–51.

[61] Heymann, B. and Grubmüller, H. (1999) Elastic properties of poly(ethylene-glycol) studied by molecular dynamics stretching simulations. *Chemical Physics Letters*, **307** (5/6), 425–432.
[62] Tasaki, K. (1996) Poly(oxyethylene)-water interactions: a molecular dynamics study. *Journal of the American Chemical Society*, **118**, 8459–8469.
[63] Tasaki, K. (1996) Conformation and dynamics of poly(oxyethylene) in benzene solution: solvent effect from molecular dynamics simulation. *Macromolecules*, **29**, 8922–8933.
[64] Rex, S., Zuckermann, M.J., Lafleur, M., and Silvius, J.R. (1998) Experimental and monte carlo simulation studies of the thermodynamics of polyethyleneglycol chains grafted to lipid bilayers. *Biophysical Journal*, **75**, 2900–2914.
[65] Alexander, S. (1977) Adsorption of chain molecules with a polar head. A scaling description. *Journal de Physique (Paris)*, **38**, 983–987.
[66] de Gennes, P.G. (1980) Conformations of polymers attached to an interface. *Macromolecules*, **13**, 1069–1075.
[67] de Gennes, P.G. (1987) Polymers at an interface: a simplified view. *Colloid and Interface Science*, **27**, 189–209.
[68] de Gennes, P.G. (1988) Model polymers at interfaces, in *Physical Basis of Cell-Cell Adhesion* (ed. P. Bongrand), CRC Press, Boca Raton, FL, pp. 39–60.
[69] Dong, H., Hyun, J.K., Durham, C., and Wheeler, R.A. (2001) Molecular dynamics simulations and structural comparisons of amorphous poly(ethylene oxide) and poly(ethylenimine) models. *Polymer*, **42**, 7809–7817.
[70] Fischer, J., Paschek, D., Geiger, A., and Sadowski, G. (2008) Modeling of aqueous poly(oxyethylene) solutions: 1. Atomistic simulations. *Journal of Physical Chemistry B*, **112**, 2388–2398.
[71] Tritopoulu, E.A. and Economou, I.G. (2006) Molecular simulation of structure and thermodynamic properties of pure tri- and tetra-ethylene glycols and their aqueous mixtures. *Fluid Phase Equilibria*, **248**, 134–146.
[72] Winger, M., de Vries, A.H., and van Gunsteren, W.F. (2009) Force-field dependence of the conformational properties of α,ω-dimethoxypolyethylene glycol. *Molecular Physics*, **107**, 1313–1321.
[73] Vorobyov, I., Anisimov, V.M., Greene, S. *et al.* (2007) Additive and classical drude polarizable force fields for linear and cyclic ethers. *Journal of Chemical Theory and Computation*, **3**, 1120–1133.
[74] Maciejewski, A., Pasenkiewicz-Gierula, M., Cramariuc, O. *et al.* (2014) Refined OPLS all-atom force field for saturated phosphatidylcholine bilayers at full hydration. *Journal of Physical Chemistry B*, **118** (17), 4571–4581.
[75] Stepniewski, M., Pasenkiewicz-Gierula, M., Róg, T. *et al.* (2011) Study of PEGylated lipid layers as a model for PEGylated liposome surfaces: molecular dynamics simulation and langmuir monolayer studies. *Langmuir*, **27** (12), 7788–7798.
[76] Borodin, O., Douglas, R., Smith, G.D. *et al.* (2003) MD simulations and experimental study of structure dynamics and thermodynamics of poly(ethylene oxide) and its oligomers. *Journal of Physical Chemistry B*, **107**, 6813–6823.
[77] Borodin, O. and Smith, G.D. (2003) Development of quantum chemistry-based force fields for poly(ethylene oxide) with many-body polarization interactions. *Journal of Physical Chemistry B*, **107**, 6801–6812.
[78] Starovoytov, O.N., Borodin, O., Bedrov, D., and Smith, G.D. (2011) Development of a polarizable force field for molecular dynamics simulations of poly(ethylene oxide) in aqueous solution. *Journal of Chemical Theory and Computation*, **7**, 1902–1915.
[79] Murtola, T., Bunker, A., Vattulainen, I *et al.* (2009) Multiscale modeling of emergent materials: biological and soft matter. *Physical Chemistry Chemical Physics*, **11**, 1869–1892.
[80] Chen, C., Depa, P., Sakai, V.G. *et al.* (2006) A comparison of united atom, explicit atom, and coarse-grained simulation models for poly(ethylene oxide). *Journal of Chemical Physics*, **124**, 234901.
[81] Bedrov, D., Ayyagari, C., and Smith, G.D. (2006) Multiscale modeling of poly(etheylene oxide)-poly(propylene oxide)-poly(ethylene oxide) triblock copolymer micelles in aqueous solution. *Journal of Chemical Theory and Computation*, **2**, 598–606.

[82] Fischer, J., Paschek, D., Geiger, A., and Sadowski, G. (2008) Modeling of aqueous poly(oxyethylene) solutions: 2. Mesoscale simulations. *Journal of Physical Chemistry B*, **112**, 13561–13571.
[83] Chen, T., Hynninen, A.P., Prud'homme, R.K. *et al.* (2008) Coarse-grained simulations of rapid assembly kinetics for polystyrene-b-poly(ethylene oxide) copolymers in aqueous solutions. *Journal of Physical Chemistry B*, **112**, 16357–16366.
[84] Juneja, A., Numata, J., Nilsson, L., and Knapp, E.W. (2010) Merging implicit with explicit solvent simulations: polyethylene glycol. *Journal of Chemical Theory and Computation*, **6**, 1871–1883.
[85] Srinivas, G., Shelley, J.C., Nielsen, S.O., Discher, D.E., and Klein, M.L. (2004) Simulation of diblock copolymer self-assembly, using a coarse-grain model. *Molecular Physics*, **108**, 8153–8160.
[86] Srinivas, G., Discher, D.E., and Klein, M.L. (2004) Self-assembly and properties of diblock copolymers by coarse-grain molecular dynamics. *Nature Materials*, **3**, 638–644.
[87] Srinivas, G. and Klein, M.L. (2004) Coarse grain molecular dynamics simulations of diblock copolymer surfactants interacting with a lipid bilayer. *Journal of Physical Chemistry*, **102**, 8153–8160.
[88] Marrink, S.J., de Vries, A.H., and Mark, A.E. (2004) Coarse grained model for semiquantitative lipid simulations. *Journal of Physical Chemistry B*, **108**, 750–760.
[89] Marrink, S.J., Risselada, H.J., Yefimov, S. *et al.* (2007) The MARTINI force field: coarse grained model for biomolecular simulations. *Journal of Physical Chemistry B*, **111**, 7812–7824.
[90] Lee, H., de Vries, A.H., Marrink, S.J., and Pastor, R.W. (2009) A course-grained model for polyethylene oxide and polyethylene glycol: conformation and hydrodynamics. *Journal of Physical Chemistry B*, **113**, 13186–13194.
[91] Rossi, G., Fuchs, P.F.J., Barnoud, J., and Monticelli, L. (2012) A course-grained MARTINI model of polyethylene glycol and of polyoxyethylene alkyl ether surfactants. *Journal of Physical Chemistry B*, **116**, 14353–14362.
[92] Yang, S.C. and Faller, R. (2012) Pressure and surface tension control self-assembled structures in mixtures of PEGylated and non-PEGylated lipids. *Langmuir*, **28**, 2275–2280.
[93] Choi, E., Mondal, J., and Yethiraj, A. (2014) Coarse-grained models for aqueous polyethylene glycol solutions. *Journal of Physical Chemistry B*, **118** (1), 323–329.
[94] Wang, Q., Keffer, D.J., and Nicholson, D.M. (2011) A coarse-grained model for polyethylene glycol polymer. *Journal of Chemical Physics*, **135**, 214903.
[95] Darvas, M., Gilányi, T., and Jedlovszky, P. (2010) Adsorption of poly(ethylene oxide) at the free water surface. A computer simulation study. *Journal of Physical Chemistry B*, **114**, 10995–11001.
[96] Darvas, M., Gulányi, T., and Jedlovszky, P. (2011) Competitive adsorption of surfactants and polymers at the free water surface. A computer simulation study of the sodium dodecyl sulfate-poly(ethylene oxide) system. *Journal of Physical Chemistry B*, **115**, 933–944.
[97] Hackett, E., Manias, E., and Giannelis, E.P. (2000) Computer simulation studies of PEO/layer silicate nanocomposites. *Chemistry of Materials*, **12**, 2161–2167.
[98] Luo, Z. and Jiang, J. (2010) Molecular dynamics and dissipative particle dynamics simulations for the miscibility of poly(ethylene oxide/poly(vinyl chloride) blends. *Polymer*, **51**, 291–299.
[99] Mu, D., Huang, X.R., Lu, Z.Y., and Sun, C.C. (2008) Computer simulation study of the compatibility of poly(ethylene oxide)/poly(methyl methacrylate) blends. *Chemical Physics*, **348**, 122–129.
[100] Yang, H., Ze-Sheng, L., Qian, H.J. *et al.* (2004) Molecular dynamics simulation studies of binary blend miscibility of poly(3-hydroxybutyrate) and poly(ethylene oxide). *Polymer*, **45**, 453–457.
[101] Lin, B., Boinske, P.T., and Halley, J.W. (1996) A molecular dynamics model of the amorphous regions of polyethylene oxide. *Journal of Chemical Physics*, **105** (4), 1668–1681.
[102] Hezaveh, S., Samanta, S., Milano, G., and Roccatano, D. (2012) Molecular dynamics simulation study of solvent effects on conformation and dynamics of polyethylene oxide and polypropylene oxide chains in water and in common organic solvents. *Journal of Chemical Physics*, **136**, 124901.

[103] Annis, B.K., Borodin, O., Smith, G.D., Benmore, C.J., Soper, A.K., and Londono, J.D. (2001) The structure of a poly(ethylene oxide) melt from neutron scattering and molecular dynamics simulations. *Journal of Chemical Physics*, **115** (23), 10998–11003.
[104] Brodeck, M., Alvarez, F., Arbe, A., Juranyi, F., Unruh, T., Holderer, O., Colmenero, J., and Richter, D. (2009) Study of the dynamics of poly(ethylene oxide) by combining molecular dynamic simulations and neutron scattering experiments. *Journal of Chemical Physics*, **130**, 094908.
[105] Smith, G.D., Yoon, D.Y., Jaffe, R.L. *et al.* (1996) Conformations and structures of poly(oxyethylene) melts from molecular dynamics simulations and small-angle neutron scattering experiments. *Macromolecules*, **29**, 3462–3469.
[106] Wu, C. (2011) Simulated glass transition of poly(ethylene oxide) bulk and film: a comparative study. *Journal of Physical Chemistry B*, **115**, 11044–11052.
[107] Sondjaja, H.R., Hatton, T.A., and Tam, K.C. (2008) Self-assembly of poly(ethylene oxide)-block-poly(acrylic acid) induced by CaCl2: mechanistic study. *Langmuir*, **24**, 8501–8506.
[108] Vogel, M. (2008) Conformational and structural relaxations of poly(ethylene oxide) and poly(propylene oxide) melts: molecular dynamics study of spatial heterogeneity, cooperativity, and correlated forward-backwards motion. *Macromolecules*, **41**, 2949–2958.
[109] Kuramochi, H., Andoh, Y., Yoshii, N., and Okazaki, S. (2009) All-atom molecular dynamics study of a spherical micelle composed of n-acetylated poly(ethylene glycol)-poly(γ-benzyl L-glutamate) block copolymers: a potential carrier of drug delivery systems for cancer. *Journal of Physical Chemistry B*, **113**, 15181–15188.
[110] Luo, Z. and Jiang, J. (2012) pH-sensitive drug loading/releasing in amphiphilic copolymer pae-peg: integrating molecular dynamics and dissipative particle dynamics simulations. *Journal of Controlled Release*, **162**, 185–193.
[111] Nawez, S., Redhead, M., Mantovani, G. *et al.* (2012) Interactions of peo-ppo-peo block copolymers with lipid membranes: a computational and experimental study linking membrane lysis with polymer structure. *Soft Matter*, **8**, 6744–6754.
[112] Kuppa, V. and Manias, E. (2003) Dynamics of poly(ethylene oxide) in nanoscale confinements: a computer simulations perspective. *Journal of Chemical Physics*, **118** (7), 3421–3429.
[113] Lee, H., Venable, R.M., Mackerell, A.D., and Pastor, R.W. (2008) Molecular dynamics studies of polyethylene glycol: hydrodynamics radius and shape anisotropy. *Biophysical Journal*, **95**, 1590–1599.
[114] Borodin, O., Bedrov, D., and Smith, G.D. (2001) A molecular dynamics study of polymer dynamics in aqueous poly(ethylene oxide) solutions. *Macromolecules*, **34**, 5687–5693.
[115] Borodin, O., Trouw, F., Bedrov, D., and Smith, G.D. (2002) Temperature dependence of water dynamics in poly(ethylene oxide)/water solutions from molecular dynamics simulations and quasielastic neutron scattering experiments. *Journal of Physical Chemistry B*, **106**, 5184–5193.
[116] Borodin, O., Bedrov, D., and Smith, G.D. (2002) Concentration dependence of water dynamics in poly(ethylene oxide)/water solutions from molecular dynamics simulations. *Journal of Physical Chemistry B*, **106**, 5194–5199.
[117] Borodin, O., Bedrov, D., and Smith, G.D. (2002) Molecular dynamics simulation of dielectric relaxation in aqueous poly(ethylene oxide) solutions. *Macromolecules*, **35**, 2410–2412.
[118] Oelmeier, S.A., Dismer, F., and Hubbuch, J. (2012) Molecular dynamics simulations on aqueous two-phase systems – single PEG-molecules in solution. *BMC Biophysics*, **5**, 14.
[119] Smith, G.D. and Bedrov, D. (2002) A molecular dynamics simulation study of the influence of hydrogen-bonding and polar interactions on hydration and conformations of a poly(ethylene oxide) oligomer in dilute aqueous solution. *Macromolecules*, **35**, 5712–5719.
[120] Uddin, N.M., Capaldi, F.M., and Farouk, B. (2011) Molecular dynamics simulations of the interactions and dispersion of carbon nanotubes in polyethylene oxide/water systems. *Polymer*, **52**, 288–296.
[121] Lee, H. (2013) Molecular dynamics studies of PEGylated single-walled carbon nanotubes: the effect of PEG size and grafting density. *Journal of Physical Chemistry C*, **117** (49), 26334–26341.
[122] Brodeck, M., Alvarez, F., Colmenero, J., and Richter, D. (2012) Signal chain dynamic structure factor of poly(ethylene oxide) in dynamically asymmetric blends with poly(methyl methacrylate). Neutron scattering and molecular dynamics simulations. *Macromolecules*, **45**, 536–542.

[123] Toth, R., Voorn, D.J., Handgraaf, J.W. *et al.* (2009) Multiscale computer simulation studies of water-based montmorillonite/polyethylene oxide) nanocomposites. *Macromolecules*, **42**, 8260–8270.
[124] Mazo, M.A., Manevitch, L., Gusarova, E.B. *et al.* (2008) Molecular dynamics simulation of thermomechanical properties of montmorillonite crystal. 3. Montmorillonite crystals with peo oligomer intercalates. *Journal of Physical Chemistry B*, **112**, 3597–3604.
[125] Suter, J.L. and Coveney, P.V. (2009) Computer simulation study of the materials properties of intercalated and exfoliated poly(ethylene)glycol clay nanocomposites. *Soft Matter*, **5**, 2239–2251.
[126] Borodin, O., Smith, G.D., Bandyopadhyaya, R., and Byutner, O. (2003) Molecular dynamics study of influence of solid interfaces on poly(ethylene oxide) structure and dynamics. *Macromolecules*, **36**, 7873–7883.
[127] Annis, B.K., Kim, M.H., Wignall, G.D. *et al.* (2000) A study of the influence of LiI on the chain conformations of poly(ethylene oxide) in the melt by small-angle neutron scattering and molecular dynamics simulations. *Macromolecules*, **33**, 7544–7548.
[128] Borodin, O. and Smith, G.D. (2000) Molecular dynamics simulations of poly(ethylene oxide)/LiI melts 2. Dynamic properties. *Macromolecules*, **33**, 2273–2283.
[129] Borodin, O. and Smith, G.D. (2000) Molecular dynamics simulations study of LiI-doped diglyme and poly(ethylene oxide) solutions. *Journal of Physical Chemistry B*, **104**, 8017–8022.
[130] Borodin, O., G.D. Smith, and Jaffe, R.L. (2001) *Ab Initio* quantum chemistry and molecular dynamics simulations studies of lipf6/poly(ethylene oxide) interactions. *Journal of Computational Chemistry*, **22** (6), 641–654.
[131] Borodin, O., Smith, G.D., and Douglas, R. (2003) Force field development and md simulations of poly(ethylene oxide)/LiBF4 polymer electrolytes. *Journal of Physical Chemistry B*, **107**, 6824–6837.
[132] Borodin, O. and Smith, G.D. (2006) Mechanism of ion transport in amorphous poly(ethylene oxide)/LiTFSI from molecular dynamics simulations. *Macromolecules*, **39**, 1620–1629.
[133] Borodin, O. and Smith, G.D. (2007) Li + transport mechanism in oligo(ethylene oxide)s compared to carbonates. *Journal of Solution Chemistry*, **36**, 803–813.
[134] Brandell, D., Priimägi, P., Kasemägi, H., and Aalboo, A. (2011) Branched polyethylene/poly(ethylene oxide) as a host matrix for Li-ion battery electrolytes: a molecular dynamics study. *Electrochimica Acta*, **57**, 228–236.
[135] Costa, L.T. and Ribeiro, M.C.C. (2006) Molecular dynamics simulation of polymer electrolytes based on poly(ethylene oxide) and ionic liquids. I. Structural properties. *Journal of Chemical Physics*, **124**, 184902.
[136] Costa, L.T. and Ribeiro, M.C.C. (2007) Molecular dynamics simulation of polymer electrolytes based on poly(ethylene oxide) and ionic liquids. II. Dynamic properties. *Journal of Chemical Physics*, **127**, 164901.
[137] Diddens, D., Heuer, A., and Borodin, O. (2010) Understanding the lithium transport within a rouse-based model for a PEO/LiTFSI polymer electrolyte. *Macromolecules*, **43**, 2028–2036.
[138] Ennari, J., Neelov, I., and Sundholm, F. (2000) Molecular dynamics simulation of the structure of PEO based solid polymer electrolytes. *Polymer*, **41**, 4057–4063.
[139] Ennari, J., Neelov, I., and Sundholm, F. (2001) Estimation of the ion conductivity of a PEO-based polyelectrolyte system by molecular modeling. *Polymer*, **42** (19), 8043–8050.
[140] Ennari, J., Pietilä, L.O., Virkkunen, V., and Sundholm, F. (2002) Molecular dynamics simulation of the structure of an ion-conducting PEO-based solid polymer electrolyte. *Polymer*, **43**, 5427–5438.
[141] Ferreira, B.A., Bernardes, A. T., Mueller-Plathe, F. *et al.* (2002) A comparison of li + transport in dimethoxyethane, poly(ethylene oxide) and poly(tetramethylene oxide) by molecular dynamics simulations. *Solid State Ionics*, **147**, 361–366.
[142] Hektor, A., Klintenberg, M.K., Aabloo, A., and Thomas, J.O. (2003) Molecular dynamics simulation of the effect of a side chain on the dynamics of the amorphous LiPF6-PEO system. *Journal of Materials Chemistry*, **13**, 214–218.

[143] Karo, J. and Brandell, D. (2009) A molecular dynamics study of the influence of side-chain length and spacing on lithium mobility in non-crystalline LiPF6 center dot PEOx; x = 10 and 30. *Solid State Ionics*, **180**, 1272–1284.
[144] Bunker, A. (2012) Poly(ethylene glycol) in drug delivery, why does it work, and can we do better? All atom molecular dynamics simulation provides some answers. *Physics Procedia*, **34**, 24–33.
[145] Huynh, L., Neale, C., Pomès, R., and Allen, C. (2012) Computational approaches to the rational design of nanoemulsions, polymeric micelles, and dendrimers for drug delivery. *Nanomedicine: Nanotechnology, Biology and Medicine*, **8**, 20–36.
[146] Lee, H. (2014) Molecular modeling of pegylated peptides, dendrimers, and single-walled carbon nanotubes for biomedical applications. *Polymers*, **6**, 776–798.
[147] Lee, H. and Larson, R.G. (2009) Molecular dynamics study of the structure of inter-particle interactions of polyethylene glycol-conjugated pamam dendrimers. *Journal of Physical Chemistry B*, **113**, 13202–13207.
[148] Lee, H. and Larson, R.G. (2011) Effects of PEGylation on the size and internal structure of dendrimers: self-penetration of long PEG chains into the dendrimer core. *Macromolecules*, **44** (7), 2291–2298.
[149] Lee, H. and Larson, R.G. (2011) Membrane pore formation induced by acetylated and polyethylene glycol-conjugated polyamidoamine dendrimers. *Journal of Physical Chemistry C*, **115**, 5316–5322.
[150] Thakkar, F.M. and Ayappa, K.G. (2010) Effect of polymer grafting on the bilayer gel to liquid-crystalline transition. *Journal of Physical Chemistry B*, **114**, 2738–2748.
[151] Lee, H. and Pastor, R.W. (2011) Coarse-grained model for PEGylated lipids: effect of PEGylation on the size and shape of self-assembled structures. *Journal of Physical Chemistry B*, **115**, 7830–7837.
[152] Cheng, L. and Cao, D. (2011) Aggregation of polymer-grafted nanoparticles in good solvents: a hierarchical model. *Journal of Chemical Physics*, **135** (12), 124703.
[153] Takada, S. (2012) Coarse-grained molecular simulations of large biomolecules. *Current Opinion in Structural Biology*, **22** (2), 133–137.
[154] Majoros, I.J., Williams, C.R., and Baker, J.R. (2008) Current dendrimer applications in cancer diagnosis and therapy. *Current Topics in Medicinal Chemistry*, **8**, 1165–1179.
[155] Tian, W.D. and Ma, Y.Q. (2013) Theoretical and computational studies of dendrimers as delivery vectors. *Chemical Society Reviews*, **42**, 705–727.
[156] Nandy, B., Maiti, P.K., and Bunker, A. (2013) Force biased molecular dynamics simulation study of effect of dendrimer generation on interaction with dna. *Journal of Chemical Theory and Computation*, **9** (1), 722–729.
[157] Tanis, I. and Karatasos, K. (2009) Molecular dynamics simulations of polyamidoamine dendrimers and their complexes with linear poly(ethylene oxide) at different pH conditions: static properties and hydrogen bonding. *Physical Chemistry Chemical Physics*, **11**, 10017–10028.
[158] Albertazzi, L., Brondi, M., Pavan, G.M., Sato, S.S., Signore, G., B. Storti, Ratto, G.M., and Beltram, F. (2011) Dendrimer-based fluorescent indicators: *in vitro* and *in vivo* applications. *PLoS ONE*, **6** (12), e28450.
[159] Karatasos, K. (2013) Self-association and complexation of the anti-cancer drug doxorubicin with PEGylated hyperbranched polyesters in an aqueous environment. *Journal of Physical Chemistry B*, **117**, 2564–2575.
[160] Pavan, G.M., Barducci, A., Albertazzi, L., and Parinello, M. (2013) Combining metadynamics simulation and experiments to characterize dendrimers in solution. *Soft Matter*, **9**, 2593–2597.
[161] Liao, A. and Parinello, M. (2002) Escaping free-energy minima. *Proceedings of the National Academy of Science of the United States of America*, **99** (20), 12562–12566.
[162] Barbier, D., Brown, D., Grillet, A.C., and Neynertz, S. (2004) Interface between end-functionalized PEO oligomers and a silica nanoparticle studied by molecular dynamics simulations. *Macromolecules*, **37**, 4695–4710.
[163] Hong, B. and Panagiotopoulos, A.Z. (2012) Molecular dynamics simulations of silica nanoparticles grafted with poly(ethylene oxide) oligomer chains. *Journal of Physical Chemistry B*, **116**, 2385–2395.

[164] Hooper, J.B., Bedrov, D., and Smith, G.D. (2008) Supramolecular self-organization in PEO-modified c60 fullerene/water solutions: influence of polymer molecular weight and nanoparticle concentration. *Langmuir*, **24**, 4550–4557.
[165] Hooper, J.B., Bedrov, D., and Smith, G.D. (2009) The influence of polymer architecture on the assembly of poly(ethylene oxide) grafted c60 fullerene clusters in aqueous solution: a molecular dynamics simulation study. *Physical Chemistry Chemical Physics*, **11**, 2034–2045.
[166] Bedrov, D., Smith, G.D., and Li, L. (2005) Molecular dynamics simulation study of the role of evenly spaced poly(ethylene oxide) tethers on the aggregation of c60 fullerenes in water. *Langmuir*, **21**, 5251–5255.
[167] Magarkar, A., Karakas, E., Stepniewski, M. *et al.* (2012) Molecular dynamics simulation of pegylated bilayer interacting with salt ions: a model of the liposome surface in the bloodstream. *Journal of Physical Chemistry B*, **116** (14), 4212–4219.
[168] Lehtinen, J., Magarkar, A., Stepniewski, M. *et al.* (2012) Analysis of cause of failure of new targeting peptide in PEGylated liposome: molecular modeling as rational design tool for nanomedicine. *European Journal of Pharmaceutical Sciences*, **46** (3), 121–130.
[169] Magarkar, A., Róg, T., and Bunker, A. (2014) Molecular dynamics simulation of PEGylated membranes with cholesterol: building toward the doxil formulation. *Journal of Physical Chemistry C*, **118** (28), 15541–15549.
[170] Stepniewski, M., Bunker, A., Pasenkiewicz-Gierula, M., Karttunen, M., and Róg, T. (2010) Effects of the lipid bilayer phase state on the water membrane interface. *Journal of Physical Chemistry B*, **114**, 11784–11792.
[171] Vuković, L., Katib, F.A., Drake, S.P., Madriga, A., Brandenburg, K.S., Král, P., and Onyuksel, H. (2011) Structure and dynamics of highly PEG-ylated sterically stabilized micelles in aqueous media. *Journal of the American Chemical Society*, **133**, 13481–13488.
[172] Yan, X., Scherphof, G.L., and Kamps, J.A.A.M. (2005) Liposome opsonization. *Journal of Liposome Research*, **15**, 109–139.
[173] Viegas, T.X., Bentley, M.D., Harris, J.M. *et al.* (2011) Polyoxazoline: chemistry, properties and applications in drug delivery. *Bioconjugate chemistry*, **22**, 976–986.
[174] Pöyry, S., Róg, T., Karttunen, M., and Vattulainen, I. (2008) Significance of cholesterol methyl groups. *Journal of Physical Chemistry B*, **112** (10), 2922–2929.
[175] Lee, H., Kim, H.R., and Park, J.C. (2014) Dynamics and stability of lipid bilayers modulated by thermosensitive polypeptides, cholesterols, and pegylated lipids. *Physical Chemistry Chemical Physics*, **16**, 3763–3770.
[176] Wu, X., Chang, H., Mello, C. *et al.* (2013) Effect of interaction with coesite silica on the conformation of cecropin P1 using explicit solvent molecular dynamics simulation. *Journal of Chemical Physics*, **138** (4), 045103.
[177] Han, E. and Lee, H. (2013) Effects of pegylation on the binding interaction of magainin 2 and tachyplesin i with lipid bilayer surface. *Lamgmuir*, **29** (46), 14214–14221.
[178] Yang, C., Lu, D., and Liu, Z. (2011) How PEGylation enhances the stability and potency of insulin: a molecular dynamics simulation. *Biochemistry*, **50**, 2585–2593.
[179] Jain, A. and Ashbaugh, H.S. (2011) Helix stabilization of poly(ethylene glycol)-peptide conjugates. *Biomacromolecules*, **12**, 2729–2734.
[180] Li, Y.C., Rissanen, S., Stepniewski, M., Cramariuc, O. *et al.* (2012) Study of interaction between PEG carrier and 3 relevant drug molecules: piroxicam, paclitaxel, and hematoporphyrin. *Journal of Physical Chemistry B*, **116**, 7334–7341.
[181] Rissanen, S., Kumorek, M., Martinez-Seara, H. *et al.* (2014) Effect of pegylation on drug entry into lipid bilayer. *Journal of Physical Chemistry B*, **118** (1), 144–151.

12

3D Structural Investigation of Solid Dosage Forms

Xianzhen Yin[1,3], Li Wu[1,4], You He[2], Zhen Guo[1,3], Xiaohong Ren[1,4], Qun Shao[3], Jingkai Gu[4], Tiqiao Xiao[2], Peter York[3], and Jiwen Zhang[1,4]

[1] *Shanghai Institute of Materia Medica, Chinese Academy of Sciences, China*
[2] *Shanghai Synchrotron Radiation Facility, Shanghai Institute of Applied Physics, Chinese Academy of Sciences, China*
[3] *Institute of Pharmaceutical Innovation, University of Bradford, UK*
[4] *Research Center for Drug Metabolism, Jilin University, China*

12.1 Structural Architectures of Solid Dosage Forms and Methods of Investigation – an Overview

Therapeutic drugs must be formulated and assembled into suitable deliverable dosage forms for convenient administration to patients. Pharmaceutics is the subject which provides the underpinning scientific knowledge concerned with the transformation of drug substance, via formulation, into dosage forms and the various manufacturing processes used to provide high quality, efficacious, and safe medicines. A key challenge, attracting increasing attention from pharmaceutical scientists, is to provide a mechanistic understanding of the behavior of drug substances and functional formulation components at all stages in the transition of a drug substance from a powdered state into final dosage forms and deliver optimal bioperformance of the medicine in patients. New approaches including predictive computational methods and advanced analytical techniques are providing valuable tools to address these challenges.

The different types of dosage forms are usually classified by the route of administration, the physical form, and the size of the dosage unit. The various dosage forms can be considered

Computational Pharmaceutics: Application of Molecular Modeling in Drug Delivery, First Edition.
Edited by Defang Ouyang and Sean C. Smith.

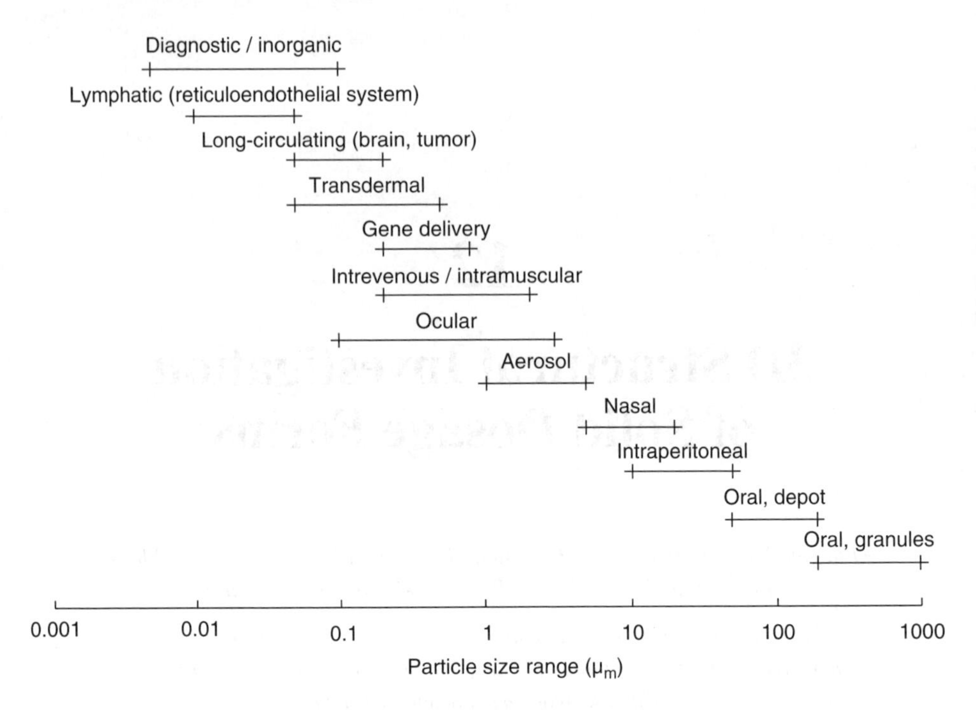

Figure 12.1 *Particle size range for different dosage forms and routes of administration [1]. Reproduced with permission from Ref. [1]. Copyright 2007 Wiley–Liss, Inc.*

across different scales of scrutiny and size (see Figure 12.1)–from the molecular to the macro level [1]. For clinical applications, the final pharmaceutical structures can be divided into numerous conventional types–solutions, emulsions, suspensions, semisolids, and solids such as pellets, capsules, and tablets. During the processing of materials into final dosage forms, other intermediate structures can be found–crystals, powders, granules, and so on. Furthermore, the internal architecture and structure of these dosage forms undergo dynamic changes due to phenomena such as hydration, swelling, and diffusion during drug release and dissolution. Of the range of marketed drug delivery systems (DDSs), solid dosage forms are the most common forms and provide the vast majority of marketed medicines.

The structure of the solid dosage unit plays an important role in the function and effectiveness of final drug products. Drug substances and other nonactive formulation ingredients exist in the majority of solid dosage forms as dry powders or crystals and solid dispersions in particle size ranging from nanocolloids to millimeter-sized granules. The physicochemical and mechanical characteristics of the powders and crystals, for example, particle size, particle size distribution, solubility, yield strength, determine the bulk properties, product performance, processability, stability, and appearance of the end product. In particular, the particle size and size distribution of the drug substance have a major impact on the critical quality and performance criteria of the final products, including drug content uniformity and drug dissolution rate [2].

Characterization of particles is an important requirement during preformulation and formulation studies since particle morphology, size, shape, and mechanical properties can influence the selection of other formulation ingredients and the selection of specific production processes. As mentioned above, drug particle size can influence a range of important performance criteria, such as powder flow and mixing, dosage unit content uniformity, dissolution rate, bioavailability, and stability, and thus has an important role in formulation, processing, and ultimately therapeutic efficiency in patients. Thus manufacturers producing a particulate-based medicine need to understand and quantify any differences between batches of particulate materials for product development and quality control purposes. For some applications routine particle size analysis provides sufficient information for sample differences to be fully rationalized but where samples exhibit similar size distributions identification of subtle variations in other relevant properties, such as particle shape or surface area may be necessary to ensure product quality and consistency.

With the structural features of solid dosage forms playing a dominant role in determining the quality and performance of medicines, it is appropriate to examine further how these structures are assembled and evaluated. The primary building blocks of all dosage forms are the molecules of drug and formulation additives. Assemblies of the molecules can be in an ordered state (i.e., crystalline) or nonordered (i.e., amorphous), and the type of molecular structure can have a profound effect on formulation design, processing route selection, and final product performance such as stability and drug dissolution. Further complexity in defining the crystalline solid state occurs through a common feature in drug molecules called polymorphism. A substance is polymorphic if the molecules take up an alternative packing or conformational arrangement in the crystal lattice. Polymorphic forms can also provide a spectrum of different physicochemical and mechanical properties which need to be understood when selecting the preferred form for formulating, processing, and preparing the final medicine [3–5]. Recent computational studies with drug substances have made major progress in predicting the range of different molecular architectures in crystals from empirical chemical formulae, thus describing the range of possible polymorphs. Such capabilities, together with other predictive computational tools in related fields, such as drug absorption and pharmacokinetics, will undoubtedly be extremely valuable in guiding and accelerating the development programs for solid dosage forms.

In formulating solid dosage forms, the molecular "building blocks" of drugs and other formulation constituents, now assembled into a particulate state, are generally compounded together into physical mixtures (e.g., powder filled hard capsules) or dose units (e.g., pellets, tablets). In doing so, the prepared structures containing a unit dose of drug are routinely assessed by a range of analytical methods. Highly sophisticated and advanced chemical techniques provide a full chemical profile of the unit dose (or a chosen number of dose units). A key *in vitro* physicochemical test is drug dissolution, and again this is carried out with a single or multiple dose units. Physical measurements, with appropriate specifications for uniformity of the final dose unit(s) are also carried out, including dimensional and mechanical testing (e.g., tablet crushing strength). However the internal architecture and microstructure, which will clearly play critical roles in determining the properties of the final unit dose, are not examined. This is true across the various scales of scrutiny relevant for solid dosage forms – from hundreds of nanometers to millimeters. This lack of investigation and characterization of internal microstructure is due to a lack of suitable *in situ* imaging techniques and methodologies to quantify the features of 3D structures.

Several attempts have been made to probe the internal structure of solid dosage forms so as to determine the structural features in relation to functions and dosage unit properties.

Various techniques have been applied, including atomic force microscopy (AFM), NMR imaging, confocal microscopy, near infrared (NIR) spectral imaging [6], and conventional microscopy (optical and electron).

Electron microscopy (SEM and TEM) is the most widely used method to observe the surface structures of pharmaceutical dosage forms and powders, such as the morphology, particle size distribution, and thermal properties of microcapsules [7]. It also provides the ability to investigate surfaces directly at nanometer to subangstrom resolution in ambient and liquid environments. However, SEM is not an analytical tool able to examine *in situ* the internal structural detail of solid dosage form without damaging sample preparation methods.

A quantitative ultrafast magnetic resonance imaging (MRI) technique, together with 19F NMR spectra and 1D 19F profiles, has been applied to study the dissolution process of commercial hydroxypropyl methylcellulose (HPMC) matrix tablets [8]. The 19F 1D-MRI profiles from the hydration layer, with a pixel resolution along the axial length of 375 μm, shows the transport activity of the drug inside the polymer matrix. The integration of the 19F spectra is consistent with the drug release profile obtained from UV measurement.

However, most of the above-mentioned methods are not true 3D approaches. Techniques such as SEM, Raman, MRI (including NMR), AFM, ultrasound, vibrational spectroscopy, and NIR spectroscopy are generally applied for 2D observation. These methods reveal only structural information about the surface or inner region just below the surface, and in order to observe internal structures, samples must be sectioned or cut which destroys their original 3D structure.

Apart from the invasive nature for these conventional methods, there are also other drawbacks of the current techniques. One drawback is the limited penetration and low resolution. Though the image analysis of 2D observations in some cases can give some quantitative information, a real *in situ* 3D structural investigation method is important to achieve a real understanding of the internal microstructure of pharmaceutical dosage forms.

X-ray computed microtomography (μCT), a powerful noninvasive investigative technique, has been applied to observe the 3D structure of various objects and has great potential for providing information on the microstructure for particulate-based solid dosage forms. In contrast to conventional techniques, the μCT technique allows noninvasive visualization of internal and microstructural details at micron resolution [9, 10]. The μCT technique has been developed for investigating the morphology and internal structure of solid dosage forms and powders. For example, the whole spatial information on a particular powder can be obtained by this method. In addition, distributions of geometric characteristics describing size, shape, and spatial arrangement of the particles can be estimated after an image analytic separation of the individual particles [11]. This powerful method has been employed to provide detailed morphological information, such as the pore shape, spatial distribution, and connectivity of porous particles which correlated with the dissolution properties of the DDSs [12, 13]. It has also been reported that the μCT combined with discrete element method (DEM) can be used to investigate particle packing in a process of pharmaceutical tablet manufacture by powder compaction [14].

In pharmaceutical research, noninvasive high resolutions are essential for extracting and rendering the finest structural information and to minimize the partial volume effect typical of medical CT scanners. Nonetheless, the major limit of laboratory μCT equipments is that they can usually accommodate only pharmaceutical samples not exceeding a few centimeters in diameter. The current "ideal" system for imaging highly (re)mineralized objects of pharmaceutical interest is synchrotron radiation X-ray computed microtomography (SR-μCT). The advanced performance of SR-μCT can be attributed to the wide X-ray energy and its high intensity, high brilliance, and high polarization. In combination with image processing and 3D reconstruction, researcher has demonstrated that SR-μCT based 3D structure is opening up a new understanding of DDSs from a structural perspective.

12.2 Synchrotron Radiation X-Ray Computed Microtomography

Computed tomography (CT) is a nondestructive technique that provides 3D images of the internal structure of an object. The basic idea of this imaging technique goes back to J. Radon, who proved in 1917 that an n-dimensional object can be reconstructed from its $(n-1)$-dimensional projections [15]. The physical principle is an interaction of ionizing radiation, such as X-rays, with matter in the energy range typically used for CT imaging, so that the so-called photo effect creates the main interaction mechanism. The photo effect attenuates the photons in proportion to the third power of the order number of the elements and inversely proportional to the third power of the photon energy. Thus the actual attenuation depends not only on the material but also on the energy spectrum of the X-ray source.

Synchrotron radiation is an electromagnetic radiation emitted from radially accelerated charged particles. It is produced in synchrotron facilities with bending magnets, undulators, or wigglers, and resembles cyclotron radiation except for being generated by the acceleration of ultrarelativistic charged particles through magnetic fields. Synchrotron radiation is artificially achieved in synchrotrons, storage rings, or naturally by fast electrons passing magnetic fields. When the circulating electron beam is deflected by the bending magnets in the storage ring, an intense flux of electromagnetic radiation is generated. Synchrotron radiation is characteristically polarized, emitted in a narrow cone, and has frequencies that can range over the entire electromagnetic spectrum [16].

SR-μCT uses the synchrotron radiation X-ray as a light source to achieve high-speed imaging, intensive strength, high spatial resolution (to submicron or nanoscale), and noninvasive fluoroscopy. SR-μCT can quantitatively evaluate and visualize the 3D DDS structure. The advanced performance of SR-μCT is due to: (i) the wide X-ray energy region (1 to >200 keV photon energy) that can be generated, (ii) the high intensity of the total power (i.e., 600 KW) which is tens of thousands of times higher than that of X-ray tubes, (iii) the reduced time for obtaining the experimental data, (iv) the high brilliance (a brilliance hundreds of times higher than that of an X-ray tube), and (v) totally polarized light; and the light of the electron orbit plane is elliptical polarization. In addition, it is a good tool to study the optical activity of biological molecules and dichromatism in magnetic materials [17].

12.3 Principles and Procedures for SR-μCT Studies

12.3.1 Preparation of Samples

In order to evaluate the structure of solid dosage forms and structural changes over time, such as occur during drug release, pharmaceutical samples should be carefully considered and prepared. Several critical factors such as X-ray absorption intensity, water content, stability of samples, containers, and size of samples can have major effects on the result of SR-μCT testing.

12.3.1.1 X-Ray Absorption Intensity

The intensity values associated with the different features of a SR-μCT image are determined by the X-ray transmission measured by an X-ray detection system, which is dependent on the material's atomic mass and the energy of X-rays [10, 18]. Different elements in pharmaceutical samples have different X-ray absorption intensities. To be able to identify the target element precisely, any other element present in the pharmaceutical samples for SR-μCT testing should be clearly distinguished from the target element. The physical nature of samples, such as bulk density, can also affect its X-ray absorption. For instance, porosity either above the system resolution limit or at subvoxel levels will result in differences in X-ray absorption that will be seen as different gray levels, and this will be in turn dependent on such variables as the distribution characteristic of the porosity, the point-spread function of the X-ray source, and the pixel size of the camera. The material interface also needs to be considered, as the degree of complexity will determine how well the detectable boundary will be defined.

12.3.1.2 Density and Density Distribution

Density sensitivity is an issue which must be taken into consideration when performing SR-μCT scans. Clearly the basic density of the material(s) being imaged should be assessed. If two adjacent materials have similar densities, they may not be differentiated unless other factors such as different hydration levels alter the base density.

12.3.1.3 Water Content

Water interferes with the imaging by SR-μCT and samples should contain as little water as possible owing to the interference of the water on the sample imaging acquisition using the SR-μCT technique. Thus samples containing considerable amounts of water must be dried prior to image acquisition. Three methods have been reported for the pretreatment of the samples to remove water in SR-μCT experiments: (i) drying in an oven, (ii) freeze-drying, and (iii) absorbing as much liquid as possible with dry filter paper, with storage at room temperature over silica gel. For methods (i) and (iii) when applied to gel matrix tablets [19], the tablets tend to shrink, and the gel formed on the surface of the tablet core collapses causing changes to the internal gel structure during the preparative process. However, for these samples, the freeze-drying method has been found to maintain the microstructure of the hydration layer with its rapid cooling process, causing minimal change to the structure of the tablet core.

12.3.1.4 Stability of Samples

Consideration also needs to be given to the stability of sample components as the energy of X-ray in SR-μCT is very high. Samples can undergo physical and chemical changes in some cases. Therefore, samples for a SR-μCT scan test are required to have good thermal stability.

12.3.1.5 Fixation of Samples

In order to capture sufficient image information, the sample stage is usually rotated 180° during the SR-μCT scan at a suitable rotational speed. Rotation of the sample stage could lead to movement or rearrangement of individual parts of samples, such as for granule samples.

Therefore, samples should be rigidly fixed. For example, when granular samples of microcrystalline cellulose and starch were imaged individually using the SR-μCT, the two particle systems were filled separately into a cylindrical container with a volume of 1 ml by a filling level $F = 66\%$ [20] to avoid overfilling perturbation and any particle movement during scanning.

12.3.1.6 Size of Samples

The first consideration regarding sample size is based upon the possible image resolution and should be chosen in accordance with the absorption coefficient of the samples. The higher the resolution, the smaller the acquisition window required. Thus, as limited by the size for the acquisition windows, the size of samples for the SR-μCT test should be controlled.

12.3.1.7 Choice of Containers

Samples for a SR-μCT scan are generally required to be fastened in certain containers like plastic tubes, capsule shells, and so on. These containers should be characterized with regard to rigidity, thickness, any weak X-ray absorption, and homogeneity of composition and structure.

12.3.1.8 Number of Samples

Different sample sizes can be chosen depending on the type and magnitude of the investigation. If the objective is focused on the macroscopical scale and dimension, such as a study on granular systems, a relatively large number of granules should be sampled, perhaps several hundred. However, if the objective is to investigate the microscopical internal structures, several or dozens of individual granules may be enough as long as the results have statistical significance. In order to recognize individual constituents in two- or multicomponent systems by SR-μCT, some diluent with weak absorption could be added to separate solid units from each other. For example, for the quantification of swelling and erosion in felodipine extended release tablets, 18 tablets were taken and divided into nine groups. Thus each group had two tablets and a standard dissolution test was carried out. At 0.5, 1.0, 2.0, 3.0, 4.0, 5.0, 6.0, 7.0, and 8.0 h, two tablets were removed from the dissolution medium at each time interval and prepared for further test [18].

12.3.1.9 Samples from Dynamic Processes

For samples taken at different time intervals from materials undergoing dynamic processes such as drug diffusion and/or dissolution, it would be advantageous to visualize the internal structure, focusing on the dynamic property of pore structure and porosity. SR-μCT can produce 3D images of materials with a voxel size of around several microns cubed, allowing the visualization of internal and microstructural details with different X-ray absorbencies. An *in vitro* dissolution test is usually used to evaluate the drug release behavior of tablets and other solid dosage forms. SR-μCT testing can be carried out to evaluate the temporal changes in the pharmaceutical microstructure during the drug release profile. Tablets containing metallic active pharmaceutical ingredients with relative high densities, such as ferrous sulfate, can be imaged directly using the SR-μCT technique [13] during dissolution testing. The dissolution medium will influence neither the imaging nor the visualization of the microstructure of the tablet core. However, for nonmetallic drugs like felodipine with similar density to the dissolution medium, the dissolution medium will interfere with the imaging [21]. Unfortunately, most drugs are not metallic or organometallic compounds. Therefore, tablets taken out from the dissolution medium must be dried prior to image acquisition. In some published research, tablets were first frozen in liquid nitrogen or ultralow temperature refrigerant before being dried using a freeze dryer [22]. In order to maintain the original shape of the tablets, swollen tablets were carefully removed using small spoons together with about 2 ml of the dissolution medium and placed individually in a 24-well plate to maintain the original shape of the swollen tablets with gelled surfaces. The 24-well plate containing the tablets in various states of hydration and erosion was then immediately placed into a refrigerator at −80 °C for 12 h. Then, the tablets were freeze-dried over a period of 24 h at −50 °C and 10 mTorr. The tablets were then kept in a dry cabinet under ambient temperature (relative humidity of 20%) for further SR-μCT tests and were not reused in dissolution testing [19]. The same method was applied to dry felodipine monolith osmotic pump tablets prior to image acquisition. However, due to the special structure of the monolith osmotic pump tablet and the fact that the volume expansion of water during the phase transition from liquid to ice when being frozen leads to the internally dissolved and suspension content of the drug in the tablet core being squeezed out at the initial freezing process, it was found that the freeze-drying procedure was not suitable for the monolith osmotic pump tablet. Furthermore, the remains of the tablet cores were mainly gel-based semisolids, and the formation of water crystals at low temperature may destroy the microstructure of the contents within the semipermeable membrane. Thus the deformation caused by any internal squeezing and possible crystal formation devalues the applicability of the freeze-drying method to the osmotic pump tablets in any internal structure study.

12.3.2 Image Acquisition and 3D Reconstruction

12.3.2.1 Image Acquisition

Accurate experimental results are based on the high-quality images acquired during the CT scans. In order to guarantee precision, it is imperative to optimize the parameters according to the objective of the experiment and the properties of the prepared samples. These parameters include:

1. *The resolution of images.* This is determined by the magnification of the lens and the pixel size of the charged coupled device (CCD). As the CCD is often fixed, an attempt at higher resolution means there will be a smaller area of acquisition window. There is a trade-off between the size of the whole sample being examined and the expected resolution of the smallest microstructure.
2. *X-ray has a range of 1 to >200 KeV.* The higher energy indicates not only the higher capacity to penetrate a sample, but also the higher possibility of radiation damage. For the optimization of X-ray energy, factors such as the elements in the sample(s) being investigated, the molecular weight, the density of materials, and the thickness of sample should be considered.
3. *The exposure time.* This determines the number of X-ray photons reaching the detector. When the exposure time increases, the flux of CCD to capture a projection will be

raised. A CT scan with an unsuitable exposure time will not provide high-quality images. A long exposure time may lead to an overexposure that would reduce the contrast between materials and increase the scan time of every sample. In the case of underexposure, the signal/noise ratio will be reduced. In addition, the radiation dose should be taken into account as a long exposure to X-ray may result in deformation of sample with radiation damage.

4. *The distance between sample and detector* (DSD). This is a key factor for inline phase contrast SR-μCT, which is especially useful in materials with low Z elements and low density. Within a certain distance range, a longer DSD means a better phase contrast is possible, which benefits distinguishing materials with small differences in density that cannot be distinguished by absorption contrast.
5. *The number of projections.* This is typically determined by the sample size and the precision of the tests. During the scan, the sample rotates 180°and projections will be captured over a number of angular orientations. A large number of projections will enhance the quality and restrain artifacts, whilst more projections will lengthen the scanning time and potentially lead to radiation damage as mentioned above.

Prepared samples need to be held in a suitable container and attached to the sample stage to prevent any unexpected movements during scans. Axis adjustment of the sample stage is necessary to make sure that there is no deviation in the horizontal direction during rotation. After penetration through the sample, the projections are magnified by diffraction-limited microscope optics and digitized by a CCD camera. The exposure time and the DSD are adjusted. For each acquisition, a certain number (e.g., 900) of projection images during the sample rotation of 180° are taken. Lightfield images (i.e., X-ray illumination on the beam-path without the sample) and darkfield images (i.e., X-ray illumination off) are also required during each acquisition, for the correction of electronic noise and variations in the X-ray source brightness (Figure 12.2).

12.3.2.2 3D Reconstruction

3D reconstruction is the process of capturing the shape and appearance of the real objects and is created from reconstructed slices of image projections of samples. The reconstruction is the reverse process of obtaining 2D images from 3D scenes. The essence of an image is a projection from a 3D scene onto a 2D plane, during which the depth is lost. The 3D

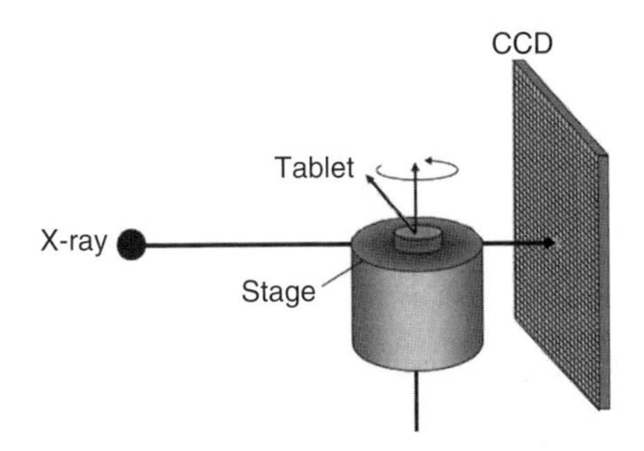

Figure 12.2 *CT scan with synchrotron radiation X-ray.*

point corresponding to a specific image point is constrained to be on the line of sight. From a single image, it is impossible to determine which point on this line corresponds to the image point. If two images are available, then the position of a 3D point can be found as the intersection of the two projection rays. Methods combining X-ray tomography, image processing, and 3D reconstructions have been developed, for example, to study the drug release kinetics of DDSs.

The projections are converted to the reconstructed slices using filtered back-projection algorithms. In order to enhance the quality of the reconstructed slices, a phase retrieval algorithm is added for phase contrast extraction. The 3D rendered data can be analyzed with commercially available software such as VGStudio Max (Volume Graphics GmbH, Germany), Amira (Visualization Sciences Group, France), and Image Pro Analyzer 3D (Media Cybernetics, Inc., USA) to obtain both qualitative and quantitative information. After segmentation, slices are all converted into black and white images by removal of the background and noise. Then 3D ISO-Surface models are constructed with segmented slices. The surface level, surface range, and the simplification parameters are adjusted to optimize the models. Then, all objects in the samples are extracted to calculate the steric parameters.

12.3.3 Model Construction and Analysis

As the model construction and analysis are computation-intensive tasks, the reconstructed slice stacks are converted into eight-bit grayscale format and cropped for accuracy and computational efficiency. Then, the resized slices of samples are processed to enhance quality and reduce noise, and are analyzed to determine the threshold gray values distinguishing different materials and microstructures. In general the most important quantitative criterion to distinguish the objects of interest is the gray value, but for some samples with ultracomplex structures or material contents the morphological information will be more suitable. All samples in the same group should be processed with consistent parameters to ensure that all results can be compared quantitatively.

The optimized 3D models present the full details of the surface morphology and the internal 3D structure. Based on the difference in gray value between materials, objects or microstructures of interest can be extracted from the 3D models by segmentation. With the ISO-Surface model, many 3D steric quantitative parameters can be calculated. The names and descriptions of these parameters are listed below:

- *Volume*: Volume of object in calibrated units;
- *Surface area*: Surface area of object in calibrated units;
- *Width*: Size of bounding box in x direction;
- *Height*: Size of bounding box in y direction;
- *Depth*: Size of bounding box in z direction;
- *Center X*: X coordinate of the center of object;
- *Center Y*: Y coordinate of the center of object;
- *Center Z*: Z coordinate of the center of object;
- *Box volume*: Volume of object's bounding box ($V = W \times H \times D$);
- *Box ratio*: Ratio between maximum and minimum size of the bounding box. ($R = Max/Min$);

- *Volume fraction*: Ratio of object's volume to box volume ($R = V_{obj}/V_{box}$);
- Diameter: Equivalent diameter of object;
- *Sphericity*: Sphericity of object, calculated as six volumes of object divided by equivalent diameter and surface area of object. For a spherical object this parameter equals 1, for all other shapes it is <1;
- *Radius (max)*: Maximum distance between an object's centroid and surface;
- *Radius (min)*: Minimum distance between an object's centroid and surface;
- *Radius ratio*: Ratio between Radius (max) and Radius (min);
- *Feret (max)*: Maximum distance between two parallel planes enclosing an object;
- *Feret (min)*: Minimum distance between two parallel planes enclosing an object;
- *Feret ratio*: Ratio between Feret (max) and Feret (min);
- *Surface deviation*: Calculated as the deviation of end points of triangle normal. The calculations are done as following: all triangle normal vectors of surface are normalized (length set to 1) and the average distance from the mean position of the endpoints to all other vectors is calculated. Uniform surface will have deviation of 0. The maximum deviation of 1.336 will have a sphere.

Combining groups of these parameters enables some specific physical structural characteristics of the samples to be deduced. These include porosity, specific surface area, and roughness, which describe the morphology of an object from a different perspective. Then correlations between all structural parameters and the properties of the sample can be explored. Sample properties to be considered include the drug content, drug release kinetics, and how the pharmaceutical processing affects the structure. In addition, multivariate analysis, data mining, and modeling methods can be introduced for the construction of statistical models.

12.4 3D Visualization and Quantitative Characterization

The 2D and 3D images for a number of solid dosage forms were acquired and used to learn more about their structures [23]. The following examples demonstrate how the structural information can be beneficial in the design and testing of dosage forms and in solving related technical problems.

X-ray microtomography provides quantitative structural information such as the shape and size of any physically different regions of a dosage form that is not accessible by other experimental techniques. For example, the technique is ideally suitable for determining the internal thickness of layers in multilayer tablets and the shape and size of the interfaces between these layers. It can be also used to elucidate the microstructure of fast-dissolving tablets such as those manufactured using lyophilization, in which the structural features reflect the size and shape of the ice crystals present in the tablet before freeze-drying. The technique is also effective in assessing the morphology and pore size distribution of pharmaceutical granules, and especially useful for studying the connectivity and shape of voids within granules made by wet granulation [12, 24]. It has also been applied in evaluation of porous matrix structures and the diffusion and dissolution processes that occur during drug release from solid dosage forms [13].

12.4.1 Internal Structure of Particles

The structural information of drug-containing granules of importance during the design stages as well as for product performance includes the distribution of drug or excipient particles, presence of any void spaces, thickness of drug or polymer layers, and shape or even roughness of the surface. Various analytical methods have been applied to obtain these structural properties. SEM is the most widely used method to observe the surface structures of particles [25–28]. Raman spectroscopy has also been successfully used to reveal the distribution of drugs and excipients [29]. However, these methods show only the structural information about the surface or inner region just below the surface of the examined particles and granules, and samples must be cut or destroyed to observe the internal structures. For example, in an investigation by SEM, samples were placed under vacuum and a coating deposited containing a heavy metal such as platinum, which can affect the surface structure of the sample [30]. Although MRI can reveal the internal structure of a drug formulation without destruction, applications in structural examination of solid systems have been limited due to its low spatial resolution. X-ray CT can elucidate the 3D structure of a formulation nondestructively due to the high penetration of X-rays, and has been applied to the structural analysis of tablets [31] and granules [32] with sizes ranging from centimeters to millimeters using an inhouse X-ray generator.

12.4.2 Dynamic Structure of Granular Systems

The mixing and segregation of granular materials are fundamental process operations with many industrial applications in the pharmaceutical, chemical, and food industries. Particulate materials are mixed in batch blenders such as tumbling bins and V-blenders. However, it is often difficult to mix particles homogeneously in this dynamic process. Moreover, composition fluctuations are frequently increased on material handling [33, 34]. Because of these inhomogeneity phenomena, if segregation occurs before tablet compaction or capsule filling, this can have a major effect on content uniformity, especially if the drug content is very low [2, 35]. In addition, control of particle blend homogeneity is important because it influences the weight variation [36], drug release [37], and dissolution properties of pharmaceutical preparations [38, 39]. For these reasons, the mixing and segregation processes of particulate materials play important roles in achieving the desired pharmaceutical product quality.

It is generally recognized that the segregation of constituent particles in a mixture is highly related to particle characteristics such as shape, size, and surface properties. For example, if particle size and shape variations exist between the individual component particles, segregation can be a major concern during powder flow into product containers [40, 41]. The characterization method for quantifying the physical properties of the powder and granular systems is therefore of particular importance for understanding the flow behavior of different materials under various process conditions.

There are several methods reported in the literature to characterize powder and granular systems, including SEM [42], NIR spectroscopy [43], and MRI [44, 45]. For instance, NIR spectroscopy has been utilized to monitor powder blend homogeneity; and the sensitivity of NIR spectra to changes in the physicochemical properties of powders has been demonstrated [41, 43]. However, it is difficult to obtain a high-resolution visualization to the internal field of multicomponent powder and granular systems. In addition, these techniques, such as SEM which probes the outer surface and not the internal structure, are not able to perform a comprehensive structural evaluation of particles [42, 46]. With these methods no quantitative characterization and statistical data were presented.

In contrast, X-ray computed tomography (μCT) has great potential for providing an improved understanding of the total (i.e., surface and bulk) structural properties of particulate materials. Unlike conventional methods, the CT technique allows noninvasive visualization of internal and microstructural details at micron and millimeter resolution [9, 10].

It is well documented that μCT has been developed for investigating the morphology and internal structure of solid dosage forms, powders, and granules. For example, the whole spatial information on a particular powder can be obtained by this method. In addition, distributions of geometric characteristics describing the size, shape, and spatial arrangement of the particles can be estimated after an image analytic separation of single particles [11]. This powerful method has been employed to provide detailed morphological information such as the pore shape, spatial distribution, and connectivity of porous particles which correlated with the dissolution properties of the DDSs [12, 13]. It has also been reported that CT combined with DEM can be used to investigate particle packing in a process of pharmaceutical tablet manufacture by powder compaction [14]. Recently, SR-μCT was used to monitor the homogeneous blending of binary mixtures noninvasively. Granular samples of microcrystalline cellulose and starch were characterized individually using SR-μCT [20]. Simultaneously, particle size distributions were investigated by calculating the frequency distribution of a statistic for testing sphericity. Then, the microcrystalline cellulose and starch granules were blended in a cylindrical container. The influences of time of rotation (TR) and time of vibration (TV) on mixture homogeneity were studied from SR-μCT data and statistical evaluation. A mixing index was also adopted to evaluate the mixture homogeneity of the particulate system. The results showed that mixture homogeneity was improved with increasing TR. Furthermore, segregation increased with longer TV when the particles exhibited different sizes and shapes. The larger starch granules of a nonspherical shape had a tendency to rise to the top of the granular bed, while the smaller microcrystalline cellulose granules which were more spherical in shape tended to migrate to the bottom of the mixture. Therefore, it was demonstrated that the SR-μCT technique can be applied to investigate the mixing and segregation of granular materials in 3D combined with a statistic method without removing any samples from the granular bed (Figure 12.3).

12.4.3 Microstructure of Monolith Osmotic Pump Tablets

Recently, advances have been reported with the structural design of osmotic pump drug delivery systems (ODDSs), from single chamber osmotic pumps to multiple chamber osmotic pumps [47]. More than 30 ODDS products have been developed and launched into the market [48]. Hundreds of patents and numerous publications have reported on the formulation aspects, clinical results, and safety aspects of ODDSs [49]. Whilst the overall structure and internal architecture of the unit are obviously the key physical characteristics of the ODDS, little research has been carried out to visualize the internal structure and the dynamical changes of the ODDS tablet core during the drug release process.

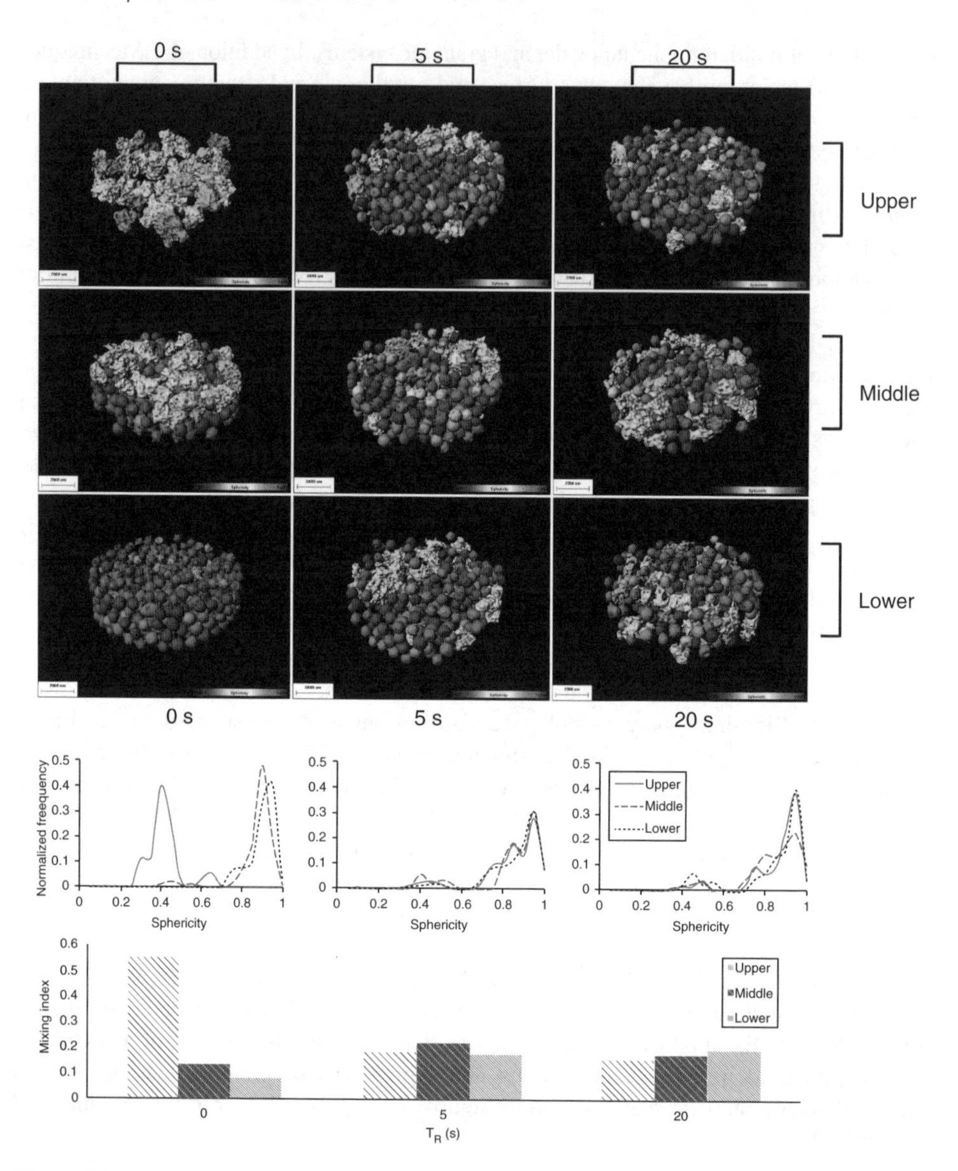

Figure 12.3 *3D reconstruction and quantification of mixing and segregation of granules [20] (see color plate section). Reproduced from Ref. [19], with permission from Elsevier.*

Among the various oral controlled DDSs available, ODDSs delivering active agents by constant inner osmotic pressure have attracted much interest since they have numerous advantages over other oral controlled DDSs. However, the drug release kinetics of the ODDS shows dependency on formulation factors, including the solubility of drugs within the tablet core, the osmotic pressure of the core component(s), the semipermeable membrane characteristics, and the delivery orifice size [50]. Microstructural investigation of the internal 3D steric data helps in understanding the drug release mechanism of monolith pump controlled release systems [31].

12.4.3.1 Visualization of the Surface Morphology and Internal 3D Structure

From the 2D monochrome X-ray CT images of felodipine monolith osmotic pump tablets (MOTSs) at different sampling times (0.5, 1.0, 3.0, 6.0, and 8.0h) during dissolution testing, the semipermeable membrane, the tablet core, and the drug delivery orifices have been clearly visualized, with small voids visible in the tablet core. It is also observed that the solid content is detached from the tablet core following erosion and swelling. The surface properties of the tablet shell maintain the original morphology at most of the sampling time points, with evidence of some collapse at the surface of the tablet shells after 8.0h dissolution testing. Most of the tablet core remains as a solid or semisolid form. Some cracks are seen at 0.5 and 1.0h, whilst at 3.0 and 6.0h, the tablet core becomes disaggregated and several voids form. The voids become increasingly larger adjacent to the drug delivery orifice. At 8.0h, the content of the tablet core is nearly empty. It is interesting to observe that the shape of the remaining tablet core varies irregularly, which is markedly different from the form and shape a formulation scientist might anticipate. It is also observed that some aggregates adhered to the internal wall of the tablet at 8.0 and 10.0h dissolution testing, which correlates in part with the accumulated release profiles and percentages of felodipine at 8.0 and 10.0h as 73.0 and 82.4%.

The reconstructed 3D tomographic images at different drug release time points demonstrate the dynamic changes in the internal 3D structure of the MOTS. The shape of the tablet core is elliptic at 0.5 and 1.0h. As a result of hydration of the tablet core, the shape becomes more irregular with several voids after 3.0h. Also, the shape change is more notable adjacent to the delivery orifices, suggesting the release of felodipine near the delivery orifices is much faster than from other locations (Figure 12.4).

12.4.3.2 Correlation between the 3D Steric Parameters and the Remaining Percentage of Drug in the Tablet Cores

The volume and surface area of the remaining core of the tablets are calculated from the reconstructed 3D images. From the SR-μCT studies with felopidine MOTS, the remaining percentages of drug in the tablet cores at 0.5, 1.0, 3.0, 6.0, and 8.0h were calculated by taking 100% minus the *in vitro* release percentage of felodipine. As a result, the 3D volumes correlate well with the remaining percentages of felodipine in the MOTS ($R=0.9988$), suggesting that the 3D parameter accurately reflects the release characteristics of the felodipine MOTS.

For MOTS, the release of felodipine is co-controlled by the osmotic pressure and drug in suspension. Initially, the pump is controlled by the osmotic mechanism. After water is imbibed, the core within the pump is covered by the liquid suspension. The remaining solid content of the tablet core (containing solid drug) is eroded into suspension. 2D images of cross-sections of felodipine MOTS acquired at 3.0 and 6.0h clearly indicate erosion of the solid content from the tablet core to the suspension. As a result of the osmotic pressure and suspension co-mechanism, the suspension is pumped out through the orifice. When considering the specific processes taking place during drug dissolution and release from MOTS systems, the detachment of the solid content from the tablet core is complex. According to Equation 12.1,

$$\frac{\mathrm{d}M}{\mathrm{d}t} = \frac{\pi C}{8}\frac{R^4}{\eta}\frac{P_1 - P_2}{h} \tag{12.1}$$

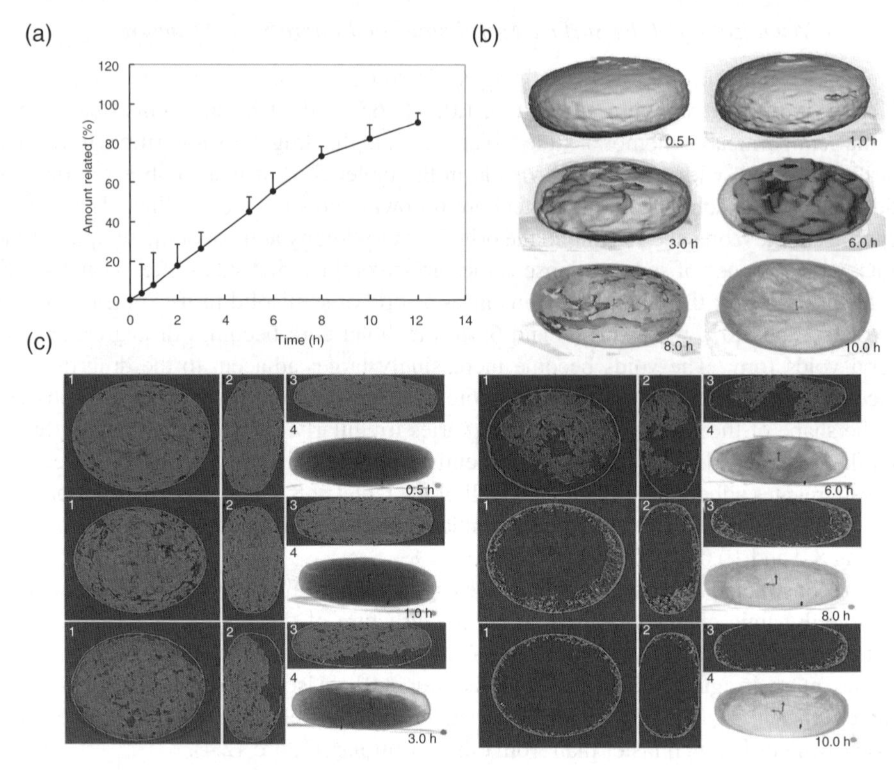

Figure 12.4 *(a) Amount of felodipine MOTS released. (b) Reconstructed 3D images of felodipine MOTS at different sampling times (yellow represents the solid moiety of the tablet core, air appears gray). (c) 2D monochrome X-ray CT images of felodipine MOTS viewing from four different aspects: 1. top, 2. front, 3. back, and 4. reconstructed image. Air appears dark, gray represents the solid moiety of the tablet core, and gray edge represents the semipermeable membrane [31] (see color plate section). Reproduced from [31], with permission from Elsevier.*

for MOTS with zero-order drug release kinetics, values of R, η, $(P_1 - P_2)$ and h are nearly constant. However, the apparent value of C is dependent on the extent of solid content detached from the tablet core to the suspension. This is the erosion controlled process which can be expressed as the following Equation 12.2:

$$\frac{\mathrm{d}M}{\mathrm{d}t} = \frac{DAC_S}{l} \tag{12.2}$$

where D, A, C_s, and l are the diffusion coefficient, the surface area of diffusion or erosion, drug solubility, and the thickness of boundary layer, respectively. For a given drug and tablet, values of D, l and C in Equation 12.2 are invariable parameters. Therefore, the concentration of drugs in suspension in Equation 12.1 is determined by A, the surface area of diffusion or erosion, namely, the surface area of the remaining tablet core. Initial assessments might expect that the surface area as well as the volume of the tablet core would decrease with the disappearance of the solid content during the drug release process. It is observed that the surface area of the tablet core during the drug release process is quite stable, only changing by a small amount although the shape of the internal solid content does change markedly from elliptical to irregular and, as discussed above, containing voids. This results in the 3D surface area values being almost constant.

In summary, values of C, R, η, $(P_1 - P_2)$ and h in Equation 12.1 are all constant during the drug release process, which demonstrates the intrinsic mechanism of the drug release kinetics of the MOTS. Thus, the 3D surface area could be regarded as a key steric parameter for the quality control of felodipine MOTS products.

12.4.4 Fractal Structure of Monolith Osmotic Pump Tablets

The quality of oral controlled DDSs is commonly assessed using conventional parameters such as the drug dissolution profile, content uniformity, related compounds, and the weight gain during coating for film-coated products. Among these parameters, the dissolution is often considered as an indicator of potential drug release *in vivo*. Clear correlations between *in vitro* and *in vivo* pharmacokinetic profiles whilst desirable are not routinely found, and formulations with similar dissolution profiles but variable *in vivo* biological activity can be found in the literature [51, 52]. These findings indicate that the dissolution profile frequently lacks specificity for sufficient understanding of the intrinsic quality of DDSs. One approach to identify such required understanding is to develop new methodologies to visualize the internal characteristics of DDSs to reveal the dynamic changes taking place in the architecture within the tablet during drug release. Thus, the conventional *in vitro* release method, which does not take into account the internal structure, can be calibrated [23]. Tomographic imaging techniques, such as μCT [13, 53], MRI [54, 55], and terahertz imaging [56] offer improved understanding of the structural characteristics of solid dosage forms and provide important information on the drug release mechanism and the quality of pharmaceutical products. Further understanding of the hydration kinetics has also been obtained through imaging techniques, including the visualization of the surface morphology and internal structure (i.e., front position or layer thickness), the rate of water penetration into polymer matrices, the polymer concentration across the formed hydro gel, and the location of the drug substance in the hydro gel during swelling [57, 58]. In addition, mathematical models describing drug release mechanisms and critical parameters for successful formulation design have been proposed [59].

It is only over the last 15 years that tomographic methods have been applied to consider the *in vitro* characterization of dosage forms. The rapid developments in the field of tomographic imaging techniques have also led to the application of nondestructive 3D imaging in pharmaceutics [23]. Most of these studies provided only qualitative information on the surface morphology and internal structure of the tablets during drug dissolution, which are irregular, complex, and continuously changing. Thus, a quantitative method or a structural parameter is required to define the complexity of the surface morphology and internal structure of tablets during drug dissolution.

Fractal analysis is a quantitative analytical method used to characterize the morphometric variability and complexity of an object in nature [60, 61]. The irregular degree of a subject can be highly abstracted into a noninteger fractal dimensional value (D_f). Fractal analysis has been applied in medical signal processing and pharmacokinetic modeling.

The fractal dimension has also been used to quantitatively evaluate the surface roughness of a poly(2-hydroxyethyl methacrylate) hydro gel based on the images acquired by a light microscope [62]. The fractal dimension exhibited a major decrease in value during swelling and was highly correlated with the swelling ratio, thus indicating that the fractal dimension provided an interesting quantitative parameter related to surface roughness. For solid pharmaceutical dosage forms, D_f values derived from surface imaging techniques have been employed to characterize the surface morphology of particles, mainly focusing on 2D static measurements, such as SEM and AFM [63–66]. Information on surface roughness and shape for particles and tablets has been gained from the values of fractal dimensions. However, there are few publications focusing on fractal analysis for 3D structures and its relationship with drug dissolution from controlled release dosage forms.

Our previous work investigated the surface morphology and the internal 3D structure of felodipine osmotic pump tablets based on slice by slice 2D imaging projections acquired via SR-μCT [1]. Thus, a 3D fractal analysis based on a box-counting method was developed to simultaneously quantify the entire shape, interior porous channels, and surface structure of felodipine osmotic pump tablets during the drug release process. The 3D volume and surface area related fractal dimensional values were then calculated and correlated with drug release kinetics. Finally, the mechanism of drug release for the felodipine controlled release system was elucidated through 3D fractal data. Results showed that values of $D_{f,surface}$ correlated well with the drug release rate. $D_{f,surface}$ was found to be an efficient fractal parameter that could be used to characterize the complex changes to the tablet core that take place during drug release. Fractal analysis, especially architectural fractal analysis, was initially employed in this work to characterize the internal structural changes taking place during drug release that cannot be measured by conventional methods. The degree of correlation between the fractal dimensions and drug release kinetics may prove to be a valuable parameter in quantifying the role of architectural changes in the mechanism of drug dissolution from controlled DDSs (Figure 12.5). Thus the fractal dimension has been shown to be a valuable quantitative indicator reflecting the drug release performance and can be regarded as a key indicator for the quality control of oral controlled DDSs [67].

12.4.5 Dynamic Structure of HPMC Matrix

Among a variety of oral controlled release DDSs, water swellable matrix systems, particularly those containing HPMC, are widely used because of the relative simplicity of formulation compositions, ease of manufacturing, and low cost, as well as acceptance by regulatory authorities and applicability to drugs with a wide range of dose and solubility [68]. Much attention has been paid to the mechanism governing drug release from hydrophilic matrices, where it has been suggested that the diffusion front at the interface between the undissolved drug and the dissolved drug in the hydration layer influences the rate of drug release from cellulose matrix systems. In addition, water penetration and/or diffusion are postulated to be the rate-limiting steps for the release of highly water-soluble drugs. For poorly water-soluble drugs, matrix erosion is considered to provide a major contribution to the mechanism of the drug release [69]. Our previous work has investigated the quantification of swelling and erosion in the controlled release of a poorly water-soluble drug using synchrotron μCT [19].

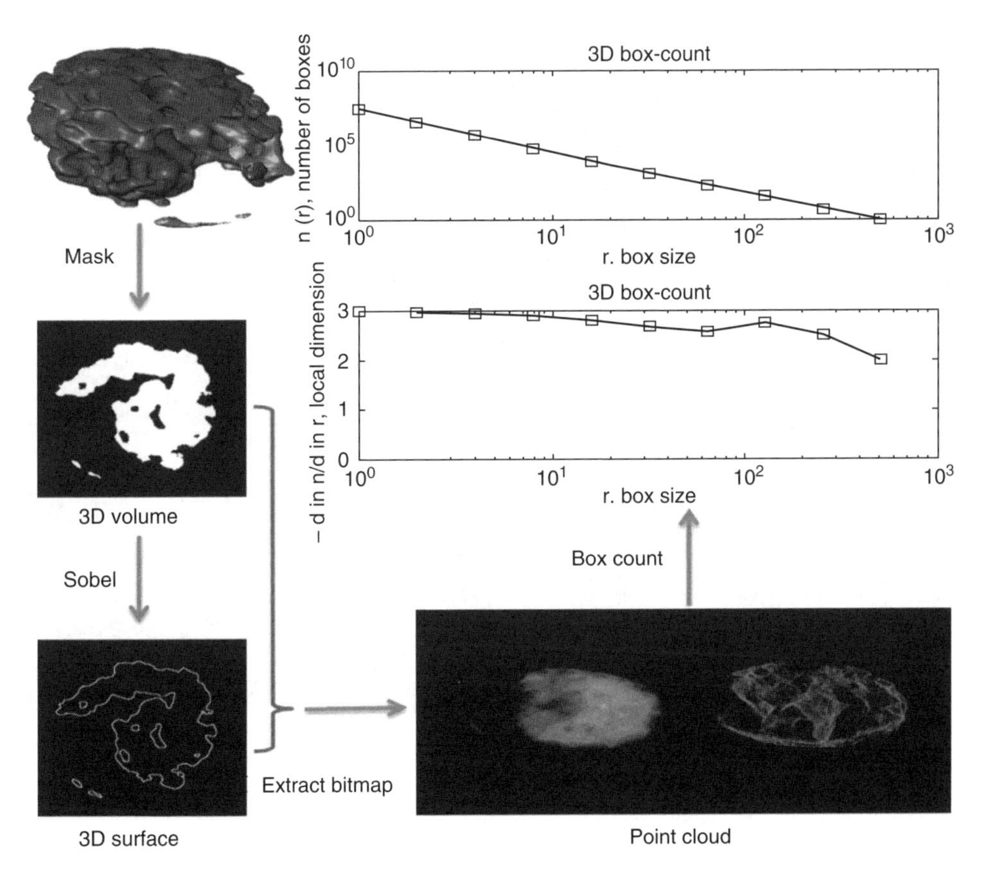

Figure 12.5 *Computing procedures for the fractal dimension values of the felodipine osmotic pump tablet core [67]. Reproduced with permission from [67]. Copyright 2013 Royal Pharmaceutical Society.*

12.4.5.1 Visualization of the Surface Morphology and the Hydration Layer of Felodipine Tablets

After SR-μCT evaluation, the hydration layer, the glassy tablet core, and the relative movement with time of the erosion and swelling fronts are clearly visible. The microstructures and changes in the core can be observed (Figure 12.6). The HPMC matrix swells following absorption of water resulting in the increase of the matrix dimensions. After 0.5 h, the hydration layer is clearly observed and grows gradually with time. After 5.0 h, the hydration layer becomes thinner. The size of the glassy core reaches a maximum at 6.0 h and then reduces as the polymer hydrates to a greater extent. After 8.0 h, the matrix is entirely hydrated with no core remaining. In the period from 1.0 to 6.0 h, the thickness of the hydration layer exhibits only minor changes, indicating the constant release rate of the felodipine HPMC matrix tablets.

Reconstructed 3D structures of felodipine HPMC matrix tablets are created with segmented slices (see Figure 12.6). The length, width, and height of the tablet all increase within the first 1.0 h as the tablet swells following its contact with the dissolution medium.

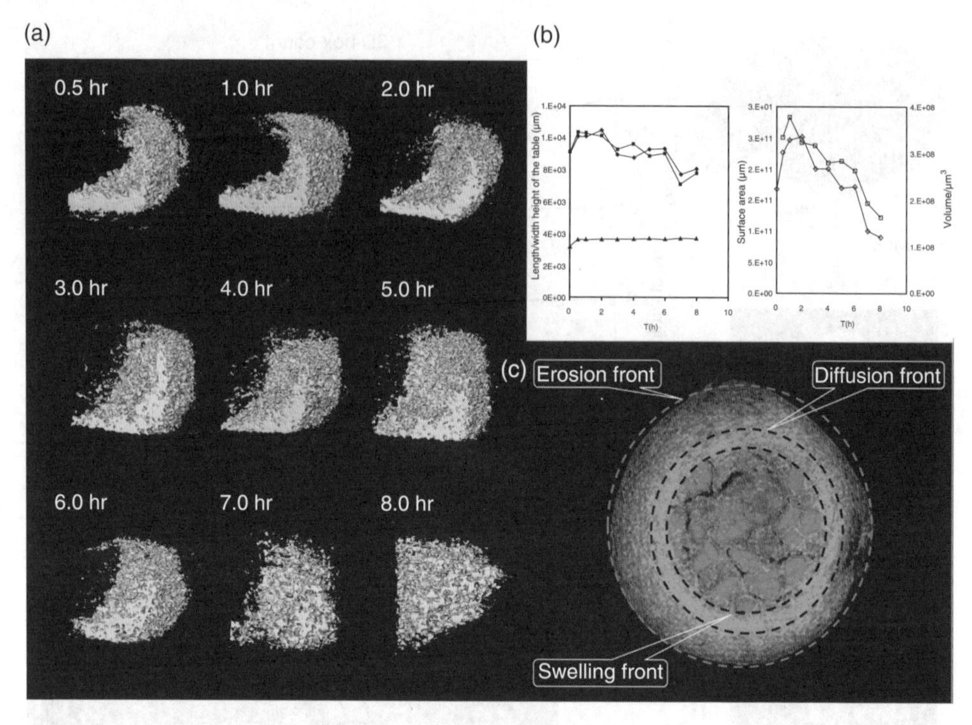

Figure 12.6 *(a) Reconstructed 3D images of the hydration layer during drug dissolution. (b) Changes to internal structures of the whole tablet during drug dissolution. Length (filled diamonds), width (filled squares), and height (filled triangles). (c) Schematic diagram of the swelling front, the diffusion front, and the erosion front of felodipine HPMC tablet during drug dissolution [19]. Reproduced from [19], with kind permission from Springer Science and Business Media.*

The length and width then start to decrease from 2.0 h onwards as the HPMC matrix erodes. The height of the tablet, however, remains the same. The volume and surface area of the tablet diminish with time. The reconstructed slice stack is converted into eight-bit grayscale format and resliced vertically. Then, the vertical slices of tablets at different time points (include the well hydrated tablets and the tablet with totally dried core) are analyzed to determine the threshold gray values to distinguish the erosion font, the diffusion front, and the swelling front. The reduction in volume of the hydration layer is slower than that of the whole tablet, indicating the drug release rate-determining influence of matrix erosion.

12.4.5.2 Correlation of the 3D Hydration Parameters with Drug Release Kinetics

In total, 23 3D steric parameters have been calculated based on extracted Iso-Surface models. The correlation between volume and cumulative release percentage has been investigated. All the volume parameters, including volume of the whole tablet, volume of the glassy core, and volume of the hydration layer, correlate to a good degree with percentages of drug release ($R^2 = 0.90, 0.94$, and 0.81, respectively). Drug release rate (dM/dt) values provide the key characteristics of DDSs. The correlation (R^2) values between surface area parameters and drug release rate are 0.45 for the surface area of the tablet and 0.80 for the surface area of the hydration layer, while the R^2 value between the surface area of the glassy core and the release rate is 0.87, indicating that the surface area of the glassy core is one of the key factors in determining constant drug release. Thus, the matrix swelling has an important and controlling effect on the drug release kinetics.

In addition, a statistical model was constructed using SPSS PASW statistics (version 18.0) for multivariate analysis, using the drug release rate at different time points as the dependent variable ($n=9$). The sample at 0h was excluded as it is not possible to calculate the value of specific surface area at 0h. Independent variable reduction was carried out taking into account the value of R^2, the equation significance, the parameter coefficient significance, and the physical meaning of the variables. Three parameters with a major influence on drug release from felodipine tablets were identified from 23 3D structural parameters by independent variable reduction and used to establish an equation to predict the drug release rate:

$$\frac{dM}{dt} = 420.4SA_{\text{hydration layer}} + 128.4SA_{\text{glassy core}} + 214.9SSA_{\text{hydration layer}} - 23.4 \quad (12.3)$$

The $SA_{\text{hydration layer}}$ is the surface area of the hydration layer, which includes the areas of pores and channels inside the hydration layer. The $SA_{\text{glassy core}}$ is the surface area of the nonhydrated glassy core, which is the interface where the glassy core swells into gel and hydration occurs. The $SSA_{\text{hydration layer}}$ is the specific surface area of the hydration layer, which reflects the magnitude of the hydration layer. All the relevant parameters in Equation 12.3 have been normalized to ensure that all parameters have equal determinant strength to drug release rate. The statistical model exhibits relative desirable predictability, with $R^2=0.96$ and significant p values (<0.001).

12.4.6 Release Behavior of Single Pellets

Multiparticulate solid dosage formulations (e.g., pellets and granules) contain numerous discrete particles that are combined into hard gelatin capsules or compressed into tablets, forming a single dose unit. Multiparticulate dosage forms have gained considerable popularity over conventional single unit products for controlled release DDSs. Conventional dissolution test of multiparticulate systems establishing a release profile clearly does not provide any information on the individual contributions of single pellets and is insufficient because the release kinetics for ensembles is a composite profile of the individual units [70–73]. An ensemble of zero-order releasing units will exhibit first-order kinetics under certain conditions [74]. The results from single pellet experiments can also be used to simulate release behavior on the dose level [75–79]. A recent study involved SR-μCT in correlating structural detail with the single pellet release characteristics of pellets of tamsulosin hydrochloride sustained release capsules (TSH; Brand name, Harnal).

12.4.6.1 Visualization and Internal 3D Parameter Calculations

The high-resolution 3D images obtained from SR-μCT refer to a model that reproduces the morphology and microstructure of the pellets; the 3D rendering model displays fully detailed structural information about every pellet from a sample capsule. Hollow pellets and solid pellets were identified and randomly selected to show morphological and microstructural information. Differences in the shape, size, morphology, and internal structure of each pellet were clearly recognized. Pellet properties, such as diameter, surface roughness, and internal void, can be extracted and used to calculate the volume and specific surface area for each pellet. The ability to visualize the microstructures of an individual pellet from the results of CT scans and 3D reconstructions is also especially valuable when investigating the drug release mechanism.

In attempting to describe the pellets, approximately 40 steric quantitative parameters were obtained from the reconstructed model. The most important ten parameters closely related to the release profile were then selected. Based on the derived data, the pellets were markedly different in size and shape from one another within the approximately 1000 pellets in each TSH sustained release capsule. The particle system was described using a frequency distribution. The size (volume, surface area, and diameter) distributions had a wide range and did not exhibit Gaussian distributions. In addition, most pellets had a value close to 1 for sphericity, while some pellets were still irregular. The specific surface area of the majority of pellets was in the range between 0.007 and 0.035 μm^{-1}. Approximately 50% of the pellets were completely solid, and the other half had voids with variable volumes and surface area distributions. The smaller pellets (with respect to surface area and volume) were more spherical. A few pellets were more ovoid in shape, a feature thought to be due to the pressure effects during the pellet formation and coating processes. The deviations in size decreased when the particles were more spherical. However, no strong correlation was observed between the void and the other parameters. The void fraction distribution and the diameter showed a random tendency without a significant correlationship (Figure 12.7).

12.4.6.2 Single Pellet Release Kinetics

The drug loading of individual pellets was distributed across a wide range and correlated to some degree ($R^2 = 0.7790$) to the pellet volume. The points on the scatter plot spread further as the diameter increased, indicating that the drug loading in small pellets with high sphericity showed a higher homogeneity. Furthermore, the drug concentration in the pellet matrix (content in unit of volume) of most pellets was approximately 1–2 ng/μm^3, with the drug loading of individual pellets ranging from <100 to 700 ng. This feature is important for understanding the drug release mechanism. However, several outliers were apparent, indicating that the manufacturing process was unable to produce totally uniform, consistent pellets. The dissolution rate of individual pellets gradually decreased during the release process, with complete drug release after 240 min for almost all of the pellets. The distribution of percentage drug at each time point was similar.

12.4.6.3 Correlation between Structural Parameters and Drug Release

The cumulative amount of drug released correlated well with pellet surface area, with particularly strong linear correlations observed over the dissolution time period of 45–120 min. Thus the surface area of the pellet represents a key steric parameter that is a primary determinant of the drug release profile. The drug release kinetics are also likely to be influenced by the diffusion of drug from the pellet. From our observations, we noted that the drug-containing pellets maintained their original shape during the drug release tests; swelling or disintegration was not observed.

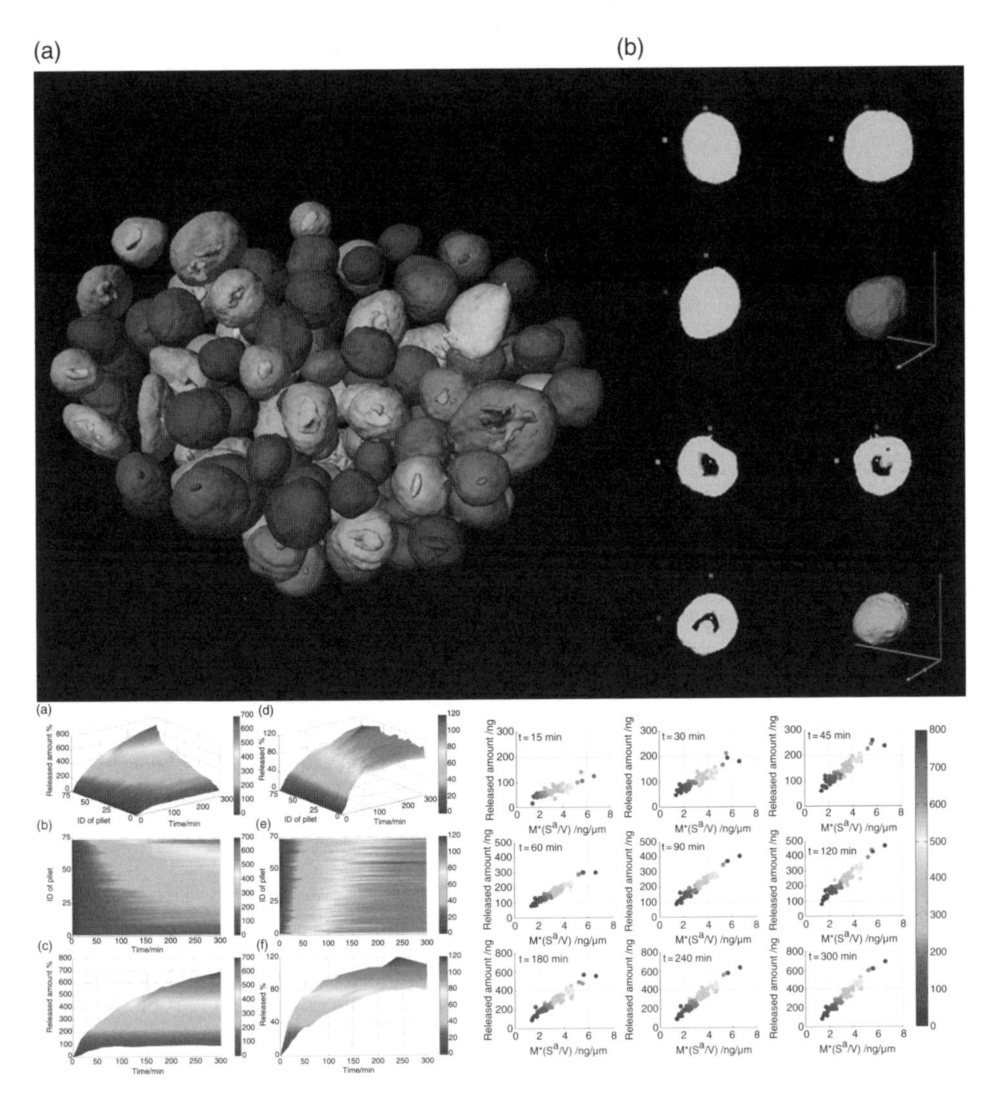

Figure 12.7 *Characterization and release behavior of pellets [78]. Reproduced from [78], with kind permission from Springer Science and Business Media.*

The drug release patterns of single pellets can be expressed as the well known relationship reported by T. Higuchi and reduced to Equation 12.4:

$$\frac{Q}{A} = \sqrt{\frac{DK}{\tau}\left(2 - KC_s\right)C_s t} \tag{12.4}$$

The release rate is controlled by the drug concentration, the dissolution speed, and how fast the dissolved drug diffuses throughout the matrix. The largest value of $M\infty/V$ for the pellets under investigation is 3.5 μg/mm^3, a value considerably lower than the solubility of TSH. Therefore, if TSH dissolves readily during the first and middle periods of release, the drug concentration in the pellet is constant and equal to M_∞/V because the surface area of the pellets hardly changes, allowing τ and the surface area to remain constant. For this case where D, K, C, and τ are constant, this equation can be reduced to Equation 12.5:

$$\frac{Q}{A} = \sqrt{\omega t} \tag{12.5}$$

where ω is the slope of the square root of the time relationships and Q can be rewritten as dM/S; A can be rewritten as M/V, allowing Equation 12.5 to be rewritten as Equation 12.6:

$$\mathrm{d}M = \frac{M}{V} S\sqrt{\omega t} \tag{12.6}$$

where M is the drug loading of the pellet, and V and S are the volume and surface area of the pellet, respectively. The cumulative amount of drug released is a linear function of the square root of time. Therefore, the level of correlation between $M\times(S/V)$ and drug release kinetics was investigated. The results confirmed a high level of correlation. At the beginning of dissolution, only the outer surface will be infiltrated with dissolution medium. After the pellet is well hydrated, the voids will also be filled with dissolution medium and the surface area of internal void will affect the dissolution and speed up the diffusion process. Then, for calculation purposes, the surface area, an important steric parameter for the release profile, is adjusted to S_a (S_a = surface area of pellet + surface area of void). The adjusted parameter $[M\times(S_a/V)]$ produced an improved correlation with higher R^2 values, demonstrating that the voids also play an important role during drug release.

12.5 Future Prospects

Much attention has been given recently in the field of designing solid DDSs to computer simulation at the molecular level and direct observation of the constituent particles of dosage forms using instruments such as electron microscopes. In related studies it is often assumed that the macro structure of conventional dosage forms and their unit constituents, such as tablets, capsules, pellets, granules, and powders has been thoroughly and sufficiently investigated for their various structures and forms. However, it is the external structure that is generally considered rather than the total microstructure, including the internal architecture of the solid system. It appears that the microstructure of dosage forms across the size range from hundreds of nanometers to millimeters has not been well investigated quantitatively and presents a missing area of study, perhaps due to the lack of a sensitive *in situ* imaging method and difficulties in quantifying 3D features. Although there are limitations to using SR-μCT, the examples highlighted above demonstrate the vitality of this approach and demonstrate how additional knowledge revealed by such studies can inform the design of a range of solid dosage forms for specific desired performance criteria. The re-visit to interrogate holistically the structure of dosage forms using this technology will enrich our 3D structure visions and provide opportunities in the creation and design of new structures for DDSs.

References

[1] Feng, Y., Grant, D.J., and Sun, C.C. (2007) Influence of crystal structure on the tableting properties of n-alkyl 4-hydroxybenzoate esters (parabens). *Journal of Pharmaceutical Sciences*, **96** (12), 3324–3333.

[2] Shekunov, B.Y., Chattopadhyay, P., Tong, H.H., and Chow, A.H. (2007) Particle size analysis in pharmaceutics: principles, methods and applications. *Pharmaceutical Research*, **24** (2), 203–227.

[3] Rohrs, B.R., Amidon, G.E., Meury, R.H. *et al.* (2006) Particle size limits to meet USP content uniformity criteria for tablets and capsules. *Journal of Pharmaceutical Sciences*, **95** (5), 1049–1059.

[4] Jain, S. (1999) Mechanical properties of powders for compaction and tableting: an overview. *Pharmaceutical Science & Technology Today*, **2** (1), 20–31.

[5] Sun, C. and Grant, D.J. (2001) Influence of crystal structure on the tableting properties of sulfamerazine polymorphs. *Pharmaceutical Research*, **18** (3), 274–280.

[6] Zhang, Q., Gladden, L., Avalle, P., and Mantle, M. (2011) In vitro quantitative ((1))H and ((19)) F nuclear magnetic resonance spectroscopy and imaging studies of fluvastatin™ in Lescol® XL tablets in a USP-IV dissolution cell. *Journal of Controlled Release: Official Journal of The Controlled Release Society*, **156** (3), 345–354.

[7] Lyon, R.C., Lester, D.S., Lewis, E.N. *et al.* (2002) Near-infrared spectral imaging for quality assurance of pharmaceutical products: analysis of tablets to assess powder blend homogeneity. *AAPS PharmSciTech*, **3** (3), E17.

[8] Xie, Y.L., Zhou, H.M., Liang, X.H. *et al.* (2010) Study on the morphology, particle size and thermal properties of vitamin A microencapsulated by starch octenylsucciniate. *Agricultural Sciences in China*, **9** (7), 1058–1064.

[9] Cooper, D.M., Turinsky, A.L., Sensen, C.W., and Hallgrimsson, B. (2003) Quantitative 3D analysis of the canal network in cortical bone by micro-computed tomography. *Anatomical Record, Part B: New Anatomist*, **274** (1), 169–179.

[10] Stock, S.R. (1999) X-ray microtomography of materials. *International Materials Reviews*, **44** (4), 141–164.

[11] Redenbach, C., Ohser-Wiedemann, R., Loffler, R. *et al.* (2012) Characterization of powders using micro computed tomography. *Particle and Particle Systems Characterization*, **28** (1/2), 3–12.

[12] Farber, L., Tardos, G., and Michaels, J.N. (2003) Use of X-ray tomography to study the porosity and morphology of granules. *Powder Technology*, **132** (2), 57–63.

[13] Young, P.M., Nguyen, K., Jones, A.S., and Traini, D. (2008) Microstructural analysis of porous composite materials: dynamic imaging of drug dissolution and diffusion through porous matrices. *American Association of Pharmaceutical Scientists*, **10** (4), 560–564.

[14] Fu, X.W., Dutt, M., Bentham, A.C. *et al.* (2006) Investigation of particle packing in model pharmaceutical powders using X-ray microtomography and discrete element method. *Powder Technology*, **167** (3), 134–140.

[15] Radon J., Parks, P.C. (translator). (1986) On the determination of functions from their integral values along certain manifolds, *IEEE Transactions on Medical Imaging*, **5** (4): 170–176.

[16] Elder, A.M., Langmuir, R.V., and Pollock, H.C. (1947) Radiation from electrons in a synchrotron. *Physical Review*, **71** (11), 829–830.

[17] Wiedemann, H. (1999) Synchrotron radiation, Particle Accelerator Physics I, Springer, Berlin, Heidelberg, pp. 300–336.

[18] Wang, Y.M., Heng, P.A., and Wahl, F.M. (2003) Image reconstructions from two orthogonal projections. *International Journal of Imaging Systems and Technology*, **13** (2), 141–145.

[19] Yin, X.Z, Li, H.Y, Guo, Z. *et al.* (2013) Quantification of swelling and erosion in the controlled release of a poorly water-soluble drug using synchrotron X-ray computed microtomography. *American Association of Pharmaceutical Scientists*, **15** (4), 1025–1034.

[20] Liu, R.H., Yin, X.Z., Li, H.Y. *et al.* (2013) Visualization and quantitative profiling of mixing and segregation of granules using synchrotron radiation X-ray microtomography and three dimensional reconstruction. *International Journal of Pharmaceutics*, **445** (1/2), 125–133.

[21] Karakosta, E., Jenneson, P. M., Sear, R. P., McDonald, P. J. *et al.* (2007) Observations of coarsening of air voids in a polymer-highly-soluble crystalline matrix during dissolution. *Physical Review E*, **74** (1), 011504.
[22] Chauve, G., Raverielle, F., and Marchessault, R.H. (2007) Comparative imaging of a slow-release starch exciplent tablet: evidence of membrane formation. *Carbohydrate Polymers*, **70** (1), 61–67.
[23] Zeitler, J.A. and Gladden, L.F. (2009) In-vitro tomography and non-destructive imaging at depth of pharmaceutical solid dosage forms. *European Journal of Pharmaceutics and Biopharmaceutics*, **71** (1), 2–22.
[24] Appoloni, C.R., Macedo, A., Fernandes, C.P., and Philippi, P.C. (2002) Characterization of porous microstructure by x-ray microtomography. *X-Ray Spectrometry*, **31** (2), 124–127.
[25] Shiny, J., Ramchander, T., Goverdhan, P. *et al.* (2013) Development and evaluation of a novel biodegradable sustained release microsphere formulation of paclitaxel intended to treat breast cancer. *International Journal of Pharmaceutical Investigation*, **3** (3), 119–125.
[26] Lin, Q., Pan, J., Lin, Q., and Liu, Q. (2013) Microwave synthesis and adsorption performance of a novel crosslinked starch microsphere. *Journal of Hazardous Materials*, **263** (2), 517–524.
[27] Islan G.A., and Castro G.R. (2014) Tailoring of alginate-gelatin microspheres properties for oral Ciprofloxacin-controlled release against Pseudomonas aeruginosa. *Drug Delivery*, **21** (8), 615–626.
[28] Caktü K., Baydemir G., Ergün B., and Yavuz H. (2014) Cholesterol removal from various samples by cholesterol-imprinted monosize microsphere-embedded cryogels. *Artificial Cells, Nanomedicine, and Biotechnology*, **42** (6), 365–375.
[29] Gordon, K.C., and McGoverin, C.M. (2011) Raman mapping of pharmaceuticals. *International Journal of Pharmaceutics*, **417** (1/2), 151–162.
[30] Suzuki, E. (2002). High-resolution scanning electron microscopy of immunogold labelled cells by the use of thin plasma coating of osmium. *Journal of Microscopy*, **208** (3): 153–157.
[31] Li, H.Y., Yin, X.Z., Ji, J.Q. *et al.* (2012) Microstructural investigation to the controlled release kinetics of monolith osmotic pump tablets via synchrotron radiation X-ray microtomography. *International Journal of Pharmaceutics*, **427** (2), 270–275.
[32] Crean, B., Parker, A., Le Roux, D. *et al.* (2010) Elucidation of the internal physical and chemical microstructure of pharmaceutical granules using X-ray micro-computed tomography, Raman microscopy and infrared spectroscopy. *European Journal of Pharmaceutics and Biopharmaceutics*, **76** (3), 498–506.
[33] Dasgupta, S., Khakhar, D.V., and Bhatia, S. K. (1991) Axial segregation of particles in a horizontal rotating cylinder. *Chemical Engineering Science*, **46** (5/6), 1513–1517.
[34] Savage, S.B., and Lun, C.K.K. (1988) Particle-size segregation in inclined chute flow of dry cohesionless granular solids. *Journal of Fluid Mechanics*, **189**, 311–335.
[35] Yalkowsky, S.H., and Bolton, S. (1990) Particle-size and content uniformity. *Pharmaceutical Research*, **7** (9), 962–966.
[36] Fan, A., Parlerla, S., Carlson, G. *et al.* (2005) Effect of particle size distribution and flow property of powder blend on tablet weight variation. *Latin American Journal of Pharmacy*, **8**, 73–78.
[37] Heng, P.W., Chan, L.W., Easterbrook, M.G., and Li, X. (2001) Investigation of the influence of mean HPMC particle size and number of polymer particles on the release of aspirin from swellable hydrophilic matrix tablets. *Journal of Controlled Release: Official Journal of The Controlled Release Society*, **76** (1/2), 39–49.
[38] Carless, J.R., and Sheak, A. (1976) Changes in the particle size distribution during tableting of sulphathiazole powder. *Journal of Pharmacy and Pharmacology*, **28** (1), 17–22.
[39] Jillavenkatesa, A., Kelly, J., Dapkunas, S.J. (2002) Some issues in particle size and size distribution characterization of powders. *Latin American Journal of Pharmacy*, **5**, 98–105.
[40] Berman, J. and Planchard, J.A. (1995) Blend uniformity and unit dose sampling. *Drug Development and Industrial Pharmacy*, **21** (11), 1257–1283.
[41] El-Hagrasy, A.S., Morris, H.R., D'Amico, F. *et al.* (2001) Near-infrared spectroscopy and imaging for the monitoring of powder bend homogeneity. *Journal of Pharmaceutical Sciences*, **90** (9), 1915.

[42] Poutiainen, S., Pajander, J., Savolainen, A. *et al.* (2011) Evolution of granule structure and drug content during fluidized bed granulation by X-Ray microtomography and confocal Raman spectroscopy. *Journal of Pharmaceutical Sciences*, **100** (12), 5254–5269.
[43] El-Hagrasy, A.S. and Drennen, J.K. (2006) A Process Analytical Technology approach to near-infrared process control of pharmaceutical powder blending. Part III: quantitative near-infrared calibration for prediction of blend homogeneity and characterization of powder mixing kinetics. *Journal of Pharmaceutical Sciences*, **95** (2), 422–434.
[44] Porion, P. (2004) Dynamics of size segregation and mixing of granular materials in a 3D-blender by NMR imaging investigation. *Powder Technology*, **141** (1/2), 55–68.
[45] Nguyen, T.T.M., Sederman, A.J., Mantle, M.D., and Gladden, L.F. (2011) Segregation in horizontal rotating cylinders using magnetic resonance imaging. *Physical Review E*, **84** (1), 011304.
[46] Akseli, I., Iyer, S., Lee, H.P., and Cuitino, A.M. (2011) A quantitative correlation of the effect of density distributions in roller-compacted ribbons on the mechanical properties of tablets using ultrasonics and X-ray tomography. *AAPS PharmSciTech*, **12** (3), 834–853.
[47] Malaterre, V., Ogorka, J., Loggia, N., and Gurny, R. (2009) Oral osmotically driven systems: 30 years of development and clinical use. *European Journal of Pharmaceutics and Biopharmaceutics*, **73** (3), 311–323.
[48] Verma, R.K., Arora, S., and Garg, S. (2004) Osmotic pumps in drug delivery. *Critical Reviews in Therapeutic Drug Carrier System*, **21** (6), 477–520.
[49] Kumar, P. and Mishra, B. (2007) An overview of recent patents on oral osmotic drug delivery systems. *Recent Patents on Drug Delivery & Formulation*, **1** (3), 236–255.
[50] Verma, R.K., Krishna, D.M., and Garg, S. (2002) Formulation aspects in the development of osmotically controlled oral drug delivery systems. *Journal of Controlled Release: Official Journal of The Controlled Release Society*, **79** (1/3), 7–27.
[51] Sawada, T., Sako, K., Fukui, M. *et al.* (2003) A new index, the core erosion ratio, of compression-coated timed-release tablets predicts the bioavailability of acetaminophen. *International Journal of Pharmaceutics*, **265** (1/2), 55–63.
[52] Sako, K., Sawada, T., Nakashima, H. *et al.* (2002) Influence of water soluble fillers in hydroxypropylmethylcellulose matrices on in vitro and in vivo drug release. *Journal of Controlled Release*, **81** (1/2), 165–172.
[53] Laity, P.R., Mantle, M.D., Gladden, L.F. and Cameron, R.E. (2010) Magnetic resonance imaging and X-ray microtomography studies of a gel-forming tablet formulation. *European Journal of Pharmaceutics and Biopharmaceutics*, **74** (1), 109–119.
[54] Metz, H. and Mader, K. (2008) Benchtop-NMR and MRI – a new analytical tool in drug delivery research. *International Journal of Pharmaceutics*, **364** (2), 170–175.
[55] Dorozynski, P.P., Kulinowski, P., Młynarczyk, A. and Stanisz G.J. (2012) Foundation review: MRI as a tool for evaluation of oral controlled release dosage forms. *Drug Discovery Today*, **17** (3/4), 110–123.
[56] Zeitler, J.A., Shen, Y., Baker, C. *et al.* (2007) Analysis of coating structures and interfaces in solid oral dosage forms by three dimensional terahertz pulsed imaging. *Journal of Pharmaceutical Science*, **96** (2), 330–340.
[57] Mantle, M.D. (2011) Quantitative magnetic resonance micro-imaging methods for pharmaceutical research. *International Journal of Pharmaceutics*, **417** (1/2), 173–195.
[58] Mikac, U., Kristl, J. and Baumgartner, S. (2011) Using quantitative magnetic resonance methods to understand better the gel-layer formation on polymer-matrix tablets. *Expert Opinion on Drug Delivery*, **8** (5), 677–692.
[59] Kimber, J.A., Kazarian, S.G. and Stepanek, F. (2011) Microstructure-based mathematical modelling and spectroscopic imaging of tablet dissolution. *Computers and Chemical Engineering*, **35** (7), 1328–1339.
[60] Lopes, R. and Betrouni, N. (2009) Fractal and multifractal analysis: a review. *Medical Image Analysis*, **13** (4), 634–649.
[61] Mandelbrot, B.B. (1985) Self-affine fractals and fractal dimension. *Physica Scripta*, **32**, 257–260.
[62] Mabilleau, G., Baslé, M.F. and Chappard, D. (2006) Evaluation of surface roughness of hydrogels by fractal texture analysis during swelling. *Langmuir*, **22** (10), 4843–4845.

[63] Jelcic, Z., Hauschild, K., Ogiermann, M. and Picker-Freyer, K.M. (2007) Evaluation of tablet formation of different lactoses by 3D modeling and fractal analysis. *Drug Development and Industrial Pharmacy*, **33** (4), 353–372.

[64] Fini, A., Ospitali, F., Zoppetti, G. and Puppini, N (2008) ATR/Raman and fractal haracterization of HPBCD/progesterone complex solid particles. *Pharmaceutical Research*, **25** (9), 2030–2040.

[65] Cavallari, C., Rodriguez, L., Albertini, B. *et al.* (2005) Thermal and fractal analysis microparticles obtained by of diclofenac/Gelucire 50/13 ultrasound-assisted atomization. *Journal of Pharmaceutical Sciences*, **94** (5), 1124–1134.

[66] Li, T. and Park, K. (1998) Fractal analysis of pharmaceutical particles by atomic force microscopy. *Pharmaceutical Research*, **15** (8), 1222–1232.

[67] Yin, X.Z., Li, H.Y., Liu, R.H. *et al.* (2013) Fractal structure determines controlled release kinetics of monolithic osmotic pump tablets. *Journal of Pharmacy and Pharmacology*, **65** (7), 953–959.

[68] Tahara, K., Yamamoto, K. and Nishihata, T. (1995) Overall mechanism behind matrix sustained-release (Sr) tablets prepared with hydroxypropyl methylcellulose-2910. *Journal of Controlled Release*, **35** (1), 59–66.

[69] Efentakis, M., Pagoni, I., Vlachou, M. and Avgoustakis, K. (2007) Dimensional changes, gel layer evolution and drug release studies in hydrophilic matrices loaded with drugs of different solubility. *International Journal of Pharmaceutics*, **339** (1/2), 66–75.

[70] Hoffman, A., Donbrow, M. and Benita, S. (1986) Direct measurements on individual microcapsule dissolution as a tool for determination of release mechanism. *Journal of Pharmacy and Pharmacology*, **38** (10), 764–766.

[71] Hoffman, A., Donbrow, M., Gross, S.T. *et al.* (1986) Fundamentals of release mechanism interpretation in multiparticulate systems-determination of substrate release from single microcapsules and relation between individual and ensemble release kinetics. *International Journal of Pharmaceutics*, **29** (2/3), 195–211.

[72] Benita, S., Babay, D., Hoffman, A. and Donbrow, M. (1988) Relation between individual and ensemble release kinetics of indomethacin from microspheres. *Pharmaceutical Research*, **5** (3), 178–182.

[73] Gross, S.T., Hoffman, A., Donbrow, M. and Benita, S. (1986) Fundamentals of release mechanism interpretation in multiparticulate systems – the prediction of the commonly observed release equations from statistical population-models for particle ensembles. *International Journal of Pharmaceutics*, **29** (2/3), 213–222.

[74] Borgquist, P., Zackrisson, G., Nilsson, B. and Axelsson, A. (2002) Simulation and parametric study of a film-coated controlled-release pharmaceutical. *Journal of Controlled Release*, **80** (1/3), 229–245.

[75] Borgquist, P., Nevsten, P., Nilsson, B. *et al.* (2004) Simulation of the release from a multiparticulate system validated by single pellet and dose release experiments. *Journal of Controlled Release*, **97** (3), 453–465.

[76] Sirotti, C., Colombo, I. and Grassi, M. (2002) Modelling of drug-release from poly-disperse microencapsulated spherical particles. *Journal of Microencapsulation*, **19** (5), 603–614.

[77] Sirotti, C., Coceani, N., Colombo, I. *et al.* (2002) Modeling of drug release from microemulsions: a peculiar case. *Journal of Membrane Science*, **204** (1/2), 401–412.

[78] Yang, S., Yin, X.Z., Wang, C.F. *et al.* (2014) Release behaviour of single pellets and internal fine 3D structural features co-define the in vitro drug release profile. *AAPS Journal*, **16** (4), 860–871.

[79] Broadbent, A.L., Fell, R.J., Codd, S.L. *et al.* (2010) Magnetic resonance imaging and relaxometry to study water transport mechanisms in a commercially available gastrointestinal therapeutic system (GITS) tablet. *International Journal of Pharmaceutics*, **397** (1/2), 27–35.

13

Physiologically Based Pharmacokinetic Modelling in Drug Delivery

Raj K. Singh Badhan

Medicines Research Unit, School of Life and Health Sciences, Aston University, UK

13.1 Introduction

The design and development of new drugs (and drug delivery systems) is a complex and costly process, with a significantly high level of attrition. It is estimated that from the 10 000 new compounds which are investigated during discovery phases only one is eventually granted approval by regulatory authorities, with an associated cost of over $1 billion [1]. These high rates of attrition and spiralling costs have sparked a resurgence in the application of pharmacokinetics modelling to better optimise drug discovery processes.

Drug metabolism and pharmacokinetics (DMPK) have an important role to play during drug discovery and development, particularly during lead optimisation and drug candidate identification. The key metrics of importance for progressing the candidate forward are its absorption, distribution, metabolism and excretion (ADME) characteristics and an early assessment of metabolic and pharmacokinetic (PK) properties [2, 3].

Over the past two decades, pharmacokinetic modelling and simulation has made a significant impact on the development process and helped reduce attrition, as a result of poor human pharmacokinetics, to about 10% [4] of all development failures, from 50% in 1990 [5]. Pharmacokinetic modelling is now well established as a tool to aid in ensuring

Computational Pharmaceutics: Application of Molecular Modeling in Drug Delivery, First Edition.
Edited by Defang Ouyang and Sean C. Smith.

efficiency and cost saving during the development stages, from target validation, ADME assessment and dose escalation to humans.

13.2 Modelling and Simulation Process

At the heart of it, any modelling process is essentially based around an investigation involving a large number of mathematical, statistical and numerical techniques, which are applied to a set of data, often in an exploratory analysis approach.

An important concept in modelling and simulation is the idea that all models are only approximations of any system they are designed to represent and that many different modelling approaches exist which differ in their strengths and limitations when applied to pharmacokinetic modelling and simulation and which often are dictated by the goals and outcomes of the investigation. Once a model is selected, driven by *a priori* knowledge, it inevitably undergoes further refinements until a convergence or goal is reached and therefore many preclinical models are considered as working models and are adapted or updated during the development process when more clinically relevant data exists.

13.3 Pharmacokinetic Principles

The clinical outcome of any therapeutic agent administered and delivered into the body is primarily governed by the pharmacokinetic properties of the active drug (and associated delivery system excipients). The clinical outcome is often associated with a concentration–response relationship, which is directed by the temporal magnitude of the therapeutic agent at the receptor active site.

As receptor active sites are often located intracellular and therapeutic agents are delivered to the systemic blood, the temporal concentration–response relationship is primarily one dictated by the concentration of the drug in the blood and the ability of the drug to partition into the intracellular active sites within tissues and organs. Thus, studying the processes which can alter the temporal blood concentrations of a therapeutics agent, will inevitably impact upon the rate and extent of the response (i.e. the clinical outcome) and pharmacokinetics, is a quantitative approach to understanding ADME processes (Figure 13.1).

13.3.1 Drug Absorption

The process of absorption, assuming the drug is complexly liberated from its carrier or delivery system, will depend significantly upon the route through which the formulation is delivered and is characterised by the appearance of the therapeutic agent within the systemic circulation.

Two commonly used routes exist to administered drugs into the body: parenteral routes [including intravenous (IV), intramuscular (IM), transdermal (TD) and subcutaneous (SC)] and enteral routes. All but IV administered drugs require an absorption phase prior to

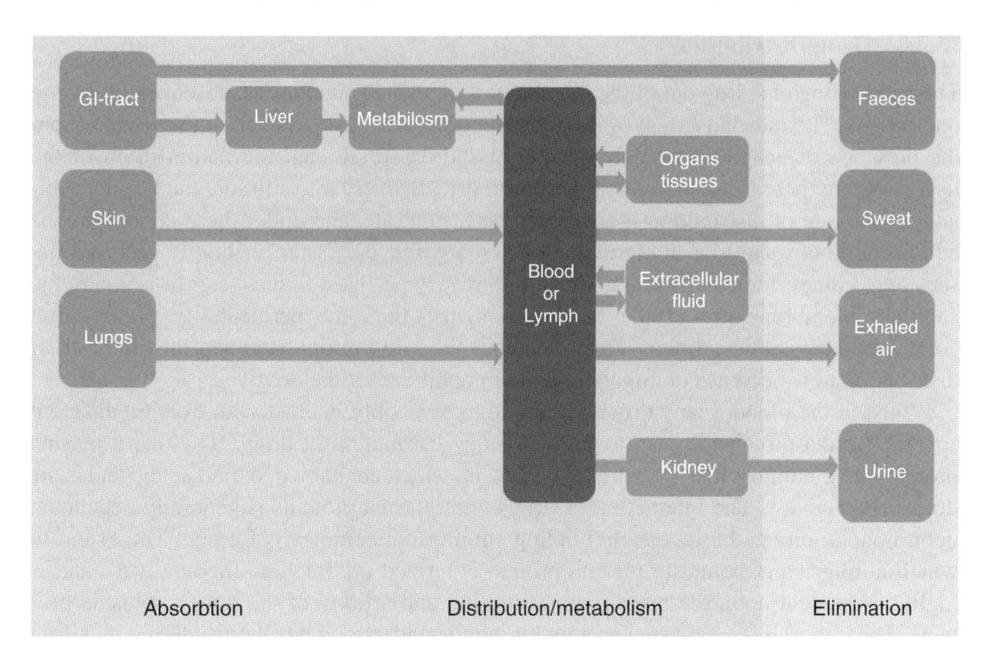

Figure 13.1 *Processes involved in absorption, distribution, metabolism and elimination (see colour plate section).*

delivery into the systemic circulation, whereby the agent will diffuse into local capillaries, which feed into the systemic circulation.

By far the most common route of administration is the oral route. Oral absorption is a complex process involving numerous steps before the therapeutic agent is able to access the systemic circulation. Before a drug can be absorption across the gastrointestinal (GI) tract mucosal wall, disintegration of the solid dosage form and dissolution to solubilise the drug particles into the gastric media must occur. Absorption across the GI wall occurs predominantly in the small intestine and is highly dependent upon the physicochemical properties of the drug (charge, molecular weight and lipophilicity), the formulation system and the physiology/anatomy of the subject.

The large surface area of the small intestine, a result of the increase in overall surface area due to the villi and microvilli of the intestinal enterocytes, makes it an ideal location for drug absorption. Despite this large surface area, drugs absorbed into intestinal enterocytes are frequently subjected to extra-hepatic metabolism (phase I and II) or active transport through membrane-localised transporters, which hinder overall flux across the intestinal lumen wall. Drugs that successfully pass through the intestinal enterocytes will enter the hepatic portal system and be subjected to metabolism in a process known as first-pass metabolism through a complement of drug metabolism enzymes, predominantly from the Cytochrome P450 family.

Generally speaking, we are interested in the rate of absorption as a measure for the speed of absorption from the dosage formulation and these are often summarised in terms such as the peak concentration (C_{max}) and time to reach peak concentration (T_{max}).

13.3.2 Drug Distribution

The partitioning of a drug out of the circulation is a prerequisite for the subsequent diffusion into tissues. The rate and extent of drug distribution is therefore related to tissue perfusion, the presence of anatomical barriers at the tissue organ site and the composition of the tissue. Where there is no anatomical barrier to drug delivery into a tissue, the key determinant of the rate of distribution is the tissue perfusion rate, and under these conditions the equilibrium between drug in blood and in tissue will occur faster for highly perfused tissues (e.g. kidney, lung and liver) compared to slowly perfused tissues (e.g. fat). Where the capillary membrane provides a barrier to distribution, the permeability of the drug compound across this cellular barrier becomes the rate-limiting step and thus the rate of distribution now becomes dominated by the lipophilicity of the drug.

Whilst in the blood, many drugs are prone to reversibly interact with over 60 different soluble plasma proteins known to influence the distribution of drugs. The major plasma proteins are albumin and α-1 acidic glycoprotein, which are known to bind acidic and basic drugs, respectively. The interaction of drugs with plasma proteins is generally dependent upon lipophilicity and a reversible binding equilibrium complex is formed. The extent to which a drug is not bound to plasma protein is termed the fraction unbound (fu_{p}) and is highly dependent upon the protein concentration and affinity of the drug to plasma proteins. The importance of plasma protein binding is understood when considering that drug diffusion out of blood capillaries and into target tissue/organs depends highly upon the bound or unbound state, with only unbound drugs being capable of passing across capillary walls and entering target tissues.

Thus, the primary effect of plasma protein binding on the pharmacokinetics of a drug is related to altering this distribution process, which is envisaged in the term volume of distribution (V_{D}), a term which can be thought of as reflecting the overall distribution of the drug in the body but which is more correctly considered as a proportionality constant between the concentration of a drug in the plasma and the amount of a drug.

As most drugs do not exert an effect in the systemic circulation, penetration into tissues/organs is essential for reaching target sites, and therefore the larger the value of V_{D}, the larger the extent of distribution of drug into tissues and the less drug there is in the blood circulation. In addition, the key factors controlling the extent of the V_{D} is essentially protein binding and the balance to which drug binds to plasma protein and nonspecific tissue binding, the equilibrium of which will dictate whether the overall balance resides in the tissue or in the blood.

13.3.3 Drug Metabolism and Elimination

The predominant sites of drug elimination from the body are the liver (metabolism and biliary excretion) and the kidneys (excretion), both of which are highly perfused tissues. The removal of drugs from the body is largely dependent upon its physicochemical properties, with highly lipophilic compounds predominantly undergoing metabolism and hydrophilic drugs compounds excreted unchanged in the kidneys.

The rate of biotransformation of a drug is an important factor in determining the availability of the drug in the systemic circulation and thus its eventual clinical efficacy. Biotransformation of drugs is the primary mechanism by which lipophilic drugs are eliminated from the body and is mediated through two pathways, phase I transformation of the parent compound to a more polar metabolite through the unmasking of functional groups and phase II conjugating reaction to form more hydrophilic end products capable of being renally excreted.

The most common phase 1 metabolic enzymes are the cytochrome P450 (CYP) enzyme family; and phase 2 enzymes consist mainly of transferases such as the UDP-glucuronosyltransferases (UGTs) and sulfotransferases (SULTs).

Drugs that have undergone phase I and II processes are subsequently eliminated from the body through either renal extraction via glomerular filtration and tubular secretion or through hepatic biliary excretion, with other routes (e.g. hair, saliva, tears) playing a smaller role, generally considered as unimportant.

13.4 Pharmacokinetic Modelling Approaches

The application of pharmacokinetic modelling within the drug development field essentially allows one to develop a quantitative description of the temporal behaviour of a compound of interest at a tissue/organ level, by identifying and defining relationships between a dose of a drug and dependent variables.

In order to understand and characterise the pharmacokinetics of a drug, it is often helpful to employ pharmacokinetic modelling using empirical or mechanistic approaches. In essence, such models are designed to represent a system (investigational species) with a primary goal of reproducing the structure and function of the system and relating this to temporal changes in drug concentration in the system.

13.4.1 Empirical (Classical Compartmental) Modelling

Classical empirical models are essentially derived from existing data describing the *in vivo* temporal concentration profile of drug blood. Many approaches exist to quantitatively describe kinetic systems but empirical and mechanistic approaches are commonly used in pharmacokinetics.

An empirical, or data-based, model aims to replicate existing observed data and therefore require very little or no *a priori* knowledge of the system under investigation but does require temporal drug concentrations in blood or plasma describing the event being modelled.

Using this *in vivo* plasma data as a starting point, a model is constructed which consists of a sufficient number of compartments needed to describe the ADME of the drug compound. Drug within each compartment is assumed to be instantly and homogenously distributed throughout.

Most empirical modelling approaches employ one-, two- or three-compartment models (Figure 13.2). In its simplest form, a one-compartment model assumes that drug is instantly and homogenously distributed within the compartment and this is often referred to as the central compartment. In a two-compartment model the assumption is made that drug can also distributed to a second compartment termed the peripheral compartment where the distribution to this compartment is defined by a rate. This can often be representative of distribution into different tissue types with the additional of third compartment also provided a defined distribution rate into a further tissue compartment.

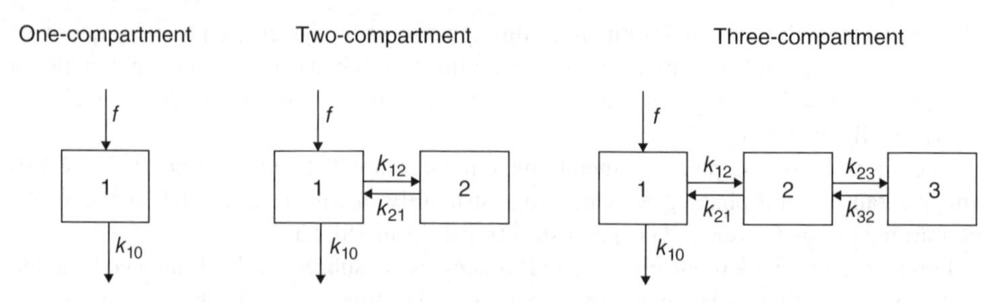

Figure 13.2 *Typical empirical compartmental models.*

Mathematically, such models are described by the sum of exponential terms defining events within each compartment (C_i: concentration in the *i*-th compartment) and inter-compartmental transfer of drug (k_i: transfer rate constant for the *i*-th compartment):

$$C_t = \sum C_i \cdot e^{-k_i \cdot t} \tag{13.1}$$

The actual choice of model is defined by *a priori* data and this approach is generally useful for interpolation of data and deriving terms such as the half life and clearance of drugs but lack refinement for extrapolation as the terms describing the kinetics within each compartment lack any true physiological meaning.

13.4.2 Noncompartmental Analysis

Prior to the advent of the personal computer and associated computational power, noncompartmental analysis (NCA) was the mainstay of all pharmacokinetics-based analysis of data. NCA is a model-independent approach to determine fundamental pharmacokinetic parameters and assumes (and requires) little or no prior knowledge of the system.

In such approaches, the time course of the drug is considered as a statistical distribution curve and the area under the plasma concentration–time curve (AUC), properly referred to as the area under the zero moment curve, is calculated. Furthermore, moment analyses has allowed the determination of key parameters such as clearance (*Cl*), mean residence time (MRT) and volume of distribution at steady state (V_{ss}). A classical example of the application of noncompartmental modelling is the bioequivalence study, where the rate and extent of drug absorption of a test and reference product evaluated by comparing the C_{max}, t_{max} and AUC.

13.4.3 Mechanistic (Physiological) Modelling

Mechanistic pharmacokinetic modelling, more often referred to as physiologically based pharmacokinetic (PBPK) modelling offers a more rationale approach to understanding the pharmacokinetics of a drug. The concept of PBPK modelling was introduced as far back at 1937 by Teorell [6], and the approach further developed and reported by Bischoff and Brown [7] and Dedrick [8].

The basis of this approach is grounded in the use of compartmental modelling approaches, but rather than using a lumping approach to reduce tissue compartments to one, two or three compartments, PBPK modelling explicitly describes all tissue and organs in the body

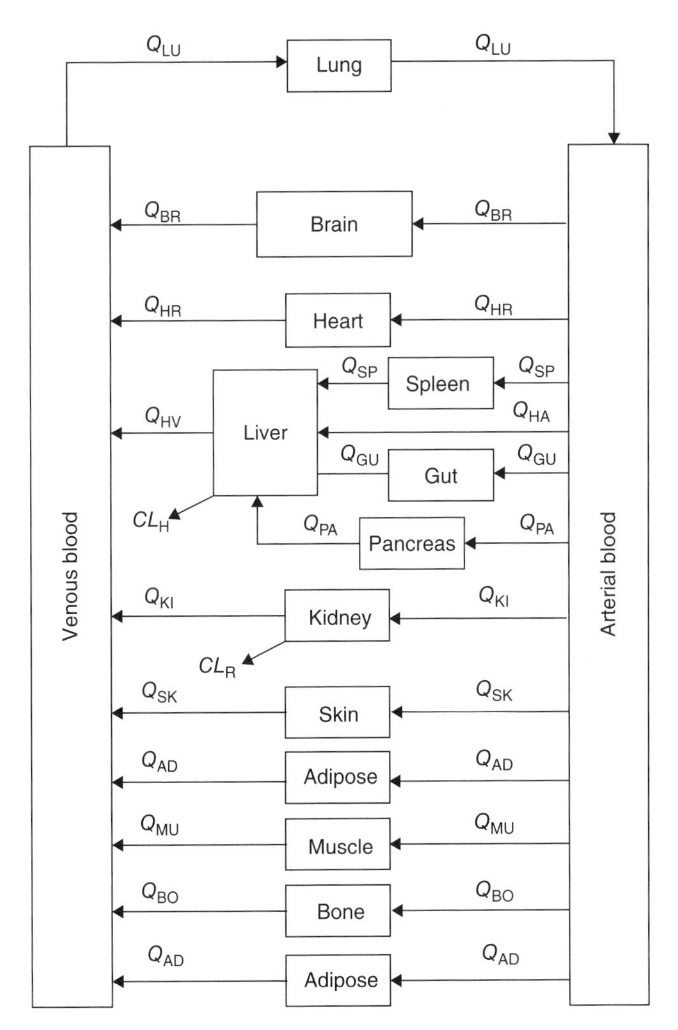

Figure 13.3 *A typical PBPK model.*

by compartments (Figure 13.3). Each compartment is assigned physiological volumes and interconnected through physiological perfusion rates. Drug entry into tissues comes from the arterial blood and returns to the heart through the venous blood. Organs with specific clearance routes (e.g. liver, kidney) include specific terms for the clearance of drugs.

Drug entry into tissue can be assumed to be limited by either blood flow (perfusion rate-limited) or ability to permeate into tissues (permeability rate-limited). For most lipophilic drugs the assumption is made that entry into tissues is solely limited by perfusion to that tissue. Whereas if *a priori* data suggests a permeability limitation to the partitioning of the drug into tissue, one can describe uptake into the tissue through a permeability-limited model which explicitly accounts for flux into the tissue through an extrapolated permeability metric. Furthermore, where metabolism is known to impact upon the ADME of the drug, the relevant metabolising enzyme activities can be incorporated into a tissue or organ within the model.

The goal of any PBPK modelling approaches is therefore creating a balance between parsimony and biological realism, development of a suitably minimal model structure, which adequately describes the pharmacokinetics of the drug to be studied.

Despite the advantages PBPK modelling has to offer over more traditional compartment/noncompartmental approaches, its application has only recently become realised with widespread application in drug development and academia, which has primarily been driven by new approaches to determine fundamental operational parameters such as the tissue to plasma partition coefficient (Kp_t) [9, 10] and approaches to carry out robust *in vitro* to *in vivo* extrapolation of drug permeability and metabolic clearance terms.

13.5 Pharmacokinetic Software for Modelling

With a recent resurgence in the application of PBPK in research and pharmaceutical industry, numerous commercial platforms are available with which to develop pharmacokinetic models. Pharmacokinetic models can be developed within mathematical and statistical commercial software such as MATLAB (MathWorks, Inc., Natick, Mass., USA) [11] using traditional mathematical and computation coding, or by using the Simbiology Toolbox available within MATLAB for a graphical user interface approach to developing PBPK models. Freely available alternatives include using the statistical program R (R Development Core Team) [12].

A number of commercial and specialist pharmacokinetic modelling software packages are also available and include GastroPlus (Simulations Plus: www.simulations-plus.com) and SimCYP (Simcyp Ltd: www.simcyp.com) and for mechanistic oral absorption and population-based modelling; Pheonix WinNonlin (Certara: www.certara.com) for PK/PD and NCA; PK-Sim (Bayer: www.systems-biology.org) for oral absorption, mechanistic and systems-based modelling.

13.6 Developing a PBPK Model for an Orally Dosed Compound

For the preclinical scientist, PBPK modelling is an invaluable tool with which to integrate a wide variety of routinely generated *in vitro* compound-specific data into systems, which will allow the prediction of whole-body temporal tissue and cellular concentrations.

The development of a PBPK model begins with a clear understanding of the question or problem in hand, which will significantly guide the design of the model structure and complexity. Thus the level of detail required from the model is primarily dictated by the kinetic events within compartments and the type of chemicals tracking within the model, such as metabolites in addition to any parent compound.

13.6.1 Conceptualisation of a PBPK Model Structure

A key component of a PBPK is visualisation of the overall elements of the model. This diagrammatic representation of the model system is helpful in this respect and often visualises key physiological elements controlling the ADME of a compound. The basis of the model is compartments for each of the key tissues/organs under investigation. The level of detail in the inclusion of compartments and tissues can be driven by an *a priori* understanding of the key tissue sites for an ADME effect. However some key elements must be considered as a minimum structural level to ensure correct consideration of ADME properties:

1. Administration site of drug – often the blood, gastrointestinal, skin or lung compartments;
2. Metabolism and excretion sites of drug – mainly liver and kidney;
3. Storage sites of drugs – often the adipose or a bone site.

To aid in the selection of which tissues to incorporate into the model, a well established generic 'whole-body' physiologically based pharmacokinetic (WB-PBPK) model is widely used in mechanistic pharmacokinetic modelling (Figure 13.3) [13–20]. The WB-PBPK model provides an excellent balance between model complexity and simplicity, which is generally considered to be biologically and physiologically plausible.

A WB-PBPK model is typically comprised of a total of 14 compartments, which include adipose, bone, brain, gut, heart, kidney, liver, lung, muscle, pancreas, skin, stomach, spleen and thymus. Each compartment has a primary assumption of being a single homogenous representation of the tissue or organs. The drug is assumed to be instantaneously and homogeneously distributed within the compartment, with perfusion providing the only limitation to the rate of uptake into the compartment. However, for dosing tissues (e.g. small intestine/gut) or target tissues (e.g. brain) it is often common to assume drug uptake into the tissue can be limited by the permeability across the cell membrane and these are often termed permeability-limited. Model complexity is often increased when there is a need to describe processes relevant to ADME, such as dissolution and solubility (for absorption), subcellular distribution within a tissue (distribution) and drug metabolism by metabolic enzymes (metabolism).

Once a model structure has been chosen, each tissue compartment is then parameterised: provided with a physiological tissue volume, allocated a calculated tissue partition coefficient and interconnected through a systemic circulation which is comprised of both arterial and venous blood supplies. Furthermore, the WB-PBPK model may act as an initial starting point for model development strategies and individual compartments may be lumped together to reduce structural complexity if there is limited to no appreciable change in simulated pharmacokinetic profiles.

13.6.2 Parameterising the Model with Model Descriptors: Systems Data

In parameterising a PBPK model, it is essential to populate the model with all relevant systems data (physiological and biological) describing the species model system and all compound-specific data (physicochemical, permeability and metabolism; Figure 13.4). Systems-based data for commonly used species (rodents, dogs, primates and humans) are readily available [21–24] and provide an established starting point for model optimisation. Furthermore, biochemical data such as enzyme and transporter abundances for certain tissues such as the GI-tract [25–31] and brain [32–35] are now becoming available to the wider scientific community.

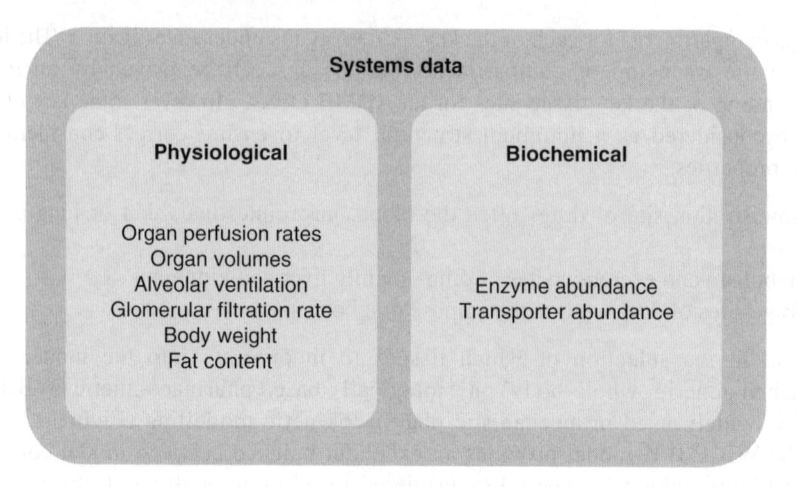

Figure 13.4 *System input parameters for a PBPK model.*

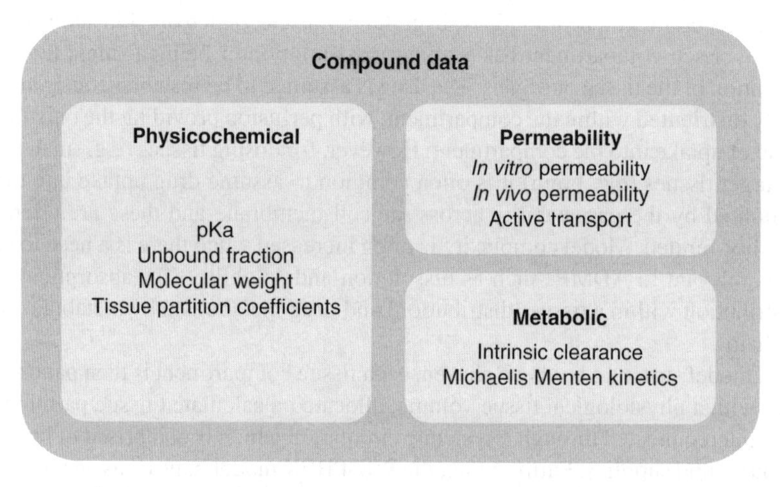

Figure 13.5 *Compound input parameters for a PBPK model.*

13.6.3 Parameterising the Model with Model Descriptors: Compound-Specific Data

Compound-specific data is often categorised as either being physicochemical or in describing specific properties of the compounds such as its *in vitro* or *in vivo* permeability across a cell membrane or its interaction with drug-metabolising enzymes (Figure 13.5).

Physicochemical data, and to some extent permeability data, can be sourced from publically available databases or predicted using specialist ADME software (GastroPlus, Simulations Plus: www.simulations-plus.com; SimCYP, Simcyp Limited: www.simcyp.com; QikProp, Schrondinger: http://www.schrodinger.com; VolSurf+, Molecular Discovery: http://www.moldiscovery.com), provided that a minimal level of compound-specific data is known (e.g. LogP and pKa).

Alternatively, certain compound-specific data has been successfully predicted using mechanistic and quantitative structure–activity relationship (QSAR) approaches and can be readily incorporated into PBPK modelling strategies. Of these, the determination of *in vivo* drug clearance (metabolic or renal), permeability and tissue partitioning are key compound-specific input parameters that are often derived from *in vitro* data and mechanistic approaches.

13.6.4 Orally Dosed Formulations

For formulations dosed orally, a prerequisite for clinical activity is the entry of the drug into the systemic circulation. The primary barrier to ensuring systemic delivery is passage across the small intestine or gut compartment in any WB-PBPK model. For any orally dosed compounds, the oral bioavailability of the drug in the systemic circulation (F_o) is dependent on the fraction absorbed across the intestinal wall (F_a), the fraction escaping the intestinal enterocytes (F_g) and the fraction escaping the liver (F_h; Figure 13.6) where the product of all three yields the overall oral bioavailability.

For compounds that are orally dosed, a detailed small intestine model can be developed to model both transit and absorption *aborally*, which can then be implemented in any WB-PBPK model replacing the single homogenous gut compartment with a more advanced oral absorption model. The section below describes elements of this development process.

13.6.5 Modelling Drug Dissolution

Unless dosed in a liquid form, the processes of disintegration and dissolution are key prerequisites to the release of drug molecules and their subsequent absorption. The formulation system acts as a carrier for drug molecules and therefore the liberation of drug molecules

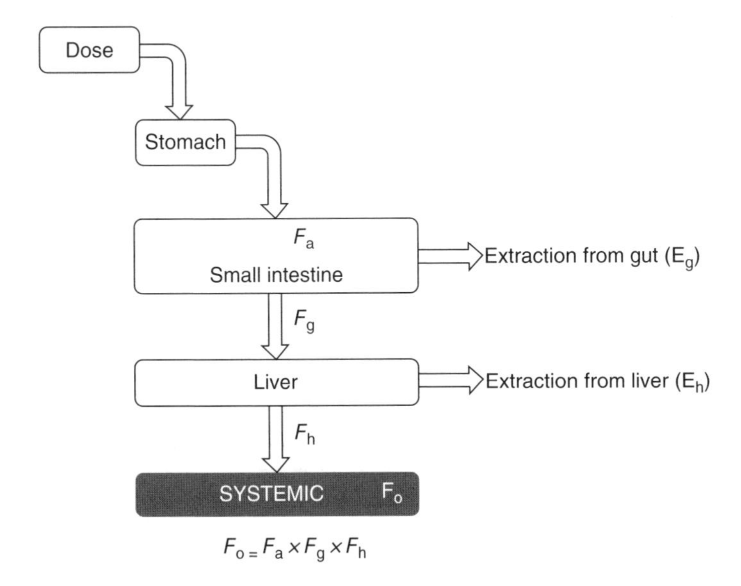

Figure 13.6 *Pathways for calculating oral absorption.*

from this carrier is essential. Both the solubility (related to the physicochemical properties of the drug) and the dissolution rate (related to the carrier system formulation) of the drug particles from the bulk formulation dictates the rate of dissolution.

Mathematical modelling of dissolution is an essential component of any oral absorption model to enable successful prediction of oral absorption when dealing with a solid dosage forms. Noyes and Whitney [36] first described a mathematical approach to model *in vitro* dissolution using:

$$\frac{dC}{dt} = k \cdot (C_s - C) \tag{13.2}$$

where C_s is the solubility concentration of the drug, C is the concentration of drug in bulk media and k is the dissolution rate constant. This was further adapted by Bruner and von Tolloczki [37] to include a description of surface area (A) of the drug particles:

$$\frac{dC}{dt} = k \cdot A \cdot (C_s - C) \tag{13.3}$$

and finally adapted by Nerst [38] and Brunner [39] to produce the Nerst–Brunner equation:

$$\frac{dC}{dt} = \frac{D \cdot A}{V \cdot h}(C_s - C) \tag{13.4}$$

where D is the diffusion coefficient, A is the surface area of the drug particles, V is the volume of the media, h is the thickness of the diffusion layer, C_s is the saturation concentration or saturation solubility of the drug particles and C is the concentration of drug in the bulb media (Figure 13.7).

This approach has its basis in the combination of both a diffusion layer model and an application of Fick's second law [40]. Many of the input parameters in Equation 13.4 are measurable, however the diffusion coefficient is often determined from the Stokes–Einstein equation [41] or the Hayduk–Laudie equations [42].

Although both the Noyes–Whitney and the Nerst–Brunner equations are widely used for *in vitro* dissolution testing, other empirical models have found favour in modelling *in vitro* dissolution, the most popular of which is the first-order Weibull equation:

$$m = 1 - \exp\left[\frac{-(t-T)^b}{a}\right] \tag{13.5}$$

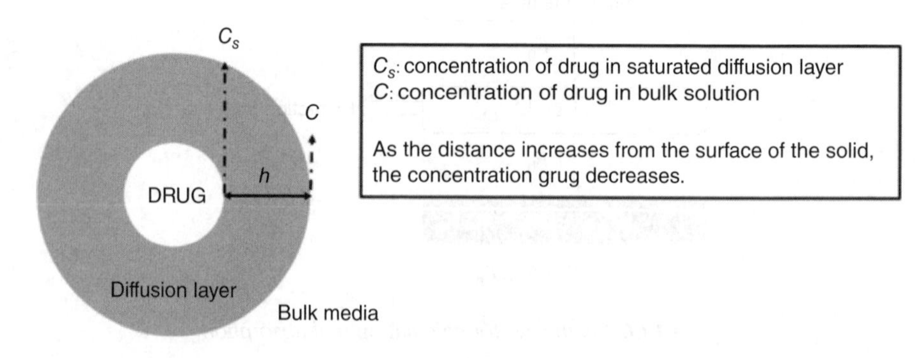
C_S
C
h
DRUG
Diffusion layer
Bulk media
C_S: concentration of drug in saturated diffusion layer
C: concentration of drug in bulk solution
As the distance increases from the surface of the solid, the concentration grug decreases.

Figure 13.7 *Diffusion layer.*

where t is time, T is a lag time, a is a scaling factor and b is a shape constant. The Weibull equation is often utilised by fitting experimental data to the model equation.

13.6.6 Modelling Drug Permeability

Having described the process of disintegration and dissolution of a drug from its dosage form, a mechanistic description of the permeability of the drug compound across the gut lumen wall is necessary.

The permeation of drug compounds across biological membranes occurs through two main pathways, namely paracellular or transcellular processes. Paracellular pathways involve the movement of drug, typically small hydrophilic compounds, through small aqueous channels within the membrane. Transcellular pathways are open to hydrophobic/lipophilic compounds of small molecular weight and involve the diffusion of drug molecules into and through the lipid bilayer.

The flux (J) of a compound across the intestinal mucosa is often described by its relationship with the intestinal effective permeability (P_{eff}), surface area (S) and a concentration gradient established across the mucosa:

$$J = P_{eff} \times S \times \Delta C \tag{13.6}$$

The intestinal effective permeability for compounds undergoing passive transcellular diffusion represents the passive passage of drug across the apical membrane of intestinal enterocytes.

The experimental determination of P_{eff} is normally carried out using an intestinal perfusion technique, which measures the loss of a perfused from an isolated segment of the jejunum [43]. In humans, the direct determination of P_{eff} has been carried out for 34 compounds (30 drugs in the intestine, two drugs and two compounds in the rectum) [43–50].

During preclinical screening, it is not routinely possible to determine P_{eff}, but rather an analogous *in vitro* permeability flux, the apparent permeability (P_{app}), is determined in cell culture based permeability studies using either human epithelial colorectal adenocarcinoma cells (Caco-2) or Madin–Darby canine kidney (MDCK-II) studies. Furthermore, the derivation of linear correlations established from *in vitro* P_{app} and literature reported *in vivo* P_{eff} are commonly used to yield an *in vivo* permeability metric.

The Caco-2 cell culture model is the most commonly used model to determine parameters to extrapolate to human P_{eff} [51], primarily because of their similar transport protein expression profile to human intestine [52], despite being colonic in origin. A limitation of this model is the variation in protein abundance compared to human intestinal cells and their widespread laboratory to laboratory variability in transport activities [53]. Similar correlations have been reported for the MDCK model and associated MDR1-transfected subclone (MDCK-MDR1) [54, 55].

Noncell-based approaches to determine an *in vitro* permeability metric derived from the use of an artificial membrane such as that employed in the parallel artificial membrane permeation assay (PAMPA) for determination of the passive permeability of compounds [56, 57].

In silico approaches have been employed to directly predict human P_{eff} and primarily employ the use of QSAR between the molecular descriptors of the compound and known human P_{eff} values (Equation 13.7 [58]) or Caco-2 derived P_{app} (Equation 13.8 [59]):

$$\mathrm{Log}\left(P_{eff}\right) = 4 - 2.546 - 0.11PSA - 0.278HBD \tag{13.7}$$

$$\mathrm{Log}\left(P_{eff}\right) = 0.6532\mathrm{Log}\left(P_{app}\right) - 0.3036 \tag{13.8}$$

Providing an *in vitro* intestinal permeability metric has been determined, the incorporation of drug flux across the intestinal mucosa can be modelled in PBPK models by the use of a permeability surface (PS) area term which often takes the units of volume/time and is analogous in units to a clearance term. The PS area metric can be calculated by correcting P_{app} or P_{eff} (units of $cm\,s^{-1}$) with the intestinal surface area, $200\,m^2$ for P_{app} or $0.66\,m^2$ for P_{eff} [60]. The surface area correction accounts for the threefold increases provided by the ridges around the lumen (values of Kerckring), a further four- totenfold expansion produced by the villi structures and a further 10- to 30-fold expansion proved by the microvilli [61]. The surface area magnification provided by the Kerkering and associated villi and microvilli are explicitly accounted for when using P_{eff} and therefore the final surface area corrections simply account for the planer tubular surface area only.

Alternative approaches to determine PS often utilise empirically correlated *in vitro* permeability to *in vivo* permeability relationships to predict P_{eff} which is subsequently scaled as described above. Furthermore, it is also possible to utilise P_{eff} to determine an absorption first-order rate constant (k_a) based on correcting for the surface/volume ratios of a cylinder:

$$k_a = \frac{2 \cdot \pi \cdot r}{\pi \cdot r^2 \cdot l} \cdot P_{eff} = \frac{2}{r} \cdot P_{eff} \tag{13.9}$$

where r and l represent the radius and length of the small intestine and the volume of a cylinder. This mechanistic approach often holds valid for drug compounds that undergo passive absorption with no active efflux (or uptake) element. However, the small intestine is known to express a wide variety of membrane-bound drug active transporter proteins which can play a significant role in altering the flux across the intestinal enterocytes. Typically the Caco-2 permeability assay will identify whether a drug is undergoing some form of active transport and subsequent studies are conducted to assess the specific transporter protein(s) responsible for the transport phenomena.

Cellular data derived from the interaction of a drug with transporter proteins should, ideally, be in the form of a Michaelis–Menten kinetic with determination of maximum velocity (V_{max}) and Michaelis constant (K_m). In reality, the determination of specific transporter elements in traditional cell culture model systems (e.g. Caco-2 or MDCK-II) is complicated by the multiplicity of transporter proteins expressed. Novel systems that may overcome this include insect-derived membrane fractions from transporter overexpressed cell lines which will allow the specific determination of K_m and V_{max} for a specific transporter pathway.

The mathematical description of active transporter compounds is similar to that for enzyme kinetics, the key feature being the exhibition of saturable kinetics. The predicted flux of the drug transporter ($J_{transporter}$) can be defined as:

$$J_{transporter} = \frac{J_{max} \cdot C_{lumen}}{K_m + C_{lumen}} \tag{13.10}$$

In the absence of kinetic data it is still possible to incorporate transporter pathways based on solely *in vitro* permeability data by including a description for the efflux ratio (ER), the ratio of drug being transported under efflux compared to passive mechanisms. Whilst lacking any concentration-dependent saturation, this term can be utilised to model the equilibrium of passive versus active transport of drug throughout the small intestine [62].

13.6.7 Modelling Drug Metabolism

The clearance of a drug from the body is an important pharmacokinetic factor that can be used to develop dosing regimens and assess the requirement for dose adjustments during clinical trials. The determination of clearance in humans is usually conducted from *in vitro* experiments and then scaled to yield an *in vivo* clearance term which can be used within the pharmacokinetic modelling strategies.

In this approach, the rate at which drug is removed from the blood by the liver (termed the intrinsic clearance) is measured *in vitro* from enzyme kinetic data such as V_{max} and K_m or from the half life from *in vitro* assays, which employ either intact cell systems (e.g. primary hepatocytes), subcellular fractions (microsomes) or transfected cell lines (MDCK and HEK cells). The choice of system to use is highly dependent upon the route of metabolism (Cytochrome P450 mediated or UGTs) and the specific isozyme(s) involved in the metabolic reaction.

A three-step approach is often utilised to scale an *in vitro* clearance to a human *in vivo* clearance (Figure 13.8). An *in vitro* intrinsic clearance (CL_{int}) is determined from substrate depletion assays. Once normalised for protein concentration and corrected for nonspecific binding, CL_{int} is now referred to as an in vitro unbound intrinsic clearance ($CL_{int,u}$) and is scaled to a human *in vivo* clearance by correcting for microsomal recovery (40 mg g^{-1} liver and 20.6 mg g^{-1} intestine) [63, 64] hepatocellualrity (120×10^6 cells g^{-1} liver) [65, 66] and finally liver weight to yield an *in vivo* intrinsic clearance ($CL_{int,u}$). This scaled term can now be used within a well defined liver model, typically the well stirred model [67, 68] after

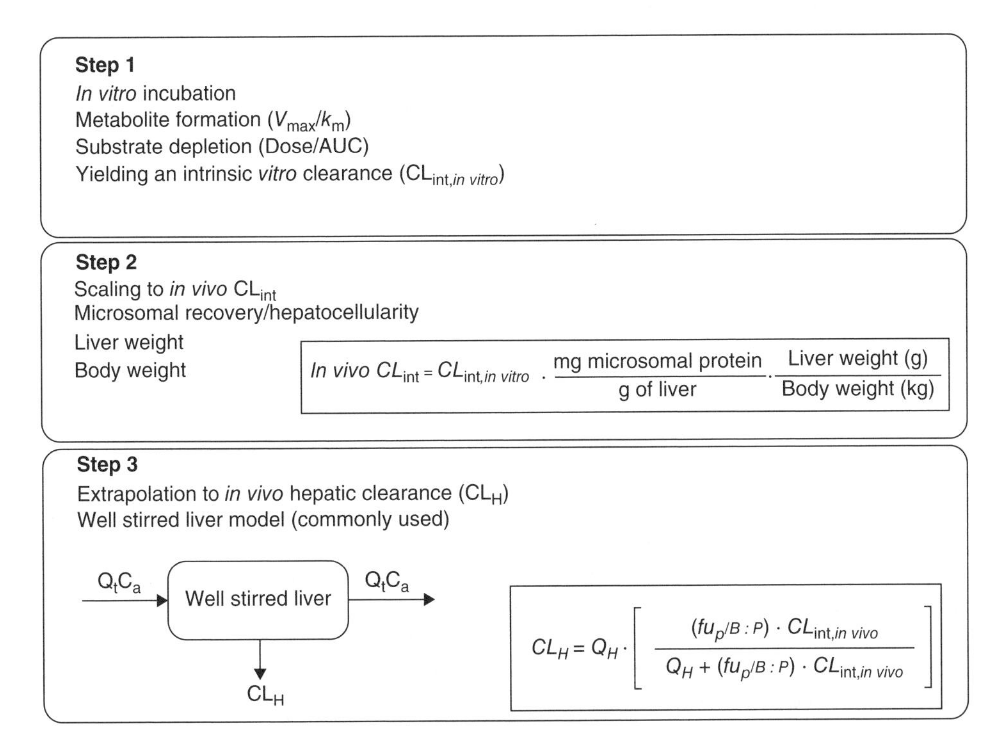

Figure 13.8 *Three-step approach to predicting* in vivo *hepatic clearance.*

correction for the blood to plasma ratio (B:P) and finally yielding a human hepatic clearance reflecting the total hepatic clearance of drug in humans.

13.6.8 Modelling Renal Clearance

Although much attention has been placed on the liver as a major clearance pathway for drugs, the kidneys are an important pathway in the final elimination of drugs from the body. Despite this, there is very little consensus on the approaches available to predict renal clearance. Current approaches are often based on extrapolation of an *in vivo* renal clearance, measured in a lower species (e.g. rodents) and using the glomerular filtration rate (GFR) to determine a human *in vivo* renal clearance [69].

More mechanistically driven models of kidney elimination, which incorporate glomerular filtration, reabsorption and secretion with the inclusion of passive and active secretion/reabsorption processes are currently lacking but progress has been made in a recent report by Neuhoff *et al.* [70] with the development of a mechanistic kidney model (Mech KiM).

13.6.9 Modelling Drug–Tissue Partitioning

The distribution of drugs from the circulatory systems into tissues is highly dependent on the perfusion rate of a drug to a tissue and the rate of uptake into the tissue. The partitioning process is influenced by the composition of the tissue, with hydrophilic drugs having a tendency to partition into organs with a high water content, such as muscle, while lipophilic drugs partition into fat-rich organs such as brain, liver and adipose.

Tissue partitioning can be described by the tissue partition coefficient (K_p) and determined from *a priori*, *in vitro* or *in vivo* data for tissue and plasma concentration at equilibrium:

$$\frac{C_{\text{tissue}}}{C_{\text{plasma}}} = K_p \tag{13.11}$$

or from *in vitro* equilibrium dialysis determination of fu_p and fu_t:

$$\frac{fu_t}{fu_p} = K_p \tag{13.12}$$

Practically, it is difficult and costly to determine the tissue affinity of drugs, but prediction approaches exist which utilise *in vitro* and *in silico* approaches to mechanistically determine K_p and this has allowed PBPK modelling, particularly during preclinical screening, to be widely used as it provides an approach to predict the partitioning of a drug into commonly used tissues within a PBPK model without the need for animal experimentation. These approaches are based on mechanistic equations first developed by Pouil and Theil [71] and Rodgers and Rowland [72–74] which are capable of predicting the steady-state tissue to plasma water unbound drug concentration ratio (K_{pu}) for moderate to strong bases and for acids, very weak bases, neutrals and zwitterion drugs. Interconversion between K_{pu} and K_p is possible by inclusion of the plasma-unbound fraction (fu_p):

$$K_{pu} = \frac{K_p}{fu_p} \tag{13.13}$$

Within PBPK modelling, it is normal to employ K_{pu} as opposed to K_p as typically it is the unbound/free drug which is able to partition into tissues.

The prediction of drug tissue partitioning requires only the pKa, LogP, blood to plasma partitioning (*Rb*) and fu_p of the drug. It is extensively based on an expression of tissue compositions and specifically accounts for the distribution of drugs into tissues (homogenously and passively) through drug–lipid nonspecific binding and reversible binding to plasma proteins and tissue proteins.

The equations developed by Rodgers and Rowland (Figure 13.9) are now widely employed and consider a tissue to be composed of neutral lipids (NLs), neutral phospholipids (NPLs)

Mechanistic tissue partition equations

One basic pka≥7

$$kpu = \left[\left(\frac{1+X.\ fiw}{1+Y}\right) + f_{EW} + \left(\frac{Ka_{AP} \cdot [AP]_T \cdot X}{1+Y}\right) + \frac{P \cdot fnl + (0.3P + 0.7) \cdot fnp}{1+Y}\right]$$

Other types

$$kpu = \left[\left(\frac{1+X.\ fiw}{1+Y}\right) + f_{EW} + (Ka_{PR} \cdot [PR]_T) + \frac{P \cdot fnl + (0.3P + 0.7) \cdot fnp}{1+Y}\right]$$

Affinity constants

Acidic phospholipids
Extracellular albumin

$$Ka_{AP} = \left[KpuBC - \left(\frac{1+Z}{1+Y}.fiw_{,BC}\right) - \left(\frac{P \cdot fnl_{,BC} + (0.3P + 0.7) \cdot fnp_{,BC}}{1+Y}\right)\right] \cdot \left(\frac{1+Y}{[AP]_{BC} \cdot Z}\right)$$

Lipoproteins

$$Ka_{PR} = \left[\frac{1}{fu} - 1 - \left(\frac{P \cdot fnl_{,p} + (0.3P + 0.7) \cdot fnp_{,p}}{1+Y}\right)\right] \cdot \left(\frac{1}{[PR]_p}\right)$$

Drug ionisation

	X	Y	Z
Monoprotic base	$10^{pKa-pH_{iw}}$	10^{pKa-pH_p}	$10^{pKa-pH_{BC}}$
Diprotic base	$10^{pKa2-pH_{iw}} + 10^{pKa1+pKa2-2pH_{iw}}$	$10^{pKa2-pH_p} + 10^{pKa1+pKa2-2pH_p}$	$10^{pKa2-pH_{BC}} + 10^{pKa1+pKa2-2ph_{BC}}$
Monoprotic acid	$10^{pH_{iw}-pKa}$	10^{pH_p-pKa}	NA
Diprotic acid	$10^{pH_{iw}-pKa1} + 10^{2pH_{iw}-pKal-pKa2}$	$10^{pH_p-pKa1} + 10^{2pH_p-pKa1-pKa2}$	NA
Zwitterion	$10^{pKa_{BASE-pH_{iw}}} + 10^{pH_{iw}-pKa_{ACID}}$	$10^{pKa_{BASE}-} + 10^{pH_p-pKa_{ACID}}$	$10^{pKa_{BASE}-pH_{BC}} + 10^{pH_{BC}-pKa_{ACID}}$
Neutral	0	0	NA

P is the n-octanol:water partition coefficient for unionised compound for all tissues except adipose (vegetable oil:water)
f is the fractional tissue volume
IW and **EW** refer to intra- and extra-cellular tissue water
NL and **NP** refer to tissue neutral lipids and neutral phospholipids
$\mathbf{AP_T}$ and $\mathbf{PR_T}$ refer to the tissue concentrations of acidic phispholipids and extra-cellular albumin (for acids and weak bases) or lipoprotein (for neutrals), respectively
$\mathbf{Ka_{AP}}$ and $\mathbf{KA_{PR}}$ are affinity constants of the drug for acidic phospholipids and either extra-cellular albumin or lipoprotein, respectively;
X,Y and **Z** account for drug ionisation
Subscripts **BC** and **P** refer to red blood cells and plasma
$\mathbf{pH_{BC}}$ is the intracellular pH of blood cells
fu is fraction unbound in plasma
$\mathbf{Kpu_{BC}}$ can be calculated from the fu and blood-to-plasma ratio

Figure 13.9 *Calculating tissue partition coefficients.*

and acidic phospholipids (APLs) with acidic drugs binding to albumin and basic drugs to α-1 acid glycoprotein (AGP) which is restricted to plasma whereas albumin is found is extracellular fluids within tissues.

Thus, the concentration of drugs in tissues was proposed as a mechanistic expression, which considered the sum of drug in extra- and intracellular waters, amounts found in intracellular regions (NL, NPL and APL) and knowledge of the tissue volumes and fractional tissue volumes of cellular components.

13.7 Developing the Model

Once a model structure has been selected and the required compound and system data obtained, the model itself is constructed using a series of linked ordinary differential equations (ODEs) to describe the kinetic events within each compartment (rate of change of drug, dC/dt).

Two choices currently exist to describe the kinetic events within each tissue, and these are defined by the rate limitations on the delivery of drug into the tissue. The first is termed the perfusion-limited model, where there is no inherent permeability barrier to the partitioning of drug into tissue (typically for highly lipophilic compounds). In this approach the rate limitation is solely dependent upon perfusion into that tissue (Figure 13.10).

The second is termed the permeability-limited model, where the partitioning of drug into the tissue is highly dependent upon the ability of the drug to enter the interstitial fluid, which is governed by a permeability barrier and cellular flux (Figure 13.10).

Each compartment is defined by a volume (V_t), which can be obtained from published physiological and anatomical sources [21–24]. Transfer of drug from one compartment to

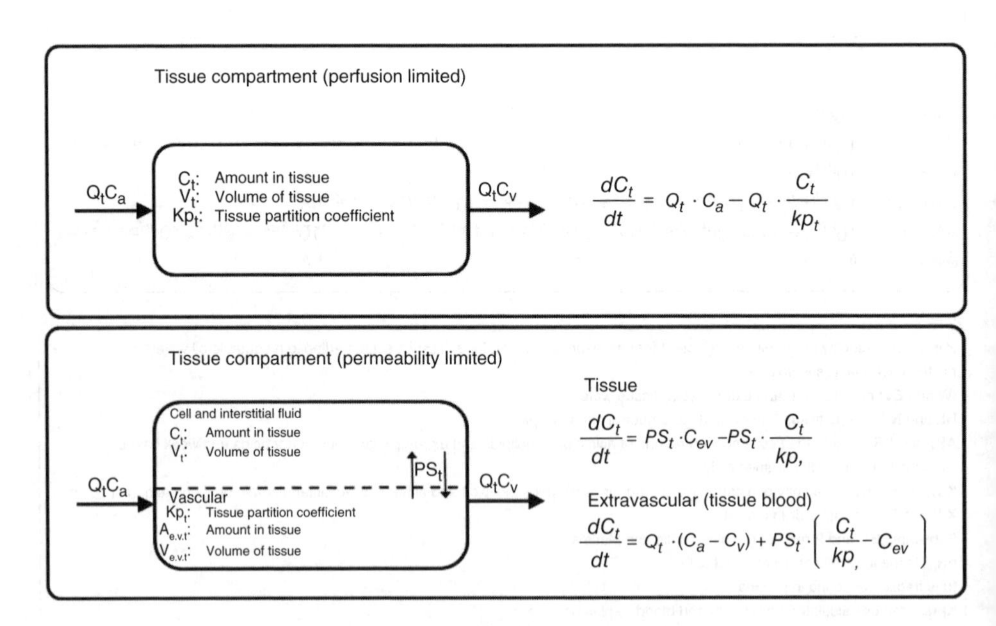

Figure 13.10 Kinetics within tissue compartments.

another is defined by a flow rate (Q_t). Partitioning of drug into tissue compartments is defined by the unbound fraction in plasma (fu_p), the tissue partition coefficient of the drug (K_{pt}) and a permeability–surface area metric (PS_t), which can determined from knowledge of the *in vitro* permeability (P_{app}) and corrected for the surface area available for absorption at the tissue site.

It is typically assumed that perfusion rate-limited kinetics apply to most drugs that are being simulated and the primary routes of elimination are the liver and kidneys.

13.7.1 Physiological Considerations

Any pharmacokinetic model, whether developed in-house or using commercial software, requires that the process of oral absorption modelling accounts for the physiology of stomach and small intestine.

13.7.1.1 Gastric Release and Small Intestine Transit Time

Gastric emptying is perhaps the first hurdle a drug formulation faces in the process of achieving systemic absorption. Gastric emptying is affected by the composition of the ingested material, which can hinder the release of the stomach content into the duodenum. Liquids are emptied from the stomach faster than solids and usually begin within 1–3 min following ingestion of a liquid and 2–3 h for digestion [75].

Upon release of the stomach content into the proximal small intestine, the propulsive and mixing movements purge the stomach contents along the three major regions of making up the small intestine, namely the duodenum, jejunum and ileum. Mean transit times along the small intestine are typically between 3 and 4 h [75, 76] and this has been supported by Yu *et al.* [77] who collected and compiled data from a total of over 400 human small intestinal transit time data from studies performed on fast/fed states with different pharmaceutical dosage forms with a mean transit time of 199 min, which is generally independent of fasted/fed states, gender, age and body weight.

13.7.1.2 Geometry and pH Variation

The small intestine can be generally considered to take the form of a conical tube with a length of approximately 280 cm and a decreasing radius from 1.75 to 1.00 cm from proximal to distal regions [78]. Knowing the lengths of the duodenum, jejunum and ileum are 22.4, 104 and 154 cm respectively [75] it is possible to then describe the transit through each segment.

The pH of the small intestine is also variable and typically has a mean of 6.63 (SD: ±0.53) in the proximal (jejujum) and 7.49 (SD: ±0.46) in the distal (ileum) regions [79]. Consideration of these pH variations would be important in dissolution-limited formulation systems. The ingestion of food also alters the pH and this should also be considered when mechanistically modelling the stomach and small intestine [stomach (pH 1.7 and 6.4) and duodenum (pH 6.2 and 6.6) reflecting fasted and fed states].

13.7.1.3 Regional Variation in Drug-Metabolising Enzymes

The gut enterocytes are the primary functional cellular unit governing the oral bioavailability or drug compounds in the small intestine. The enterocytes are terminally differentiated columnar cells which form a thin layer lining the villi of the small intestine. The enterocytes contain phase I and II metabolising enzymes and express membrane transporter proteins governing the entry of nutrients into the enterocytes but also controlling the distribution of xenobiotics into the systemic circulation.

The predominate phase I enzyme family within enterocytes are the CYP isozyme family and there are known to be 57 human isoenzymes, in 18 families and 43 subfamilies [80]. The most prominent and abundance CYP isozyme in the small intestine is the CYP3A family (CYP3A4, 3A5 and 3A7) and makes up at least 50–70% of the total P450 content in the small intestine [26]. Protein abundance data for whole and regional intestinal segments are available with total CYP3A content of 65.7, 11.46, 35.8 and 16.92 nmol in the duodenum, jejunum and ileum, respectively [81].

The abundance of CYP isozymes is vitally important in being able to allocate a regional contribution to drug metabolism *aborally* in mechanistic modelling and is important in correcting an *in vitro* derived metabolic clearance to a *in vivo* regional clearance term reflecting drug metabolism in each region of the small intestine.

13.7.1.4 Regional Variation in Transporter Protein Abundance

The small intestine has a prominent expression of a wide range of transporter proteins (influx and efflux) with the purpose of transporting anions, cations and other nutrients as well as preventing the intracellular accumulation of xenobiotics [82]. The three prominent groups of transporter proteins are ATP-binding cassette (ABC) transporters, ion channel transporters and solute carriers (SLCs).

ABC transporters are expressed in the small intestine and include a family of 49 related genes in humans which encode the expression of the membrane bound proteins [83–85]. Prominent members include P-glycoprotein (P-gp, encoded by the gene *ABCB1*), breast cancer resistance protein (BCRP, encoded by the gene *ABCG2*) and multidrug resistance protein 1 (MRP1, encoded by the gene *ABCC1*).

The expression and regional abundance of membrane-bound transporter proteins is not readily available due to the complexity in specifically isolating these proteins from membrane fractions. As an alternative to this, many groups have examined mRNA expression as way to indirectly quantify transporter expression.

P-glycoprotein is considered a highly important xenobiotic efflux transporter protein. Reports of mRNA expression in human intestinal segments have found expression to increase *aborally* [31, 86, 87] but with large interindividual variability. Further studies by Troutman and Thakker [88] have demonstrated contradictory results showing jejunum expression of P-glycoprotein exceeding that of the ileum (using only one human small intestine), and this was further supported by results from Englund *et al.* [52] who demonstrated higher jejunum P-glycoprotein expression mRNA (2.35 ± 0.114 SEM) than ileum mRNA expression (1.95 ± 0.119 SEM). Interestingly, in a novel method used to quantify ABC transporters using S-tag/S-protein systems, Tucker *et al.* [89] found that the expression of BCRP (305 fmol cm^{-2} duodenum surface area ±248 SD) was higher than that of P-glycoprotein (275 fmol cm^{-2} duodenum surface area ±205 SD) in 12 duodenum samples, which also agreed with the results reported by Englund *et al.* [52].

What is evident from these results is that there is a large interindividual variability in the expression of drug transporter proteins, which may significantly alter oral bioavailability in a population and consideration should be made for variability during the modelling approach.

An understanding of expression profiling of drug transporter proteins along the small intestine is important when utilising *in vitro* derived transport kinetic input data in mechanistic models, as the contribution of the transporter in distinct regions of the small intestine can be factored into the determination of an overall *in vivo* transport flux *aborally* through correction of regional flux for this variation in transport expression

Without viable transporter protein expression data, it is possible to utilise a relative expression approach based on normalising the transporter expression to CYP3A expression and propagating this along the small intestine based on the distribution pattern of the transporter [62].

13.7.2 Constructing the Small Intestine PBPK Model

The basis for developing a small intestine mechanistic oral absorption model was first conceptualised in the compartmental absorption–transit (CAT) model [77, 90]. This model describes a structural mechanistic model of the small intestine, which incorporated the processes of drug transit along the small intestine with linear absorption for drugs in solution or instantaneous dissolution-based formulations.

The model split the small intestine into seven compartments, which were sequentially linked, and drug transfer between compartments was modelled by linear transfer kinetics. Each volume was parameterised with a defined volume and transit time (based on their lengths) and, when linked with a well stirred liver model, was able to predict plasma concentration profiles for drugs.

Following on from the CAT model the advanced compartmental absorption–transit (ACAT) model [91] incorporated a description of drug metabolism and transporter kinetics (nonlinear), descriptions of different drug states which included undissolved and dissolved states (i.e. a description of dissolution) and further incorporated pH/geometrical variations to provide a detailed model for the prediction of oral absorption from drug formulations (Figure 13.11).

Based on this underlying structure, a number of other models have also been developed to predict drug absorption and distribution through the small intestine. These include the GITA [92], GRASS [93] and ADAM [94] models.

Implementation of oral absorption models can be done in any mathematically focussed software capable of solving ODEs, for example Matlab [11], R [12], or by using specialised commercially available pharmacokinetics software such as GastroPlus (Simulations Plus: www.simulations-plus.com), PK-Sim (Bayer Technolology Services: http://www.systems-biology.com) or SimCYP (Simcyp Ltd: www.simcyp.com).

Once the model structure has been defined, disintegration, dissolution, transit and absorption need to be linked together by a series of ODEs.

13.7.2.1 Stomach and Gastric Emptying

The release of stomach contents into the small intestine is governed by the gastric emptying rate (k_s) for both undissolved (solid) and dissolved drug particles.

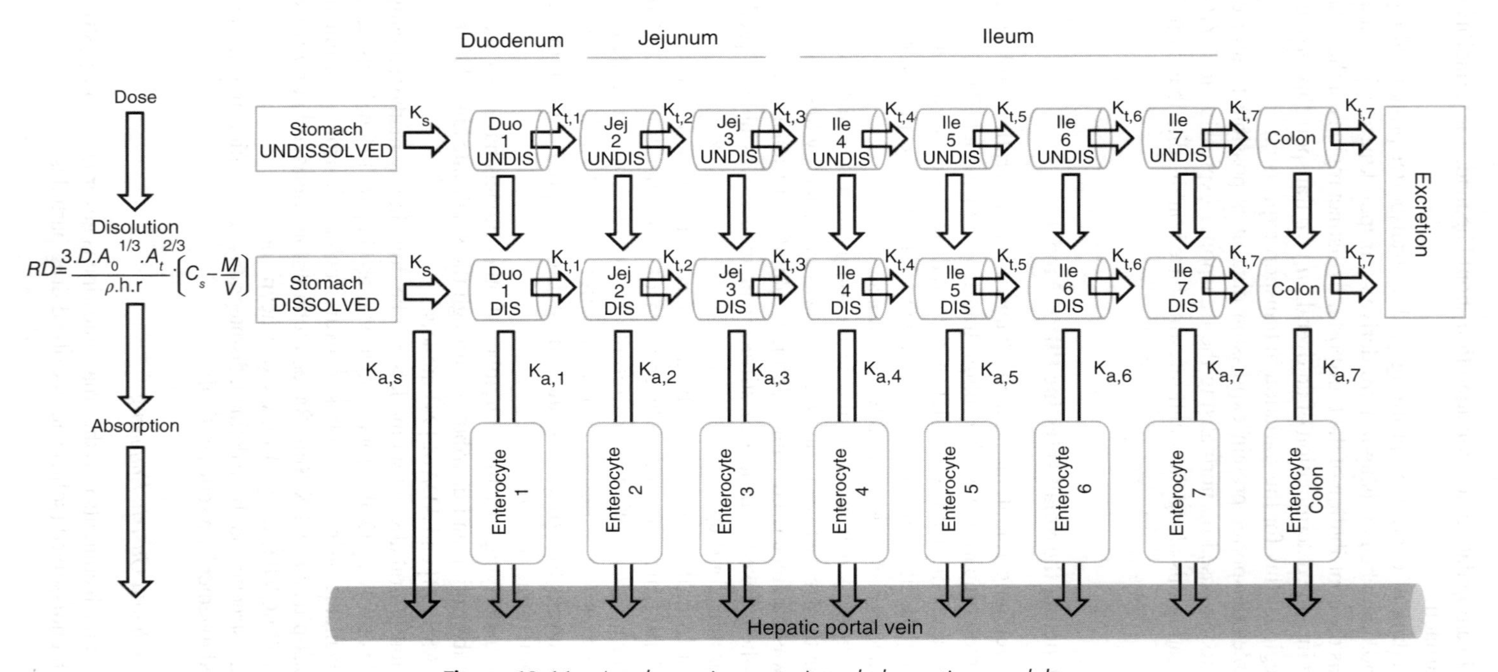

Figure 13.11 *An absorption–transit oral absorption model.*

13.7.2.2 Dissolution

The dissolution of solid material can occur in both stomach and small intestine. It is governed by the Nerst–Brunner Equation 13.4 and incorporated into the model through a rate of dissolution term (*RD*):

$$\frac{dA_{ud}}{dt} = -k_s \times A_{ud} - RD \tag{13.14}$$

where $A_{ud,s}$ refers to the mass of undissolved drug. The first term reflects the loss of undissolved drug into the next compartment (duodenum) governed by the gastric emptying rate (k_s) and the second rate reflects the undissolved drug, which has undergone dissolution (thereafter referred to as 'dissolved').

For any drug that has undergone dissolution in the stomach, the rate of change of dissolved drug ($A_{d,s}$):

$$\frac{dA_d}{dt} = -k_s \times A_{d,s} - k_{a,s} \times A_{d,s} \tag{13.15}$$

is governed by the rate lost due to gastric emptying (first term) and the amount of drug absorbed from the stomach (second term) which is governed by the absorption rate constant ($k_{a,s}$).

13.7.2.3 Small Intestine: First Compartment (Duodenum)

The release of stomach content into the proximal small intestine will first deposit solid or dissolved drug into the duodenum compartment, where the process of dissolution, transit and absorption occur are possible:

$$\frac{dA_{ud,1}}{dt} = k_s \times A_{ud,s} - k_{t,1} \times A_{ud,1} - RD \tag{13.16}$$

Gastric emptying releases the formulation into the first compartment of the small intestine (duodenum; first term) and here the process of transit *aborally* (second term) and dissolution of solid particles (third term) occurs.

For drug which has undergone dissolution, the processes of transit (third term) and absorption (fourth term) predominate *aborally*:

$$\frac{dA_{d,1}}{dt} = k_s \times A_{d,s} + RD - k_{t,1} \times A_{d,1} - k_{a,1} \times A_{d,1} \tag{13.17}$$

13.7.2.4 Small Intestine: Subsequent Compartments (n = 2–7)

For subsequent compartments, the appearance of solid or dissolved drug (first term) from the previous compartment ($n-1$) is followed by transit along subsequent compartments ($n=2$–7; second term) and dissolution (RD):

$$\frac{dA_{ud}}{dt} = k_{t,n-1} \times A_{ud,n-1} - k_{t,n} \times A_{ud,n} - RD \tag{13.18}$$

or absorption into the intestinal enterocytes (fourth term):

$$\frac{dA_d}{dt} = k_{t,n-1} \times A_{d,n-1} - k_{t,n} \times A_{d,n} + RD - k_{a,n} \times A_{d,n} \tag{13.19}$$

13.7.2.5 Colon

For the colon, dissolution is modelled for the undissolved drug:

$$\frac{dA_{ud}}{dt} = k_{t,n-1} \times A_{ud,n-1} - RD \tag{13.20}$$

and absorption for the dissolved drug:

$$\frac{dA_d}{dt} = k_{t,n-1} \times A_{d,n-1} + RD - k_{a,n} \times A_{d,n} \tag{13.21}$$

13.8 Summary

The role of pharmacokinetic modelling, and particular physiologically based pharmacokinetics, during preclinical drug development is important in that it allows for the development of a framework to integrate a wide range of routinely generated *in vitro* drug-/compound-specific properties into a system capable of predicting both oral bioavailability and tissue distribution with reasonable accuracy and confidence.

Clearly, successful predictions rely on the generation of good quality and reliable *in vitro* data for input and there is a need therefore to develop more realistic *in vitro* environments for measuring parameters such as dissolution, solubility and permeability metrics.

The development of commercially available models to predict oral drug absorption highlights the potential applications of such mechanistic modelling approaches to drug delivery and the pharmaceutical scientist with a view to reduce the attrition rate of compounds possessing poor pharmacokinetic properties later during the discovery/development phases.

References

[1] PharmaResearch (2010) Analysis for Pharmaceutical Research and Manufacturers of America, Washington D.C.

[2] Alavijeh, M.S.; Palmer, A.M., The pivotal role of drug metabolism and pharmacokinetics in the discovery and development of new medicines. *IDrugs: The Investigational Drugs Journal* 2004, **7**, 755–763.

[3] Davis, A.M.; Riley, R.J., Predictive admet studies, the challenges and the opportunities. *Current Opinion in Chemical Biology* 2004, **8**, 378–386.

[4] Kola, I.; Landis, J., Can the pharmaceutical industry reduce attrition rates?. *Nature Reviews Drug Discovery* 2004, **3**, 711–715.

[5] Prentis, R.A.; Lis, Y.; Walker, S.R., Pharmaceutical innovation by the seven UK-owned pharmaceutical companies (1964–1985). *British Journal of Clinical Pharmacology* 1988, **25**, 387–396.

[6] Teorell, T., Studies on the diffusion effect upon ionic distribution: II. Experiments on ionic accumulation. *Journal of General Physiology* 1937, **21**, 107–122.

[7] Bischoff, K.B. and Brown, R.G. (1966) Drug distribution in mammals. Chemical Engineering Progress Symposium.

[8] Dedrick, R.L., Animal scale-up. *Journal of Pharmacokinetics and Biopharmaceutics* 1973, **1**, 435–461.
[9] Poulin, P.; Theil, F.P., Prediction of pharmacokinetics prior to in vivo studies. II. Generic physiologically based pharmacokinetic models of drug disposition. *Journal of Pharmaceutical Sciences* 2002, **91**, 1358–1370.
[10] Rodgers, T.; Rowland, M., Mechanistic approaches to volume of distribution predictions: understanding the processes. *Pharmaceutical Research* 2007, **24**, 918–933.
[11] MathWorks Inc. (2003) Matlab, The MathWorks Inc., Natick, MA.
[12] R Development Core Team *R: A Language and Environment for Statistical Computing*, R Foundation for Statistical Computing: Vienna, 2010.
[13] Yoon, M.; Kedderis, G.L.; Yan, G.Z.; Clewell, H.J., III, Use of in vitro data in developing a physiologically based pharmacokinetic model: carbaryl as a case study. *Toxicology* 2014, doi: 10.1016/j.tox.2014.05.006.
[14] Wu, B., Use of physiologically based pharmacokinetic models to evaluate the impact of intestinal glucuronide hydrolysis on the pharmacokinetics of aglycone. *Journal of Pharmaceutical Sciences* 2012, **101**, 1281–1301.
[15] Schaller, S.; Willmann, S.; Lippert, J.; Schaupp, L.; Pieber, T.R.; Schuppert, A.; Eissing, T., A generic integrated physiologically based whole-body model of the glucose-insulin-glucagon regulatory system. *CPT: Pharmacometrics and Systems Pharmacology* 2013, **2**, e65.
[16] Nestorov, I., Whole body pharmacokinetic models. *Clinical Pharmacokinetics* 2003, **42**, 883–908.
[17] Lyons, M.A.; Reisfeld, B.; Yang, R.S.; Lenaerts, A.J., A physiologically based pharmacokinetic model of rifampin in mice. *Antimicrobial Agents and Chemotherapy* 2013, **57**, 1763–1771.
[18] Karlsson, F.H.; Bouchene, S.; Hilgendorf, C.; Dolgos, H.; Peters, S.A., Utility of in vitro systems and preclinical data for the prediction of human intestinal first-pass metabolism during drug discovery and preclinical development. *Drug Metabolism and Disposition: The Biological Fate of Chemicals* 2013, **41**, 2033–2046.
[19] De Buck, S.S.; Mackie, C.E., Physiologically based approaches towards the prediction of pharmacokinetics: in vitro–in vivo extrapolation. *Expert Opinion on Drug Metabolism & Toxicology* 2007, **3**, 865–878.
[20] Andrew, M.A., Hebert, M.F., and Vicini, P. (2008) Physiologically based pharmacokinetic model of midazolam disposition during pregnancy. Conference Proceedings: Annual International Conference of the IEEE Engineering in Medicine and Biology Society, pp. 5454–5457.
[21] Abduljalil, K.; Furness, P.; Johnson, T.N.; Rostami-Hodjegan, A.; Soltani, H., Anatomical, physiological and metabolic changes with gestational age during normal pregnancy: a database for parameters required in physiologically based pharmacokinetic modelling. *Clinical Pharmacokinetics* 2012, **51**, 365–396.
[22] Price, P.S.; Conolly, R.B.; Chaisson, C.F.; Gross, E.A.; Young, J.S.; Mathis, E.T.; Tedder, D.R., Modeling interindividual variation in physiological factors used in pbpk models of humans. *Critical Reviews in Toxicology* 2003, **33**, 469–503.
[23] Thompson, C.M.; Johns, D.O.; Sonawane, B.; Barton, H.A.; Hattis, D.; Tardif, R.; Krishnan, K., Database for physiologically based pharmacokinetic (PBPK) modeling: physiological data for healthy and health-impaired elderly. *Journal of Toxicology and Environmental Health, Part B: Critical Reviews* 2009, **12**, 1–24.
[24] Brown, R.P.; Delp, M.D.; Lindstedt, S.L.; Rhomberg, L.R.; Beliles, R.P., Physiological parameter values for physiologically based pharmacokinetic models. *Toxicology and Industrial Health* 1997, **13**, 407–484.
[25] Bruyere, A.; Decleves, X.; Bouzom, F.; Ball, K.; Marques, C.; Treton, X.; Pocard, M.; Valleur, P.; Bouhnik, Y.; Panis, Y., *et al.*, Effect of variations in the amounts of P-glycoprotein (ABCB1), BCRP (ABCG2) and CYP3A4 along the human small intestine on PBPK models for predicting intestinal first pass. *Molecular Pharmaceutics* 2010, **7**, 1596–1607.
[26] Canaparo, R.; Finnstrom, N.; Serpe, L.; Nordmark, A.; Muntoni, E.; Eandi, M.; Rane, A.; Zara, G.P., Expression of CYP3A isoforms and P-glycoprotein in human stomach, jejunum and ileum. *Clinical and Experimental Pharmacology and Physiology* 2007, **34**, 1138–1144.

[27] Canaparo, R.; Nordmark, A.; Finnstrom, N.; Lundgren, S.; Seidegard, J.; Jeppsson, B.; Edwards, R.J.; Boobis, A.R.; Rane, A., Expression of Cytochromes P450 3a and P-glycoprotein in human large intestine in paired tumour and normal samples. *Basic and Clinical Pharmacology and Toxicology* 2007, **100**, 240–248.

[28] Granvil, C.P.; Yu, A.M.; Elizondo, G.; Akiyama, T.E.; Cheung, C.; Feigenbaum, L.; Krausz, K.W.; Gonzalez, F.J., Expression of the human CYP3A4 gene in the small intestine of transgenic mice: in vitro metabolism and pharmacokinetics of midazolam. *Drug Metabolism and Disposition: The Biological Fate of Chemicals* 2003, **31**, 548–558.

[29] MacLean, C.; Moenning, U.; Reichel, A.; Fricker, G., Closing the gaps: a full scan of the intestinal expression of P-glycoprotein, breast cancer resistance protein, and multidrug resistance-associated protein 2 in male and female rats. *Drug Metabolism and Disposition: The Biological Fate of Chemicals* 2008, **36**, 1249–1254.

[30] Madani, S.; Paine, M.F.; Lewis, L.; Thummel, K.E.; Shen, D.D., Comparison of CYP2D6 content and metoprolol oxidation between microsomes isolated from human livers and small intestines. *Pharmaceutical Research* 1999, **16**, 1199–1205.

[31] Mouly, S.; Paine, M.F., P-glycoprotein increases from proximal to distal regions of human small intestine. *Pharmaceutical Research* 2003, **20**, 1595–1599.

[32] Agarwal, S.; Uchida, Y.; Mittapalli, R.K.; Sane, R.; Terasaki, T.; Elmquist, W.F., Quantitative proteomics of transporter expression in brain capillary endothelial cells isolated from P-glycoprotein (P-gp), breast cancer resistance protein (Bcrp), and P-gp/Bcrp knockout mice. *Drug Metabolism and Disposition: The Biological Fate of Chemicals* 2012, **40**, 1164–1169.

[33] Ohtsuki, S.; Ikeda, C.; Uchida, Y.; Sakamoto, Y.; Miller, F.; Glacial, F.; Decleves, X.; Scherrmann, J.M.; Couraud, P.O.; Kubo, Y., *et al.*, Quantitative targeted absolute proteomic analysis of transporters, receptors and junction proteins for validation of human cerebral microvascular endothelial cell line hCMEC/D3 as a human blood–brain barrier model. *Molecular Pharmaceutics* 2013, **10**, 289–296.

[34] Shawahna, R.; Uchida, Y.; Decleves, X.; Ohtsuki, S.; Yousif, S.; Dauchy, S.; Jacob, A.; Chassoux, F.; Daumas-Duport, C.; Couraud, P.O., *et al.*, Transcriptomic and quantitative proteomic analysis of transporters and drug metabolizing enzymes in freshly isolated human brain microvessels. *Molecular Pharmaceutics* 2011, **8**, 1332–1341.

[35] Uchida, Y.; Ohtsuki, S.; Katsukura, Y.; Ikeda, C.; Suzuki, T.; Kamiie, J.; Terasaki, T., Quantitative targeted absolute proteomics of human blood–brain barrier transporters and receptors. *Journal of Neurochemistry* 2011, **117**, 333–345.

[36] Noyes, A.A.; Whitney, W.R., The rate of solution of solid substances in their own solutions. *Journal of the American Chemical Society* 1897, **19**, 930–934.

[37] Bruner, L.; Tolloczko, S., Über die auflösungsgeschwindigkeit fester körper. *Zeitschrift für Physikalische Chemie* 1900, **35**, 283–290.

[38] Nernst, W., Theorie der reaktionsgeschwindigkeit in heterogenen systemen. *Zeitschrift für Physikalische Chemie* 1904, **47**, 52–55.

[39] Brunner, E., Reaktionsgeschwindigkeit in heterogenen systemen. *Zeitschrift für Physikalische Chemie* 1904, **47**, 56–102.

[40] Fick, A., Poggendorff's annalen. *Journal of the American Mathematical Society* 1855, **94** 59–86.

[41] Einstein, A., Determination of diffusion coefficients for non-electrolytes. *Anuuleu der Physik* 1905, **17**, 549–560.

[42] Hayduk, W.; Laudie, H., Prediction of diffusion coefficients for nonelectrolytes in dilute aqueous solutions. *American Institute of Chemical Engineers* 1974, **20**, 611–615.

[43] Lennernas, H.; Ahrenstedt, O.; Hallgren, R.; Knutson, L.; Ryde, M.; Paalzow, L.K., Regional jejunal perfusion, a new in vivo approach to study oral drug absorption in man. *Pharmaceutical Research* 1992, **9**, 1243–1251.

[44] Lennernas, H., Human intestinal permeability. *Journal of Pharmaceutical Sciences* 1998, **87**, 403–410.

[45] Lennernas, H., Intestinal permeability and its relevance for absorption and elimination. *Xenobiotica; the Fate of Foreign Compounds in Biological Systems* 2007, **37**, 1015–1051.

[46] Lennernas, H., Modeling gastrointestinal drug absorption requires more in vivo biopharmaceutical data: experience from in vivo dissolution and permeability studies in humans. *Current Drug Metabolism* 2007, **8**, 645–657.
[47] Lennernas, H.; Ahrenstedt, O.; Ungell, A.L., Intestinal drug absorption during induced net water absorption in man; a mechanistic study using antipyrine, atenolol and enalaprilat. *British Journal of Clinical Pharmacology* 1994, **37**, 589–596.
[48] Lennernas, H.; Fagerholm, U.; Raab, Y.; Gerdin, B.; Hallgren, R., Regional rectal perfusion: a new in vivo approach to study rectal drug absorption in man. *Pharmaceutical Research* 1995, **12**, 426–432.
[49] Lennernas, H.; Gjellan, K.; Hallgren, R.; Graffner, C., The influence of caprate on rectal absorption of phenoxymethylpenicillin: experience from an in-vivo perfusion in humans. *Journal of Pharmacy and Pharmacology* 2002, **54**, 499–508.
[50] Lennernas, H.; Knutson, L.; Knutson, T.; Hussain, A.; Lesko, L.; Salmonson, T.; Amidon, G.L., The effect of amiloride on the in vivo effective permeability of amoxicillin in human jejunum: experience from a regional perfusion technique. *European Journal of Pharmaceutical Sciences: Official Journal of the European Federation for Pharmaceutical Sciences* 2002, **15**, 271–277.
[51] Artursson, P.; Palm, K.; Luthman, K., Caco-2 monolayers in experimental and theoretical predictions of drug transport. *Advanced Drug Delivery Reviews* 2001, **46**, 27–43.
[52] Englund, G.; Rorsman, F.; Ronnblom, A.; Karlbom, U.; Lazorova, L.; Grasjo, J.; Kindmark, A.; Artursson, P., Regional levels of drug transporters along the human intestinal tract: co-expression of ABC and SLC transporters and comparison with Caco-2 cells. *European Journal of Pharmaceutical Sciences: Official Journal of the European Federation for Pharmaceutical Sciences* 2006, **29**, 269–277.
[53] Taipalensuu, J.; Tornblom, H.; Lindberg, G.; Einarsson, C.; Sjoqvist, F.; Melhus, H.; Garberg, P.; Sjostrom, B.; Lundgren, B.; Artursson, P., Correlation of gene expression of ten drug efflux proteins of the ATP-binding cassette transporter family in normal human jejunum and in human intestinal epithelial Caco-2 cell monolayers. *Journal of Pharmacology and Experimental Therapeutics* 2001, **299**, 164–170.
[54] Irvine, J.D.; Takahashi, L.; Lockhart, K.; Cheong, J.; Tolan, J.W.; Selick, H.E.; Grove, J.R., MDCK (Madin–Darby canine kidney) cells: a tool for membrane permeability screening. *Journal of Pharmaceutical Sciences* 1999, **88**, 28–33.
[55] Volpe, D.A., Variability in Caco-2 and MDCK cell-based intestinal permeability assays. *Journal of Pharmaceutical Sciences* 2008, **97**, 712–725.
[56] Faller, B., Artificial membrane assays to assess permeability. *Current Drug Metabolism* 2008, **9**, 886–892.
[57] Sugano, K.; Nabuchi, Y.; Machida, M.; Aso, Y., Prediction of human intestinal permeability using artificial membrane permeability. *International Journal of Pharmaceutics* 2003, **257**, 245–251.
[58] Winiwarter, S.; Bonham, N.M.; Ax, F.; Hallberg, A.; Lennernas, H.; Karlen, A., Correlation of human jejunal permeability (in vivo) of drugs with experimentally and theoretically derived parameters. A multivariate data analysis approach. *Journal of Medicinal Chemistry* 1998, **41**, 4939–4949.
[59] Sun, D.; Lennernas, H.; Welage, L.S.; Barnett, J.L.; Landowski, C.P.; Foster, D.; Fleisher, D.; Lee, K.D.; Amidon, G.L., Comparison of human duodenum and Caco-2 gene expression profiles for 12,000 gene sequences tags and correlation with permeability of 26 drugs. *Pharmaceutical Research* 2002, **19**, 1400–1416.
[60] Yang, J.; Jamei, M.; Yeo, K.R.; Tucker, G.T.; Rostami-Hodjegan, A., Prediction of intestinal first-pass drug metabolism. *Current Drug Metabolism* 2007, **8**, 676–684.
[61] Avdeef, A., Physicochemical profiling (solubility, permeability and charge state). *Current Topics in Medicinal Chemistry* 2001, **1**, 277–351.
[62] Badhan, R.; Penny, J.; Galetin, A.; Houston, J.B., Methodology for development of a physiological model incorporating CYP3A and P-glycoprotein for the prediction of intestinal drug absorption. *Journal of Pharmaceutical Sciences* 2009, **98**, 2180–2197.
[63] Barter, Z.E.; Bayliss, M.K.; Beaune, P.H.; Boobis, A.R.; Carlile, D.J.; Edwards, R.J.; Houston, J.B.; Lake, B.G.; Lipscomb, J.C.; Pelkonen, O.R., *et al.*, Scaling factors for the extrapolation of

in vivo metabolic drug clearance from in vitro data: reaching a consensus on values of human microsomal protein and hepatocellularity per gram of liver. *Current Drug Metabolism* 2007, **8**, 33–45.
[64] Cubitt, H.E.; Houston, J.B.; Galetin, A., Relative importance of intestinal and hepatic glucuronidation-impact on the prediction of drug clearance. *Pharmaceutical Research* 2009, **26**, 1073–1083.
[65] Bayliss, M.K.; Bell, J.A.; Jenner, W.N.; Park, G.R.; Wilson, K., Utility of hepatocytes to model species differences in the metabolism of loxtidine and to predict pharmacokinetic parameters in rat, dog and man. *Xenobiotica: The Fate of Foreign Compounds in Biological Systems* 1999, **29**, 253–268.
[66] Bayliss, M.K.; Bell, J.A.; Jenner, W.N.; Wilson, K., Prediction of intrinsic clearance of loxtidine from kinetic studies in rat, dog and human hepatocytes. *Biochemical Society Transactions* 1990, **18**, 1198–1199.
[67] Rowland, M.; Benet, L.; Graham, G., Clearance concepts in pharmacokinetics. *Journal of Pharmacokinetics and Biopharmaceutics* 1973, **1**, 123–136.
[68] Gillette, J.R., Factors affecting drug metabolism. *Annals of the New York Academy of Sciences* 1971, **179**, 43–66.
[69] Lin, J.H., Applications and limitations of interspecies scaling and in vitro extrapolation in pharmacokinetics. *Drug Metabolism and Disposition: The Biological Fate of Chemicals* 1998, **26**, 1202–1212.
[70] Neuhoff, S., Gaohua, L., Burt, H., Jamei, M., Li, L., Tucker, G., Rostami-Hodjegan, A., Accounting for transporters in renal clearance: towards a mechanistic kidney model (mech kim). In *Accounting for Transporters in Renal Clearance: Towards a Mechanistic Kidney Model (Mech Kim)*, Sugiyama, Y., Steffansen, B., Ed. Springer New York: 2013; pp 155–177.
[71] Poulin, P.; Theil, F.P., A priori prediction of tissue:Plasma partition coefficients of drugs to facilitate the use of physiologically-based pharmacokinetic models in drug discovery. *Journal of Pharmaceutical Sciences* 2000, **89**, 16–35.
[72] Rodgers, T.; Leahy, D.; Rowland, M., Physiologically based pharmacokinetic modeling 1: predicting the tissue distribution of moderate-to-strong bases. *Journal of Pharmaceutical Sciences* 2005, **94**, 1259–1276.
[73] Rodgers, T.; Leahy, D.; Rowland, M., Tissue distribution of basic drugs: accounting for enantiomeric, compound and regional differences amongst beta-blocking drugs in rat. *Journal of Pharmaceutical Sciences* 2005, **94**, 1237–1248.
[74] Rodgers, T.; Rowland, M., Physiologically based pharmacokinetic modelling 2: predicting the tissue distribution of acids, very weak bases, neutrals and zwitterions. *Journal of Pharmaceutical Sciences* 2006, **95**, 1238–1257.
[75] International Commission on Radiological Protection, Basic anatomical and physiological data for use in radiological protection: reference values. A report of age- and gender-related differences in the anatomical and physiological characteristics of reference individuals. ICRP Publication 89. *Annals of the ICRP* 2002, **32**, 5–265.
[76] International Commission on Radiological Protection, Human alimentary tract model for radiological protection. ICRP publication 100. A report of the international commission on radiological protection. *Annals of the ICRP* 2006, **36**, 25–327, iii.
[77] Yu, L.X.; Crison, J.R.; Amidon, G.L., Compartmental transit and dispersion model analysis of small intestinal transit flow in humans. *International Journal of Pharmaceutics* 1996, **140**, 111–118.
[78] Willmann, S.; Schmitt, W.; Keldenich, J.; Lippert, J.; Dressman, J.B., A physiological model for the estimation of the fraction dose absorbed in humans. *Journal of Medicinal Chemistry* 2004, **47**, 4022–4031.
[79] Evans, D.F.; Pye, G.; Bramley, R.; Clark, A.G.; Dyson, T.J.; Hardcastle, J.D., Measurement of gastrointestinal pH profiles in normal ambulant human subjects. *Gut* 1988, **29**, 1035–1041.
[80] Preissner, S.; Kroll, K.; Dunkel, M.; Senger, C.; Goldsobel, G.; Kuzman, D.; Guenther, S.; Winnenburg, R.; Schroeder, M.; Preissner, R., Supercyp: a comprehensive database on Cytochrome P450 enzymes including a tool for analysis of CYP-drug interactions. *Nucleic Acids Research* 2010, **38**, D237-D243.

[81] Paine, M.F.; Khalighi, M.; Fisher, J.M.; Shen, D.D.; Kunze, K.L.; Marsh, C.L.; Perkins, J.D.; Thummel, K.E., Characterization of interintestinal and intraintestinal variations in human CYP3A-dependent metabolism. *The Journal of Pharmacology and Experimental Therapeutics* 1997, **283**, 1552–1562.

[82] Takano, M.; Yumoto, R.; Murakami, T., Expression and function of efflux drug transporters in the intestine. *Pharmacology and Therapeutics* 2006, **109**, 137–161.

[83] Shugarts, S.; Benet, L.Z., The role of transporters in the pharmacokinetics of orally administered drugs. *Pharmaceutical Research* 2009, 2039–2054, **26**.

[84] International Transporter, C.; Giacomini, K.M.; Huang, S.M.; Tweedie, D.J.; Benet, L.Z.; Brouwer, K.L.; Chu, X.; Dahlin, A.; Evers, R.; Fischer, V., *et al.*, Membrane transporters in drug development. *Nature Reviews Drug Discovery* 2010, **9**, 215–236.

[85] He, L.; Vasiliou, K.; Nebert, D.W., Analysis and update of the human solute carrier (SLC) gene superfamily. *Human Genomics* 2009, **3**, 195–206.

[86] Lin, J.H.; Yamazaki, M., Role of P-glycoprotein in pharmacokinetics: clinical implications. *Clinical Pharmacokinetics* 2003, **42**, 59–98.

[87] Lau, Y.Y.; Huang, Y.; Frassetto, L.; Benet, L.Z., Effect of oatp1b transporter inhibition on the pharmacokinetics of atorvastatin in healthy volunteers. *Clinical Pharmacology and Therapeutics* 2007, **81**, 194–204.

[88] Troutman, M.D.; Thakker, D.R., Novel experimental parameters to quantify the modulation of absorptive and secretory transport of compounds by P-glycoprotein in cell culture models of intestinal epithelium. *Pharmaceutical Research* 2003, **20**, 1210–1224.

[89] Tucker, T.G.; Milne, A.M.; Fournel-Gigleux, S.; Fenner, K.S.; Coughtrie, M.W., Absolute immunoquantification of the expression of ABC transporters P-glycoprotein, breast cancer resistance protein and multidrug resistance-associated protein 2 in human liver and duodenum. *Biochemical Pharmacology* 2012, **83**, 279–285.

[90] Yu, L.X.; Lipka, E.; Crison, J.R.; Amidon, G.L., Transport approaches to the biopharmaceutical design of oral drug delivery systems: prediction of intestinal absorption. *Advanced Drug Delivery Reviews* 1996, **19**, 359–376.

[91] Agoram, B.; Woltosz, W.S.; Bolger, M.B., Predicting the impact of physiological and biochemical processes on oral drug bioavailability. *Advanced Drug Delivery Reviews* 2001, **50** Suppl. 1, S41-S67.

[92] Sawamoto, T.; Haruta, S.; Kurosaki, Y.; Higaki, K.; Kimura, T., Prediction of the plasma concentration profiles of orally administered drugs in rats on the basis of gastrointestinal transit kinetics and absorbability. *The Journal of Pharmacy and Pharmacology* 1997, **49**, 450–457.

[93] Grass, G.M., Simulation models to predict oral drug absorption from in vitro data. *Advanced Drug Delivery Reviews* 1997, **23**, 199–219.

[94] Jamei, M. and Yang, J. (2007) A novel physiologically-based mechanistic model for predicting oral drug absorption: the advanced dissolution, absorption, and metabolism (adam) model. 4th World Conference on Drug Absorption, Transport and Delivery.

Index

Note: Page numbers in *italics* refer to Figures; those in **bold** to Tables.

Computational Pharmaceutics: Application of Molecular Modeling in Drug Delivery, First Edition.
Edited by Defang Ouyang and Sean C. Smith.